Ferroelectric Semiconductors

Ferroelectric Semiconductors

V. M. Fridkin
Academy of Sciences of the USSR
Moscow, USSR

Translated from Russian

CONSULTANTS BUREAU · NEW YORK AND LONDON

Library of Congress Cataloging in Publication Data

Fridkin, Vladimir Mikhaĭlovich.
  Ferroelectric semiconductors.

  Translation of Segnetoèlektriki-poluprovodniki.
  Bibliography: p.
  Includes index.
  1. Ferroelectricity. 2. Ferroelectric crystals. 3. Semiconductors. I. Title.
QC596.5.F7413   537.6'22   79-14561
ISBN 0-306-10957-3

The original Russian text, published by Nauka Press in Moscow in 1974, has been corrected by the author for the present edition. This translation is published under an agreement with the Copyright Agency of the USSR (VAAP).

СЕГНЕТОЭЛЕКТРИКИ— ПОЛУПРОВОДНИКИ
В. М. ФРИДКИН

SEGNETOELEKTRIKI-POLUPROVODNIKI
V. M. Fridkin

© 1980 Consultants Bureau, New York
A Division of Plenum Publishing Corporation
227 West 17th Street, New York, N.Y. 10011

All rights reserved

No part of this book may be reproduced, stored in a retrieval system, or transmitted, in any form or by any means, electronic, mechanical, photocopying, microfilming, recording, or otherwise, without written permission from the Publisher

Printed in the United States of America

Preface to the English Edition

In preparing the book for translation into English, practically no changes have been made either in format or in content except for a small number of brief corrections and comments. The material and the bibliography devoted to the anomalous photovoltaic effect in ferroelectrics and the most recent investigations of the photorefractive effect, which is closely related to our subject, have been left out; the recently published comprehensive book by M. E. Lines and A. M. Glass, entitled *Principles and Applications of Ferroelectrics and Related Materials* (Clarendon Press, Oxford, 1977), covers the principal aspects of these phenomena. Nevertheless the material on photoferroelectric phenomena (including the photovoltaic and photorefractive effects) is extensive enough to fill a separate volume.

The author would like to thank his American colleagues Professors George Taylor and Issai Lefkowitz, co-editors of the international journal *Ferroelectrics*, for their cooperation and initiative in the publication of the English translation of this book. It is hoped that the English edition of this book will prove interesting and useful to specialists in ferroelectricity as well as to a wider circle of Western readers.

Institute of Crystallography, V. Fridkin
Academy of Sciences of the USSR, Moscow

Preface

Since Kurchatov's discovery of the ferroelectric properties of crystals of Rochelle salt, ferroelectricity has been traditionally considered one of the divisions of the physics of dielectrics. Thanks to the work of Ginzburg, Cochran, Smolenskii, Khokhlov, and others, the physics of ferroelectrics and nonlinear crystals has become in recent years an important branch of solid state physics, a branch in which modern fields such as lattice dynamics, resonance methods of investigation, nonlinear optics and electro-optics, etc., have found application and further development. One could also refer to the recent intensive development of the more traditional fields such as the thermodynamics of phase transitions, atomic structure, crystallographic studies, and crystal physics of ferroelectrics. The latter division touches fields begun classically such as the study of domain structure, the mechanism of polarization reversal, and nonlinear dielectric properties.

A fundamentally new stage in the investigation of ferroelectrics is associated with the discovery of the ferroelectric properties of barium titanate by Vul in 1944 [1, 2]. Barium titanate was the first ferroelectric semiconductor with a not too wide forbidden band ($E_g \simeq 3$ eV), so it appeared possible to use it to investigate electron conductivity and other transport phenomena, the mechanism of carrier scattering, nonequilibrium conductivity and luminescence, intrinsic and impurity optical absorption, band structure, and other semiconductor properties. After the discovery of the ferroelectric properties of barium titanate, the semiconductor properties of other ferroelectrics with the perovskite structure began to be investigated. In 1962, Merz and his colleagues [3, 4] discovered the ferroelectric properties of the relatively high-sensitivity photoconductors of the type $A^V B^{VI} C^{VII}$ (for example, SbSI with $E_g \simeq 2$ eV). The class of ferroelectric semiconductors now includes a large number of compounds, among which are both wide-band materials such as lithium niobate, barium and strontium niobate, and compounds of the type $A_2^V B_3^{VI}$, as well as narrow-band semiconductors, for example, compounds of the type $A^{IV} B^{IV}$.

Ferroelectric semiconductors, in particular those with a wide forbidden band and low carrier mobility, are not favorable objects of study if one approaches them from the usual position of semiconductor physics. The indeterminacy of the band structure, the presence in the forbidden band of a complex system of local levels of unknown nature (and the impossibility of alloying related to this), and the large Maxwell relaxation time (which renders difficult the investigation of the kinetics of nonequilibrium electron processes) are a few of the reasons these materials are of little promise in studying elementary processes in semiconductors.

What then is the reason for the great and ever-increasing interest in ferroelectric semiconductors—in investigating and obtaining new materials? It is related first of all to the existence in these materials of ferroelectric and semiconductor properties, and also to the presence of phase transitions of these semiconductors. The contribution of the free energy of the electron subsystem to the free energy of the lattice and the interaction of electrons with the "soft" ferroelectric mode lead to a series of fundamentally new physical phenomena in ferroelectric semiconductors. One of these phenomena is the effect of photoactive illumination and the corresponding nonequilibrium carriers on the Curie temperature, spontaneous polarization, and other macroscopic properties of ferroelectrics. This phenomenon has already had important practical application in holography and optical memory systems (damage effect). The application of ferroelectric semiconductors and, in particular, photoelectric phenomena in ferroelectrics will undoubtedly expand in relation to the development of electrooptics, nonlinear optics, and related disciplines.

The investigation of ferroelectric semiconductors is also timely because of the contribution it makes to the study of the nature of ferroelectricity itself. Thus, for example, the development in recent years of the vibron model of ferroelectrics based on the application of the pseudo-Jahn–Teller effect has led to an active investigation of the electron–phonon interaction both in narrow-band as well as wide-band ferroelectrics, previously considered as typical dielectrics. It is within the framework of this theory that the interband electron–phonon interaction is found to be responsible for anharmonicity and the appearance of the soft mode. Another important example is the screening of spontaneous polarization, which both in narrow-band as well as in wide-band ferroelectrics has an electron–hole nature, since it is caused either by a sharp curvature of the bands at the surface or by the participation of electron surface states.

Thus, the application of the basic ideas and methods of semiconductor physics to ferroelectrics has been very fruitful in understanding the mechanism of ferroelectricity. From this point of view, the division of ferroelectrics into dielectrics and semiconductors is very restrictive and arbitrary in a number of cases.

Thus, the effect of the electron subsystem on the generation of ferroelectricity and the ferroelectric properties is one of the aspects of the subject under consideration. Another no less important aspect is the inverse effect of the dielectric nonlinearity, and the effects of screening and phase transitions on electron processes in ferroelectrics. In this sense, ferroelectrics form a new class of nonlinear semiconductors, a class in which a majority of the phenomena known for electron semiconductors display specific features; moreover, a number of new phenomena are observed (positive temperature coefficient of resistance, a new type of domain instability, "intrinsic" field effect, etc.).

Both of the above-mentioned aspects have been reflected in this book. The author would like to make here some remarks and reservations with respect to its content. The book is written in the form of a systematic consideration of the phenomenological and partially microscopic theory of ferroelectric semiconductors and the classification of groups of phenomena which result from this analysis. The author has tried to avoid a description of the properties of individual compounds, their classification into a series of ferroelectrics, etc. The goal of the book has been to give as complete and systematic a discussion as possible of the theory and experimental results related to the study of the electron processes in ferroelectrics. Thus, the book gives a description of the phenomena, and not a description of the properties of individual or even typical ferroelectric semiconductors. The author has also tried to avoid digressions into related divisions of ferroelectric physics, referring the reader when necessary to the literature. This is justified in view of the recently published monograph of Smolenskii and his co-workers [5] which encompasses all the basic divisions of ferroelectric physics, the book of Vaks on the microscopic theory of ferroelectrics [6], and a number of other very complete monographs, for example [7 - 15].

A decisive contribution to the study of the semiconductor properties of ferroelectrics has been made by a series of Soviet investigators, including the theoretical works of Bonch-Bruevich, Larkin, Guro, Ivanchik, Pasynkov, Kristofel', Bersuker, Sandomirskii, Bursian, Chenskii, Selyuk, and others. In using these works, the author has tried to give as complete and systematic a discussion as possible. Nonetheless, the author accepts responsibility for not having avoided bias by sometimes allotting a disproportionate amount of attention to work done by his colleagues. However, the subject itself is still far from being thoroughly studied, and thus the author was faced with the problem of considering the series of features of electron phenomena in ferroelectrics, instead of a discussion of the subject as a whole.

This monograph generalizes the investigations of the author carried out with his colleagues and students at the A. V. Shubnikov Institute of Crystallography of the Academy of Sciences of the USSR over the past decade. A significant place in the book has been assigned to the results of investigations of photoelectric phenomena in ferroelectrics, which was carried out under the direction of the author in recent years by a group of students and co-workers of the Faculty of Semiconductor Physics of Rostov State University, now directed by

A. A. Grekov. The All-Union seminars on the physics of ferroelectric semiconductors, which are held regularly by the faculty, stimulated this work in many respects. The style of presentation of the book was largely dictated by the lectures read by the author in various years at Rostov State University, at the Karl Marx University (Leipzig, GDR), and at Dejon University (France).

The author considers it his duty to thank Academician B. M. Vul, Corresponding Members of the Academy of Sciences of the USSR G. A. Smolenskii and B. K. Vainshtein, and Doctor of Physical and Mathematical Sciences L. A. Shuvalov for their interest in this subject and for their attention and support. The author also would like to express his appreciation to Professor L. M. Belyaev and A. N. Lobachev, the directors of the laboratories at the Institute of Crystallography of the Academy of Sciences of the USSR and the first in our country to develop a series of ferroelectric semiconductor crystals (including $A^V B^{VI} C^{VII}$) and the first to investigate their physical properties. The author is grateful to all the colleagues working with him on this subject for many years, in particular to A. A. Grekov, and also to T. R. Volk and K. A. Verkhovskii for carefully reviewing the manuscript of the book. At the request of the author, Section 3.8 was written by B. V. Selyuk and Section 4.6 by A. G. Khasabov, to whom the author expresses his appreciation.

In conclusion, the author expresses the hope that this work will stimulate to some degree the further development of the physics of ferroelectric semiconductors and that future success can be expected in this area, whose beginning was based on the discovery of the ferroelectric properties of barium titanate. B. M. Vul [16] recently wrote: "It was difficult to foresee the ferroelectric properties of barium titanate, but after they were discovered, it was easy to hope that they would have a promising future. The past 25 years have justified this hope."

We hope that the study and practical application of ferroelectric semiconductors as a separate class of compounds will be useful both for solid state physics and for its applications.

Moscow, 1975　　　　　　　　　　　　　　　　　　　　　　　　　　V. M. Fridkin

Contents

1. **Thermodynamics of Ferroelectric Semiconductors** 1
    1.1. Free Energy of a Ferroelectric Semiconductor.............. 3
    1.2. First- and Second-Order Phase Transitions................. 5
    1.3. Effect of Electrons on the Phase Transition. Photoferroelectric Phenomena.. 9
    1.4. Anomalies of the Width of the Forbidden Band in the Region of Phase Transitions. Properties of Electrical Absorption......... 11
    1.5. Thermodynamic Interpretation of the Width of the Forbidden Band.. 14
    1.6. Electrical Conductivity of Ferroelectric Semiconductors near the Curie Point.. 17

2. **Microscopic Theory of Ferroelectric Semiconductors**............ 23
    2.1. Ferroelectric Phase Transition and the "Soft" Mode of Vibrations. 23
    2.2. Effect of Screening on the "Soft" Mode of Vibrations......... 28
    2.3. Ferroelectric Phase Transition and the Interband Electron-Phonon Interaction... 31
    2.4. Pseudo-Jahn–Teller Effect in Wide-Band and Impurity Ferroelectrics.. 38

3. **Screening of Spontaneous Polarization** 41
    3.1. Single-Domain Crystal in the Absence of Surface Levels....... 42
    3.2. The Debye Length as a Parameter of Screening Length in a Ferroelectric... 47
    3.3. "Intrinsic" Field Effect with the Contact of a Single-Domain Ferroelectric and Semiconductor........................ 52
    3.4. Screening in a Ferroelectric Capacitor. Opposing Domains 56
    3.5. Screening of Spontaneous Polarization in the Presence of a Surface Layer 59
    3.6. Role of Surface Levels in the Screening of Spontaneous Polarization.. 64
    3.7. Effect of Surface Levels on the Schottky Barrier in Ferroelectrics. 70

3.8. Effect of Screening and Surface Levels on the Domain Structure.. 74
3.9. Screening and Periodic Structure of Interphase Boundaries in a Ferroelectric.. 92

4. **Optical Absorption and the Band Structure of Ferroelectrics**......... 99
   4.1. Temperature Dependence of the Width of the Forbidden Band near First- and Second-Order Phase Transitions 100
   4.2. Intrinsic Optical Absorption of Single Crystals of $BaTiO_3$ 118
   4.3. Intrinsic Optical Absorption of Ferroelectrics of the Groups $A^V B^{VI} C^{VII}$ and $A_2^V B_3^{VI}$ 125
   4.4. Polarization Fluctuations and the Intrinsic Absorption Edge of Ferroelectrics 140
   4.5. Effect of an Electric Field on the Intrinsic Absorption Edge of Ferroelectrics 143
   4.6. Band Structure of Ferroelectrics 153

5. **Photoferroelectric Phenomena and Photostimulated Phase Transitions** . 163
   5.1. Thermodynamics of Photoferroelectric Phenomena 164
   5.2. Photostimulated Shift of the Phase-Transition Temperature and the Photohysteresis Effect 168
   5.3. Photoferroelectric Phenomena in $BaTiO_3$ 181
   5.4. Photodeformation Effect..................................... 184
   5.5. Effect of the Photoinduced Change in Birefringence 187
   5.6. Photostimulated Phase Transitions in Nonferroelectric Materials .. 202

6. **Screening Phenomena** 209
   6.1. Experimental Observation of the "Intrinsic" Field Effect....... 209
   6.2. Electroluminescence of Ferroelectrics 213
   6.3. Opposing Domains in SbSI................................... 215
   6.4. Effect of Nonequilibrium Carriers on the Screening of Interphase Boundaries in SbSI... 221
   6.5. Effect of Screening on Polarization Reversal Processes......... 229
   6.6. Photodomain Effect 238
   6.7. Screening and Short-Circuit Photocurrents 251

7. **Ferroelectric Photoelectrets** 255
   7.1. Formation of Photoelectrets with Screening of Ferroelectric Polarization by Nonequilibrium Carriers..................... 255
   7.2. Polarization of a Ferroelectric under the Effect of the Internal Field of the Photoelectret 263
   7.3. Electron Mechanism of the Effect of Radiation on Ferroelectric Polarization Reversal....................................... 270

8. **Ferroelectrics as Nonlinear Semiconductors** 279
   8.1. Electrical Conductivity and Photoconductivity of
       Ferroelectrics near the Phase Transition 280
   8.2. Effect of Screening of Spontaneous Polarization on the
       Photoconductivity of Ferroelectrics. 288
   8.3. Some Photoelectric Phenomena in SbSI 294
   8.4. Currents Limited by the Space Charge in Ferroelectrics 299
   8.5. Induction Waves and Space Charge in Ferroelectrics 304

References 309

1
Thermodynamics of Ferroelectric Semiconductors

One of the features of ferroelectric semiconductors is the effect of electrons on the fundamental thermodynamic functions and their behavior near the Curie temperature. As will be shown below, the effect of nonequilibrium electrons on the phase transitions in ferroelectrics is of greatest interest from the experimental point of view. In the literature such phase transitions have been called *photostimulated*, and phenomena related to them in ferroelectrics, *photoferroelectric* [17-24]. Photostimulated phase transitions and the effect of electrons on phase transitions generally are not characteristic of ferroelectrics only, but are typical for all semiconductors experiencing phase transitions. Investigation of ferroelectric semiconductors offers in this respect only the advantage that the photoferroelectric phenomena can be related to the basic phenomenological parameters of ferroelectrics such as the Curie-Weiss constant, spontaneous polarization and heat capacity discontinuity, as well as to other independently measured parameters.

The existence of photostimulated phase transitions in solids and, in particular, in ferroelectrics is of fundamental interest. In considering the mechanism of phase transitions, one usually neglects the contribution of the electron subsystem (or the electron free energy) in the total free energy of the crystal and thereby assumes that the mechanism of phase transitions is not related to electron excitations in the crystal. This assumption is based on the fact that at temperatures above the Debye temperature, the contribution of the electron heat capacity to the total heat capacity of the crystal can be neglected. We will consider this in somewhat more detail. Denoting the free energy by F and the internal energy by E, we write the Gibbs-Helmholtz equation

$$F = E + T\frac{\partial F}{\partial T}. \qquad (1.1)$$

After integration, this equation can be represented in the form

$$F = -T \int_0^T \frac{E}{T^2} dT. \qquad (1.2)$$

The internal energy E, by definition, can be expressed in terms of the heat capacity C_V in the following manner:

$$E(T) = E(0) + \int_0^T C_V(T) \, dT, \qquad (1.3)$$

from which, by substituting (1.3) into (1.2), we obtain the final expression for the free energy:

$$F = E(0) + T \int_0^T \frac{dT}{T^2} \int_0^T C_V(\tau) \, d\tau. \qquad (1.4)$$

It is seen from (1.4) that the free energy of the crystal is completely determined by the temperature dependence of the heat capacity. It is known from solid state physics that for a nondegenerate semiconductor or dielectric at temperatures above the Debye temperature, the ratio of the electron heat capacity C_V^{el} to the lattice heat capacity C_V^L is

$$\frac{C_V^{el}}{C_V^L} = \frac{N_c}{n_0} \ll 1, \qquad (1.5)$$

where $N_c = (2\pi m^* kT/h^2)^{3/2}$ is the density of states of electrons or holes in the band, m^* is the effective mass of the electron or hole, and n_0 is the number of atoms per cubic centimeter. For metals at temperatures above the Debye temperature, we have

$$\frac{C_V^{el}}{C_V^L} \simeq \frac{kT}{E_F} \ll 1, \qquad (1.6)$$

where the Fermi energy E_F is of the order of several electron volts, while $kT \simeq 0.025$ eV at room temperature. Thus, both for metals and for semiconductors at temperatures above the Debye temperature, the contribution of electrons to the free energy of the crystal can be neglected. The contribution of the electrons can become significant only near $T = 0$, where C_V tends to zero as $\sim T^3$.

Nonetheless, it is found that besides the region of absolute zero, the vicinity of the temperature of phase transitions is also a peculiar temperature region, where the contribution of the electron subsystem to the free energy of the crystal can be significant. The physical meaning of this assertion involves the fact that close to the temperature of a phase transition, the electrons can make a significant contribution not to the heat capacity itself, but to its anomalous part at the phase transition. This conclusion was first reached in [25] for ferroelectric phase transitions within the framework of the phenomenological theory of Landau, Ginzburg, and Devonshire, and is developed in [26, 27]. The effect of electron excitations on phase transitions of a different nature was investigated in a number of subsequent works. We will dwell in this chapter on an analysis of the photoferroelectric phenomena and the effect of electrons on the properties of ferroelectrics on the basis of the phenomenological theory of ferroelectric phase transitions.

1.1. Free Energy of a Ferroelectric Semiconductor

The presence of a relatively high concentration of carriers in a ferroelectric semiconductor makes it necessary to consider the free energy of the electron subsystem in the expression for the free energy of the crystal near the phase-transition temperature. We will then assume that the free energy of the electron subsystem F_2 is small everywhere (except at the vicinity of the Curie point) as compared to the free energy of the lattice F_1, and that the phase transition itself is related to instability of lattice vibrations (in the opposite case, we are dealing with a purely electron phase transition, as, for example, in vanadium oxides). We will also consider the ordinary ferroelectric for which, according to Landau [28], the free energy of the lattice F_1 near the Curie point can be expanded in a series in even powers of the spontaneous polarization P. We will consider separately in Section 1.4 the case of singular ferroelectrics, for which the spontaneous polarization P is not an expansion parameter. Finally, we will consider the thermodynamics of ferroelectric semiconductors neglecting the anisotropy of the dielectric properties by assuming that the polarization of the crystal occurs only in the direction of the spontaneous polarization axis (the c axis). Following [25-27], we represent the free energy of the crystal F as the sum of the free energy of the lattice in the paraelectric region F_0 and in the ferroelectric region F_1 and the free energy of the electron subsystem F_2:

$$F = F_0 + F_1 + F_2, \qquad (1.7)$$

where

$$F_0(T) = F(P=0,\ \sigma_h=0,\ N_i=0), \qquad (1.8)$$

$$F_1 = \frac{1}{2}\alpha P^2 + \frac{1}{4}\beta P^4 + \frac{1}{6}\gamma P^6 - \frac{1}{2}\sum_i\sum_h s_{ih}\sigma_i\sigma_h - P^2\sum_h v_h\sigma_h, \qquad (1.9)$$

$$F_2 = \sum_i N_i E_i(T, P, \sigma_h). \qquad (1.10)$$

Here, α, β, and γ are the components in the expansion of the free energy in the polarization P [15, 29, 30], σ_k are the components of the mechanical stress tensor, $S_{ik} = S_{ik}^P = \partial^2 F_1 / \partial \sigma_i \partial \sigma_k$ are the components of the elastic compliance tensor, and $\nu_k = \partial^2 F_1 / \partial P^2 \partial \sigma_k$ are the components of the electrostriction tensor. In expression (1.10) for the free energy of the electron subsystem, E_i and N_i are the energy of the levels and the corresponding concentrations of electrons (holes) in the crystal. For the time being, we will neglect the configuration part of the free energy in equation (1.10), while we assume the polarization to be uniform in equation (1.9) and neglect the correlation term. In the general case, equation (1.9) must be supplemented by the Poisson equation (cf. Chapter 3).

To be specific, we will consider an n-type ferroelectric semiconductor, whose band energy diagram is presented in Figure 1.1. Let there exist in a crystal having a width of the forbidden band E_g and a concentration of free electrons n, one type of electron-trapping levels (with energy u_1, concentration M, and electron concentration in these levels N) and one type of hole-trapping levels (with energy u_2 and hole concentration in the levels p). Neglecting the contribution of the recombination levels and assuming that $N, p \gg n$, as usually occurs in high-resistance semiconductors, we can represent the free energy F_2 in the form

$$F_2 = nE_g + N(E_g - u_1) - pu_2 \simeq N(E_g - u_1 - u_2) = N\tilde{E}, \quad (1.11)$$

where the condition of electrical neutrality $p = n + N$ is used and the notation $\tilde{E} = E_g - u_1 - u_2$ is introduced. Thus, N in (1.11) should be understood as the concentration of electrons (holes) in the trapping levels, which exceeds by many orders of magnitude the concentration of free electrons (holes), while $\tilde{E}$ should be understood as the width of the forbidden band with an accuracy to the energies of the trapping levels. Assuming further than the function F_2 must be invariant with respect to the same symmetry transformation as the function F_1, we expand the energy $\tilde{E}$ near the phase transition temperature in a series in P and σ_k:

$$\tilde{E}(T, P, \sigma_k) = \tilde{E}_0(T) + \frac{1}{2}(\tilde{E}_P^{II})_0 P^2 + \frac{1}{4}(\tilde{E}_P^{IV})_0 P^4 +$$
$$+ \frac{1}{6}(\tilde{E}_P^{VI})_0 P^6 + \sum_h \tilde{E}_h' \sigma_h + \frac{1}{2}\sum_h \sum_i \tilde{E}_{hi}'' \sigma_h \sigma_i + P^2 \sum_h \tilde{E}_h''' \sigma_h, \quad (1.12)$$

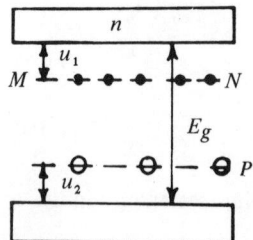

Figure 1.1. Band diagram of a ferroelectric semiconductor.

where the following notation is assumed:

$$(\widetilde{E}_P^{II})_0 = \left(\frac{\partial^2 \widetilde{E}}{\partial P^2}\right)_0 = a, \quad (\widetilde{E}_P^{IV})_0 = \left(\frac{\partial^4 \widetilde{E}}{\partial P^4}\right)_0 = b,$$

$$(\widetilde{E}_P^{VI})_0 = \left(\frac{\partial^6 \widetilde{E}}{\partial P^6}\right)_0 = c,$$

$$\widetilde{E}'_k = \left(\frac{\partial \widetilde{E}}{\partial \sigma_k}\right)_0, \quad \widetilde{E}''_{ki} = \left(\frac{\partial^2 \widetilde{E}}{\partial \sigma_k \partial \sigma_i}\right)_0, \quad \widetilde{E}'''_k = \left(\frac{\partial^3 \widetilde{E}}{\partial P^2 \partial \sigma_k}\right). \quad (1.13)$$

Combining (1.8), (1.9), (1.11), and (1.13), we obtain the final expression for the free energy of the ferroelectric semiconductor:

$$F(T, P, \sigma_k, N) = F_{0N} + \frac{1}{2}\alpha_N P^2 + \frac{1}{4}\beta_N P^4 + \frac{1}{6}\gamma_N P^6 +$$

$$+ N \sum_k \widetilde{E}'_k \sigma_k - \frac{1}{2}\sum_i \sum_k s_{Nik}\sigma_i \sigma_k - P^2 \sum_h v_{Nk}\sigma_k. \quad (1.14)$$

Renormalization of the coefficients in (1.9) leads to the following relations:

$$F_{0N} = F_0 + N\widetilde{E}_0, \quad \alpha_N = \alpha + aN, \quad \beta_N = \beta + bN,$$
$$\gamma_N = \gamma + cN, \quad v_{Nk} = v_k - \widetilde{E}'''_k N, \quad s_{Nik} = s_{ik} - \widetilde{E}''_{ik}N. \quad (1.15)$$

Equation (1.14) must be supplemented by the two equations of state

$$\frac{\partial F}{\partial P} = \mathscr{E} = \alpha_N P + \beta_N P^3 + \gamma_N P^5 - 2P \sum_h v_{Nk}\sigma_k, \quad (1.16)$$

$$\frac{\partial F}{\partial \sigma_k} = -u_k = N\widetilde{E}'_k - \frac{1}{2}\sum_i s_{Nik}\sigma_i - P^2 v_{Nk}, \quad (1.17)$$

where $\mathscr{E}$ is the electric field; u_k are the components of the deformation tensor.

1.2. First- and Second-Order Phase Transitions

Before turning to an analysis of equations (1.14), (1.16), and (1.17) and the effect of electrons on the phase transition and the ferroelectric properties, we recall the principal consequence of the thermodynamic theory of ferroelectrics [15, 28–30]. For this, it is sufficient to analyze equations (1.14), (1.16), and (1.17) for $N = 0$. For simplicity, we will also assume $\sigma_k = \mathscr{E} = 0$. We will then consider separately the cases of first- and second-order phase transitions.

According to [15, 29], in the expression for the free energy (1.14) for a second-order phase transition, it is sufficient to retain the first two terms in the expansion in the polarization:

$$F = F_0 + \frac{1}{2}\alpha P^2 + \frac{1}{4}\beta P^4. \quad (1.18)$$

The equilibrium condition corresponds to the minimum of the free energy, i.e.,

$$\frac{\partial F}{\partial P} = \alpha P + \beta P^3 = 0, \tag{1.19}$$

$$\frac{\partial^2 F}{\partial P^2} > 0. \tag{1.20}$$

It follows from (1.19) that the paraelectric phase corresponds to the solution $P = 0$, while the ferroelectric phase corresponds to

$$P^2 = -\alpha/\beta.$$

It follows from condition (1.20)

$$\alpha + 3\beta P^2 > 0 \tag{1.21}$$

that the paraelectric region ($P = 0$) corresponds to $\alpha > 0$, while in the ferroelectric region ($P^2 = -\alpha/\beta$) α is negative. By denoting the phase-transition temperature (the Curie point) by $T = T_0$, expanding α in a series near $T = T_0$, and retaining the linear term, one can represent α in the form

$$\alpha = \alpha'_T (T - T_0) = \frac{2\pi}{C}(T - T_0), \tag{1.22}$$

where the Curie–Weiss constant is denoted by C. According to (1.21), $\beta > 0$ everywhere below the Curie point. We will assume that β does not depend on the temperature. Thus the temperature dependence of the spontaneous polarization near the Curie point is given by the relation

$$P^2 = \frac{\alpha'_T}{\beta}(T_0 - T), \quad T < T_0. \tag{1.23}$$

We determine the behavior of the entropy S and heat capacity C_p near the Curie point:

$$S = -\frac{\partial F}{\partial T} = S_0 + \frac{(\alpha'_T)^2}{\beta}(T - T_0), \tag{1.24}$$

where $S_0 = -\partial F_0/\partial T$ is the entropy of the paraelectric phase. One can determine from (1.24) the discontinuity of the heat capacity $C_p = T(\partial S/\partial T)_p$ for a phase transition of second order:

$$\Delta C_p = \frac{T_0}{\beta}(\alpha'_T)^2. \tag{1.25}$$

In accordance with the general nature of second-order phase transitions, a finite discontinuity of the heat capacity occurs at the Curie point, while the latent heat of transition equals zero.

Assuming that $\mathscr{E} \neq 0$ in (1.16), one can determine the temperature dependence of the dielectric constant ϵ near the Curie temperature [29]. In the paraelectric region one has

$$\epsilon \simeq \frac{2\pi}{\alpha}, \quad T > T_0, \quad (1.26)$$

in the ferroelectric region one has

$$\epsilon \simeq -\frac{\pi}{\alpha}, \quad T < T_0. \quad (1.27)$$

By combining (1.26) and (1.22), one usually represents the temperature dependence of ϵ in the paraelectric phase in the form

$$\epsilon \simeq \frac{C}{T - T_0}. \quad (1.28)$$

Relation (1.28) is called the Curie-Weiss law.

Figure 1.2 represents schematically the dependences of the free energy on the polarization $F = F(P)$, the temperature dependence of the square of the spontaneous polarization, and the temperature dependence of the reciprocal dielectric constant near the second-order ferroelectric phase transition. It should be noted, as is seen from (1.26) and (1.27), that the ratio of the slopes $1/\epsilon = 1/\epsilon(T)$ in the ferro- and paraelectric phases equals 2 (the "two" law).

The first-order ferroelectric phase transition is described with the help of the expansion (1.14), in which the expansion coefficients in the paraelectric region have different signs. In this case, it is necessary to take into account all three terms of the expansion in P. We assume that $\beta < 0$, $\gamma > 0$, and α satisfies (1.22) in the expansion (1.14). As follows from (1.16), for $\mathscr{E} = 0$ in this case, the free energy F has two equal minima corresponding to the same temperature $T = T_1$ and two different values of P. One of these minima corresponds to $P = 0$, and the other to $P = P_0 \neq 0$. Thus, the temperature $T = T_1$ is the temperature of the first-order phase transition, for which the crystal abruptly changes from one stable state to the other. The temperature dependence $P = P(T)$ in the ferroelectric phase is determined from the condition of the minimum of the free energy (1.16):

$$\alpha P + \beta P^3 + \gamma P^5 = 0, \quad (1.29)$$

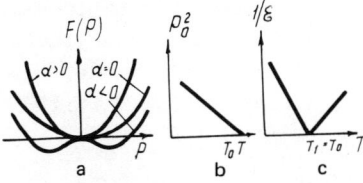

Figure 1.2. Dependences $F(P)$, $P_0^2(T)$, and $1/\epsilon(T)$ near the temperature of the second-order phase transition.

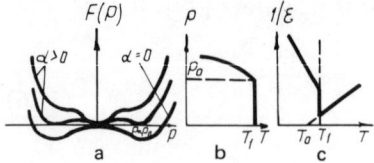

Figure 1.3. Dependences $F(P)$, $P(T)$, and $1/\epsilon(T)$ near the temperature of the first-order phase transition.

from which

$$P^2 = -\frac{\beta}{2\gamma}\left(1 + \sqrt{1 - \frac{4\alpha\gamma}{\beta^2}}\right). \tag{1.30}$$

From the equality of the free energies F at $P = 0$ and $P = P_0$, the combined solution of (1.29) and (1.31),

$$\frac{1}{2}\alpha P^2 + \frac{1}{4}\beta P^4 + \frac{1}{6}\gamma P^6 = 0, \tag{1.31}$$

determines the values P_0 and α at the point of the phase transition $T = T_1$:

$$P_0^2 = P^2(T_1) = \frac{3}{4}\left(-\frac{\beta}{\gamma}\right), \tag{1.32}$$

$$\alpha(T_1) = \frac{3}{16}\frac{\beta^2}{\gamma}. \tag{1.33}$$

Figure 1.3 illustrates for a first-order phase transition the dependence of the free energy on polarization $F = F(P)$, the temperature dependence of the spontaneous polarization $P = P(T)$, and the temperature dependence of the reciprocal dielectric constant.

Calculation of the entropy from equation (1.24) shows that, as a first-order phase transition, the phase transition under consideration is accompanied by the evolution of latent heat, where the entropy discontinuity

$$\Delta S = \frac{2\pi}{C} P_0^2. \tag{1.34}$$

Another feature of the phase transition of first order is temperature hysteresis. It is seen from Figure 1.3a that a transition of the crystal from the equilibrium state for $P = 0$ to the equilibrium state for $P = P_0$ is related to overcoming an energy barrier, whose height is proportional to $|\beta|^3/\gamma^2$. The existence of this barrier causes the temperature hysteresis of the phase transition. The magnitude of the temperature hysteresis ΔT_h can be evaluated as the difference of the phase-transition temperature T_1 and the Curie–Weiss temperature T_0 and is determined from (1.33) and (1.22):

$$\Delta T_h \simeq T_1 - T_0 = \frac{1}{8\pi} C \frac{\beta^2}{\gamma}. \tag{1.35}$$

1.3. Effect of Electrons on the Phase Transition. Photoferroelectric Phenomena

We now consider equations (1.14)–(1.17) for $N \neq 0$. Consideration of the free energy of the electron subsystem is expressed in the renormalization of the coefficients of the expansion of the free energy in the parameter P, which in turn leads to a change in the thermodynamic functions and the nature of the phase transition. We will consider in succession a series of effects, which are a result of the renormalization relations (1.15).

1.3.1 Shift of the Curie Point. It follows from relation (1.15) for α_N and (1.22) that the electrons shift the Curie temperature. Denoting the Curie temperature in the presence of electrons with a concentration N by T_{0N} and the corresponding Curie temperature in the absence of electrons by T_0, we have

$$T_{0N} - T_0 = \Delta T_N = -\frac{C}{2\pi} aN, \qquad (1.36)$$

where C is the Curie–Weiss constant and $a = (\partial^2 \tilde{E}/\partial P^2)$ is the coefficient in the expansion of the energy $\tilde{E}$ in the polarization (1.12). It follows from the condition of the minimum of the free energy $(\partial^2 F/\partial P^2)_0 = Na > 0$ that $a > 0$ and, consequently, the electrons shift the Curie point toward lower temperatures. The shift of the Curie point is proportional to the concentration of electrons N. As has already been noted in Section 1.1, N in high-resistance ferroelectric semiconductors should be understood as the concentration of electrons (holes) in the traps. When all the traps are filled ($N = M$), saturation must be observed in the dependence of the Curie point shift ΔT_N on N. Equation (1.36) then has meaning for not too large values of ΔT_N for which the Landau–Ginzburg–Devonshire expansion remains valid.

1.3.2. Change in Spontaneous Polarization. The change in spontaneous polarization due to the electrons can be obtained from (1.30) and (1.15):

$$P_N^2 = P^2(T)\left[1 + \frac{bNP^2(T)}{\beta P^2(T) + 2\alpha} + \frac{cN\alpha P^4(T)}{\beta P^2(T) + 2\alpha} - \frac{cN}{\gamma}\right]. \qquad (1.37)$$

The change in the discontinuity of $P_0 = P(T_1)$ at the point of the first-order phase transition follows from (1.32) and (1.15):

$$P_{0N}^2 = P_0^2\left[1 + \frac{bN}{\beta} - \frac{cN}{\gamma}\right]. \qquad (1.38)$$

It is assumed that $b > 0$ and $c > 0$; then, as follows from (1.37) and (1.38), the electrons decrease the spontaneous polarization over the entire temperature interval of the ferroelectric phase, including the phase-transition point itself.

1.3.3. Change in the Temperature Hysteresis. First-order phase transitions are characterized by temperature hysteresis, whose microscopic mechanism is related to the nucleation process. The magnitude of the temperature hysteresis is proportional to the formation energy of the critical nucleation center. For ferroelectric phase transitions, this energy, as has already been indicated, is proportional to $|\beta|^3/\gamma$. If it is assumed that $b > 0$ and $c > 0$, then the electrons decrease the temperature hysteresis correspondingly by the energy $|\beta|^3/\gamma$. The change in temperature hysteresis due to electrons can be evaluated quantitatively with the help of (1.35) and (1.15);

$$\Delta T_{hN} \simeq \frac{3}{32\pi} C \frac{\beta^2}{\gamma} \left[1 + \frac{2bN}{\beta} - \frac{cN}{\gamma}\right]. \quad (1.39)$$

It is seen from (1.39) that $\Delta T_{hN} < \Delta T_h$ for $b > 0$ and $c > 0$.

1.3.4. Change in Spontaneous Deformation. The effect of the electrons on the spontaneous deformation follows from the equation of state (1.17). After setting $\sigma_i = 0$, we have

$$u_{kN} = v_{kN} P_0^2 - N\widetilde{E}_k' = u_k \left[1 + \frac{\Delta P_{0N}^2}{P_0^2} - \frac{\widetilde{E}_k''' N}{v_k}\right] - N\widetilde{E}_k', \quad (1.40)$$

where $u_k = v_k P_0^2$, $\Delta P_{0N}^2 = (P_{0N} - P_0)^2$.

One can evaluate by analogy the effect of the electrons on the change in volume of a unit cell v at the phase transition:

$$\frac{\Delta v_N}{v} = \frac{\Delta v}{v}\left[1 + \frac{\Delta P_{0N}^2}{P_0^2} - N\sum_k \frac{\widetilde{E}_k'''}{v_k} - N\sum_k \frac{\widetilde{E}_k'}{v_k P_0^2}\right]. \quad (1.41)$$

1.3.5. Change in the Dielectric and Piezoelectric Properties. By combining (1.33), (1.35), and (1.26), it is not difficult to obtain the known relation relating the dielectric constant at the point of the first-order phase transitions with the temperature hysteresis:

$$\varepsilon(T_1) \Delta T_h \simeq C. \quad (1.42)$$

The effect of the electrons on the dielectric constant follows directly from (1.42). If it is assumed that the electrons do not affect the Curie–Weiss constant C and decrease the temperature hysteresis, then $\varepsilon(T_1)$ increases, i.e., $\varepsilon_N(T_1) > \varepsilon(T_1)$. The effect of the electrons on the piezoelectric module is described by the equation of state (1.17) for $\mathscr{E} \neq 0$:

$$\left(\frac{\partial u_k}{\partial \mathscr{E}}\right)_N = d_{kN} = \frac{\varepsilon_N}{2\pi} v_{Nk} P_N. \quad (1.43)$$

1.3.6. Change in the Latent Heat and Heat Capacity Discontinuity. The dependences of the entropy discontinuity ΔS_N and heat capacity discontinuity ΔC_N on the electron concentration N is obtained from (1.34) and (1.25):

$$\Delta S_N = \frac{2\pi}{C} P_0^2 \left[1 + \frac{bN}{1+\beta} - \frac{cN}{\gamma}\right], \quad (1.44)$$

$$\Delta C_N = \frac{4\pi^2 T_0}{C^2 \beta} \left[1 - \frac{bN}{\beta}\right]. \quad (1.45)$$

Other relations relating the characteristics of the phase transitions with the electron concentration can also be obtained from the equations of state (1.16) and (1.17). We note one general feature of these results. An increase in electron concentration for the assumptions ($b > 0, c > 0$) leads to $|\beta_N| < |\beta|$ and $\gamma_N > \gamma$, which changes the nature of the phase transition, moving it toward the critical Curie point [28, 29]. This corresponds to a decrease in the temperature hysteresis, a decrease in the latent heat of transition, etc.

An increase in the concentration of electrons (holes) in the band n due to nonequilibrium carriers occurs with illumination of the ferroelectric semiconductor in the spectral region of its natural photosensitivity. This in turn leads to an increase in the carrier concentration in the traps N. The relation between n and N in each individual case is determined by the trap parameters, their energy, and capture cross section. When the "natural" light is turned off, the traps empty thanks to transition of the electrons (holes) in the band and recombination. Thus, with illumination of a ferroelectric semiconductor, one must observe a shift in the Curie point and a change in the physical properties of the ferroelectric caused by the change in concentration of the nonequilibrium carriers. These phenomena have been called *photoferroelectric*. The thermodynamics of ferroelectric semiconductors, as was shown above, describe the photoferroelectric phenomena, but do not reveal their microscopic mechanism. We will discuss the latter in the second chapter.

1.4. Anomalies of the Width of the Forbidden Band in the Region of Phase Transitions. Properties of Electrical Absorption

We now turn to an analysis of relation (1.12). For simplicity, we identify the energy $\tilde{E}$ with the width of the forbidden band E_g (we accordingly neglect the dependence of the trap energies u_1 and u_2 on the polarization P). In a number of cases, for example, in the analysis of electrical conductivity and photoconductivity of ferroelectrics (cf. Section 1.5), we will show that this neglect is inadmissible. We will note now only that if N means the concentration of free electrons, then the equality $\tilde{E} = E_g$ is strictly satisfied.

In the absence of mechanical stresses, the relation (1.12) for E_g has the form

$$E_g = E_{g0} + \frac{1}{2} aP^2 + \frac{1}{4} aP^4 + \frac{1}{6} cP^6, \qquad (1.46)$$

where E_{g0} is the width of the forbidden band in the paraelectric phase. It follows from (1.46) that the first-order phase transition from the para- to the ferroelectric phase is accompanied by a discontinuity ΔE_g:

$$\Delta E_g \simeq \frac{a}{2} P_0^2 \simeq \frac{a}{4\pi} C \Delta S. \qquad (1.47)$$

For a second-order phase transition, $\Delta E_g = 0$. However, there must occur a discontinuity of the coefficients of the width of the forbidden band $(dE_g/dT)_p$ and $(dE_g/dp)_T$. Actually, by differentiating (1.46) with respect to temperature and pressure and using (1.23) and (1.24), we have

$$\Delta\left(\frac{dE_g}{dT}\right)_p = \frac{a}{2}\frac{\alpha'_T}{\beta} = \frac{a}{2}\frac{\Delta C_p}{T_0 \alpha'_T}, \tag{1.48}$$

$$\Delta\left(\frac{dE_g}{dp}\right)_T = -a\frac{\alpha'_T}{\beta}\frac{dT_0}{dp}, \tag{1.49}$$

$$\frac{\Delta\left(\frac{dE_g}{dp}\right)_T}{\Delta\left(\frac{dE_g}{dT}\right)_p} = -\frac{dT_0}{dp}, \tag{1.50}$$

where dT_0/dp is a constant characterizing the shift of the Curie point with pressure. Thus, for a second-order phase transition, the coefficients of the change in width of the forbidden band with temperature and pressure experience a finite discontinuity, the sign and magnitude of which are determined, respectively, by the sign and magnitude of the constant $a > 0$.

In the presence of an external electric field $\mathscr{E}$, applied in the direction of spontaneous polarization, P in (1.46) should be taken as the sum of the spontaneous and induced polarizations. Keeping the quadratic term in (1.46), we have

$$E_g \simeq E_{g0} + \frac{1}{2}a\left(P + \frac{1}{4\pi}\varepsilon\mathscr{E}\right)^2. \tag{1.51}$$

In the ferroelectric region, keeping the linear term in $\mathscr{E}$, we have

$$\Delta E_g^{\mathscr{E}} = \Delta E_g^0 + \frac{1}{4\pi}a\varepsilon\mathscr{E}P. \tag{1.52}$$

Here, $\Delta E_g^{\mathscr{E}}$ is the change in width of the forbidden band of the ferroelectric under the effect of the electric field; $\Delta E_g^0 = \frac{1}{2}aP^2$ is the spontaneous change in E_g at the phase transition, which was discussed above; and ϵ is the dielectric constant. In the paraelectric region, as follows from (1.51), a quadratic dependence of $\Delta E_g^{\mathscr{E}}$ on the field $\mathscr{E}$ obtains. Since $a > 0$, then, according to (1.52), the field increases the width of the forbidden band of the ferroelectric and accordingly shifts the edge of the intrinsic optical absorption toward shorter wavelengths. Thus, the sign of this effect is opposite the sign of the Franz-Keldysh effect, and its magnitude is an order of magnitude greater than was shown by the first observations of Kern and Harbeke [31, 32] performed for the ferroelectric SbSI.

It also follows from (1.52) that the temperature dependence of the Kern-Harbeke effect is determined by the temperature dependence of the dielectric constant and the spontaneous polarization and differs significantly for first- and second-order phase transitions. Actually, for a first-order phase transition, by substituting $\epsilon \sim (T - T_0)^{-1}$ and $P = P_0$ into (1.52), we have $\Delta E_g^{\mathcal{E}} \sim (T - T_0)^{-1}$. For a second-order phase transition, $P \sim (T - T_0)^{1/2}$ and $\epsilon \sim (T - T_0)^{-1}$ and, consequently, $\Delta E_g^{\mathcal{E}} \sim (T - T_0)^{-1/2}$. Thus, although $\Delta E_g^{\mathcal{E}}$ has a maximum at the Curie point in both cases, this maximum must be flatter for the second-order phase transition.

Investigation of the intrinsic optical absorption of a ferroelectric close to the phase transition and the simultaneous change in the temperature dependence of the spontaneous polarization, according to (1.46), make it possible to evaluate the constants a, b, and c. As was shown in Section 1.3, these constants determine the magnitude of a number of photoferroelectric phenomena. Thus, parallel investigation of the intrinsic optical absorption, electrical absorption, and photoferroelectric phenomena makes it possible to evaluate all the phenomenological parameters of the ferroelectric semiconductor which determine its thermodynamic functions. Chapters 4 and 5 are devoted to this problem.

In concluding this section, we will briefly discuss singular ferroelectrics. As was already stated at the beginning of this chapter, the effect of electrons (in particular, nonequilibrium) on the phase transition and the thermodynamic functions of the crystal close to the phase transition is a general phenomenon characteristic of all semiconductors experiencing a phase transition. The so-called singular ferroelectrics are one of the cases convenient for analysis [33, 34]. Like ordinary ferroelectrics, they have spontaneous polarization in the ferroelectric phase, but it is not a parameter of the phase transition. This means that the free energy of the crystal close to the temperature of the phase transition is expanded in a series in parameters having another physical meaning: for example, purely structural. In accordance with the expression for the free energy obtained for singular ferroelectrics in [33], the width of the forbidden band E_g can be represented in the form

$$E_g = E_{g0} + \frac{a_1}{2}(\eta^2 + \xi^2) + \frac{a_2}{4}(\eta^2 + \xi^2)^2 + \frac{a_3}{2}(\eta\xi)^2 +$$
$$+ \frac{a_4}{6}(\eta^2 + \xi^2)^3 + \frac{a_5}{2}(\eta^2 + \xi^2)(\eta\xi)^2 + \frac{a_6}{2}P^2 + a_7\eta\xi P, \quad (1.53)$$

where η and ξ are the parameters of the phase transition and P is the polarization (without considering stresses). According to [33], the equilibrium values of η, ξ, and P are

$$\eta = \pm \xi, \quad P = \pm a\chi\eta^2, \quad (1.54)$$

$$\eta^2 = \frac{-\beta + \sqrt{\beta^2 - 4\gamma\alpha}}{2\gamma}, \quad \beta < 0 \text{ first-order phase transition}, \quad (1.55)$$

$$\eta^2 = -\frac{\alpha}{\beta}, \quad \beta > 0 \text{ second-order phase transition}, \quad (1.56)$$

where $\alpha = \alpha'_T(T - T_0)$, β, γ, and α are coefficients in the expansion of the free energy (cf. [33], and χ is the dielectric susceptibility. Substituting (1.54) into (1.53), we can express E_g as a function of the polarization P:

$$E_g = E_{g0} \pm a_1^* P + a_2^* P^2 + a_3^* P^3,$$
$$a_1^* = a_1 (a\chi)^{-1},$$
$$a_2^* = a_2 (a\chi)^{-2} + \frac{a_3}{2}(a\chi)^{-2} + \frac{a_6}{2} + a_i (a\chi)^{-1},$$
$$a_3^* = \frac{4}{3} a_4 (a\chi)^{-3} + a_5 (a\chi)^{-3}. \tag{1.57}$$

It is seen by comparing (1.57) with (1.46) that the dependence $E_g(P)$ must include the linear term for the singular ferroelectrics. The behavior of $E_g(T)$ at the Curie point is analogous to that observed for ordinary ferroelectrics: the discontinuity $\Delta E_g \simeq a_1 (|\beta|/\gamma)$ for the first-order phase transition and a discontinuity $\Delta(dE_g/dT) = a_1(\alpha'_T/\beta)$ for the second-order phase transition. Keeping the quadratic term in (1.57), we have, by comparing with (1.52),

$$\Delta E_g^\mathscr{E} = \Delta E_g^0 + a_1 a^{-1} \mathscr{E} + \frac{1}{2\pi} a_2^* \varepsilon \mathscr{E} P. \tag{1.58}$$

Since the value of ϵ and the temperature dependence $\epsilon = \epsilon(T)$ are insignificant for singular ferroelectrics [33, 34], the magnitude of the electrical absorption effect $\Delta E_g^\mathscr{E}$ and its temperature dependence must also be small.

Consideration of the free energy of the electron subsystem $F_2 \sim NE_g$ in the expression for the free energy of a singular ferroelectric leads to renormalization relations analogous to (1.15). This in turn determines the effect of the electrons on the phase transition, in particular, the shift of the Curie point toward lower temperatures, and conversely (as for ordinary ferroelectrics).

1.5. Thermodynamic Interpretation of the Width of the Forbidden Band

We will now show that relations (1.47), (1.48), and (1.49) are a consequence of the general relation between the width of the forbidden band of the crystal and the heat capacity in the Debye approximation.

The temperature coefficient of the width of the forbidden band is determined in the following manner:

$$\left(\frac{\partial E_g}{\partial T}\right)_p = \left(\frac{\partial E_g}{\partial T}\right)_V - \frac{\beta^*}{\chi^*}\left(\frac{\partial E_g}{\partial p}\right)_T, \tag{1.59}$$

where $\beta^* = (1/V)(\partial V/\partial T)_P$ is the temperature coefficient of volume expansion and $\chi^* = -(1/V)(\partial V/\partial p)_T$ is the compressibility. Thus a change in E_g with temperature is determined by two terms, the first of which takes into account the change in E_g with temperature at constant volume and is caused by the electron–phonon interaction, while the second is related to the temperature expansion of the crystal.

In describing above the expressions for the free energy of the electron subsystem (1.10) and (1.11), we started from the fact that the width of the forbidden band is the free energy in the calculation of one electron–hole pair. This definition was rigorously justified in the works of James and Brooks [35, 36]. Keyes, in developing these thermodynamic ideas [37], used the Slater model of the energy levels in the crystal [38]. According to Keyes, the energy required for optical excitation of an electron from the state j in the valence band to the state l in the conduction band is determined by the expression

$$E_{g_0} = \omega(V) + 3\bar{\lambda}kT + p_l^2/2m_c^* + p_j^2/2m_v^*, \qquad (1.60)$$

$$v_\alpha[V(l,j)] = v_\alpha(V)\left[1 + \lambda_\alpha \frac{n_i}{N}\right], \qquad (1.61)$$

where p is the quasimomentum of the electron, ω is the magnitude of the width of the forbidden band at $T = 0$, V is the volume of the crystal, m_c^*, m_v^* are the effective masses of the electrons and holes, $\bar{\lambda}$ is the average value of the parameters λ_α, which are coefficients in the expansion (1.61) of the lattice vibration frequencies v_α in n_i, α is the parameter normalizing the lattice vibrations, n_i is the total number of internal carriers, N is the total number of atoms in the crystal, and l and j are numbers defining the state of the electrons and holes. The term $E_{act} = \omega(V) + 3\bar{\lambda}kT$ in expression (1.60) is the minimum energy required to produce an electron–hole pair, i.e., the free activation energy. The other two terms in (1.60) are the kinetic energy of the electrons and holes at the edge of the conduction band and valence band.

By using classical statistics for small values of n_i and considering lattice vibrations above the Debye temperature in the classical approximation, Keyes found an expression for the free energy of the crystal and its derivatives. We present here the expressions for the entropy S and the chemical potential ζ:

$$S = -k\ln(n_i^2/N_cN_vV^2) + 3k + 3k\bar{\lambda} + 2kT\beta^* - \beta^*(d\omega/d\ln V), \qquad (1.62)$$

$$\zeta = kT\ln(n_i^2/N_cN_vV^2) + \omega + 3kT\bar{\lambda}, \qquad (1.63)$$

where β^* is the thermal expansion coefficient and N_c, N_v are the effective densities of states in the bands.

The entropy defined by (1.62) consists of a configuration term S_1, the activation entropy at constant volume $S_2 = 3k\bar{\lambda}$, and the term S_3 related to thermal expansion. It is obvious from the definition of $\bar{\lambda}$ that the term S_2 appears as a result of the effect of electron excitation on the lattice vibration frequencies.

In the quantum-mechanical consideration, the term $3k\bar{\lambda}$ is replaced by an expression of the form

$$S_2 = -\frac{k}{3N} \sum_{\alpha=1}^{3N} \lambda_\alpha \left(\frac{h\nu_\alpha}{kT}\right)^2 \frac{\exp(h\nu_\alpha/kT)}{[\exp(h\nu_\alpha/kT)-1]^2}. \tag{1.64}$$

Expression (1.64) resembles the formula for the specific heat capacity of the solid C_V, having the form

$$C_V = kN \frac{(h\nu/kT)^2 \exp(h\nu/kT)}{[\exp(h\nu/kT)-1]^2}. \tag{1.65}$$

Comparing (1.64) with (1.65) and setting

$$\lambda_\alpha = \bar{\lambda} = \sum_{\alpha=1}^{3N} \frac{\lambda_\alpha}{3N}, \tag{1.66}$$

we have

$$S_2 = -\bar{\lambda} C_V/N. \tag{1.67}$$

Since the activation entropy is $S_2 = -(\partial E_{act}/\partial T)_V$, while the free activation energy is the width of the forbidden band, then the value of $(\partial E_g/\partial T)_V$ is related linearly to the heat capacity of the crystal C_V according to (1.67).

The contribution of the thermal expansion of the crystal determined by the term S_3 also depends linearly on C_V, since

$$\beta^* = \gamma_G \frac{\chi^*}{V} C_V, \tag{1.68}$$

where γ_G is the Grüneisen constant. Hence, according to (1.68), the second term on the right side of equation (1.59) is also proportional to C_V and, consequently, the linear dependence

$$\left(\frac{\partial E_g}{\partial T}\right)_p \sim C_V \tag{1.69}$$

occurs in the most general case.

Thus, the anomalies in the width of the forbidden band in the region of ferroelectric phase transitions described by relations (1.47), (1.48), and (1.49) and obtained in Section 1.4 from the thermodynamic theory of ferroelectrics are a consequence of the general Keyes relation (1.69). Hence it follows that other nonferroelectric phase transitions are also accompanied by a discontinuity in the width of the forbidden band or its coefficients.

Experimental verification of (1.69) can go in two directions. First, one can investigate the temperature dependence $(\partial E_g/\partial T)_p$ in the region of temperatures close to absolute zero where, according to Debye, $C_V \propto (T/\theta)^3$; θ is the Debye temperature. These measurements were performed for germanium and silicon in [37] and showed good agreement with (1.69). For both semiconductors, the dependence $E_g = E_g(T/\theta)$ was the same, satisfying (1.69). Second, it is advisable to investigate the anomalies of E_g and its derivatives in the region of the first- and second-order phase transitions. This second way is a most natural and convenient method for experimental verification of (1.69). The results of these measurements as applied to ferroelectrics will be discussed in Chapter 4.

1.6. Electrical Conductivity of Ferroelectric Semiconductors near the Curie Point

In Section 1.1, in expressions (1.10) and (1.11) for the free energy of the electron subsystem F_2 we neglected the configuration part of the energy, which takes into account all possible ways of distributing N electrons with two opposing spins over M trapping levels. We turn to the band diagram of an n-type ferroelectric semiconductor with a concentration of donor levels N having an activation energy E (Figure 1.4). According to [39], the free energy of the electron subsystem in the approximation of Fermi–Dirac statistics has the form

$$F_2 = nE - nkT\left[\ln\left(\frac{2\pi m^* kT}{h^2}\right)^{3/2} + \ln\frac{1}{n} + \ln 2\right] -$$

$$- kT \ln \frac{N!}{n!(N-n)!} - kT(N-n)\ln 2, \qquad (1.70)$$

where n is the concentration of free electrons in the band and m^* is the effective mass. The first term nE here is analogous to the term $N\tilde{E}$ in (1.11), the second term corresponds to the kinetic energy of the free electrons in the band, while the remaining terms are the configuration part of the free energy of the electrons with consideration of spin. It is seen from (1.70) that for weak ionization of the donors, i.e., for $n \ll N$, the configuration part of the energy can be neglected. Just as for near filling of the trapping levels $N \simeq M$ (on the energy diagram of

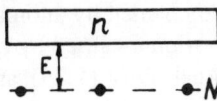

Figure 1.4. Band diagram for the electrical conductivity of a ferroelectric semiconductor with donor levels.

Figure 1.1), one can neglect the contribution of the configuration energy to the free energy F_2, as was done in writing (1.10) and (1.11). In the general case, for arbitrary filling of the levels, one should take into account the configuration term in (1.10) and (1.11).

Far from the Curie point in the ferroelectric region $F_1 \gg F_2$, the electrons also do not make a significant contribution to the free energy. On the other hand, in the paraelectric region, where $P = 0$, the free energy $F = F_0 + F_2$ and, consequently, the equilibrium concentration of free electrons n is determined from the condition

$$(\partial F/\partial n)_T = (\partial F_2/\partial n)_T = 0. \tag{1.71}$$

We obtain from (1.70) and (1.71) in the approximation $n \ll N$ the expression for the concentration of free electrons for the paraelectric region:

$$n(E_0) = 2(NN_c)^{1/2} \exp(-E_0/2kT), \tag{1.72}$$

where E_0 is the activation energy of the donor centers in the paraelectric region and N_s is the density of states in the band:

$$N_s = (2\pi mkT/h^2)^{3/2}. \tag{1.73}$$

In the ferroelectric region near the phase transition, F_2 gives a significant contribution to the free energy of the crystal, where we obtain from the condition $(\partial F/\partial n)_T = 0$ an expression analogous to (1.72) for the concentration of free electrons $n = n(E)$:

$$n(E) = 2(NN_c)^{1/2} \exp(-E/2kT), \tag{1.74}$$

where E is the activation energy of the donors in the ferroelectric region. By analogy to (1.12) and (1.46), the energy E can be expanded in a series in the polarization:

$$E = E_0 + \frac{1}{2} a_1 P^2 + \frac{1}{4} b_1 P^4 + \frac{1}{6} c_1 P^6. \tag{1.75}$$

The solution of (1.74) determines the dependence $n = n(P)$ in the vicinity of the Curie point. By comparing (1.72), (1.74), and (1.75), we are led to the conclusion that the phase transition in the ferroelectric semiconductor is accompanied by anomalies in the temperature dependence of the carrier concentration n caused by a change in the donor activation energy near the Curie point. Analysis of (1.75), analogous to that carried out in Section 1.4, shows that a discontinuity in the activation energy ΔE occurs for a first-order phase transition at the point of the phase transition, while the activation energy undergoes a continuous change in the vicinity of the Curie point for second-order phase transitions. These features lead in turn to the appearance of anomalies in the temperature dependence of the carrier concentration near the Curie point. The anomalies in the concentration n near the Curie point can also have another cause. They can, in particular, be caused by changes in the effective mass of the carriers (and accordingly the density of states) and the concentration of donor centers. In principle, in the region of the phase transition there can arise centers of another nature, for example, levels related to nucleation centers of the new phase (the Kanzig region). In the general case, the generation of nucleation centers of the new phase or fluctuations related to local changes in the transition parameter can be stabilized by electron capture and the formation of new electron combinations (phasons and fluctuons, cf. Section 5.6) near the phase transition. The formation of these electronic states can in turn cause anomalies in the carrier concentration n in the region of the phase transition.

As will be shown in the following chapter, the microscopic nature of the effect of the electron subsystem on the phase transition in ferroelectrics consists of the interaction of soft optical phonons (soft ferroelectric mode [5]) with free or localized electrons. Thus, it is natural to suppose that the anomalies noted above in the behavior of the activation energy of the donor centers near the Curie point, the possible change in concentration, and the nature of the centers themselves are caused by the generation of a coupled state of quasiparticles— optical phonons and electrons [40]. In this case, the observation of the coupled state only near the Curie point is due to the appearance of the known energy threshold of decay [40], which in the case of a ferroelectric semiconductor is caused by the temperature dependence of the energy of transverse optical phonons near the Curie point (soft mode).

As is known, the electrical conductivity σ is related to the concentration of free carriers n in the following manner:

$$\sigma = qn\mu, \qquad (1.76)$$

where q is the elementary charge, and μ is the mobility. Thus, possible anomalies in the temperature dependence $n = n(T)$ cause the peculiarities in the electrical conductivity of the ferroelectric semiconductor near the Curie point. Another possible cause of the anomalies in the electrical conductivity near the Curie point [this follows from (1.76)] is the behavior of the mobility near the phase transition. We will discuss this and other possible mechanisms for the anomalies in the electrical conductivity in ferroelectric semiconductors in Chapter 8.

In conclusion we note the characteristics of multiminimum ferroelectric semiconductors. It is known that the band structure of a number of ferroelectric semiconductors, for example, with the perovskite structure [5], is characterized by a conduction band having six minima located along the principal axes of symmetry (cf. Figure 1.5). In the cubic paraelectric region, these minima are degenerate. Below the temperature of the phase transition for $T < T_1$, the

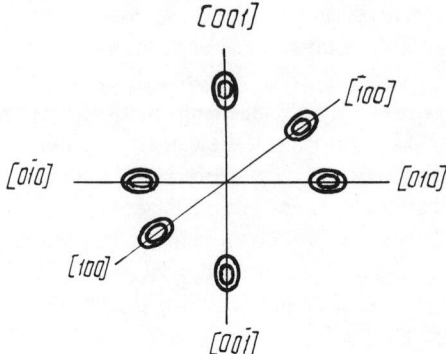

Figure 1.5. Energy minima of the conduction band for $BaTiO_3$ crystals.

degeneracy is reduced because of the tetragonal distortion of the lattice. The energy of the minima located in the direction of spontaneous polarization [001] is less than the other minima (located along the [100] and [010] axes) by an amount $V(P, \sigma_k)$. This leads to a redistribution of the conduction electrons among the minima.

Let the electrical conductivity be impurity in nature and caused by donors which are completely ionized and which have a concentration N and activation energy E. We denote the concentration of electrons in the minima of the conduction band by n_z, and, accordingly, $n_x = n_y$. Then the free energy of the electron subsystem can be written by analogy to (1.11):

$$F_2 = 2n_x E_g + n_z(E_g - V) - NE. \tag{1.77}$$

By using the redistribution condition

$$n_x = n_y = n_z \exp(-V/kT) \tag{1.78}$$

and the electrical neutrality condition

$$2n_x + n_z = N, \tag{1.79}$$

we obtain the following expression for the free energy:

$$F_2 = N\widetilde{E} - N \frac{V}{2\exp(-V/kT) + 1}, \tag{1.80}$$

where

$$\widetilde{E}(P, \sigma_h) = E_g - E. \tag{1.81}$$

Expression (1.80) for the free energy of the electron subsystem coincides with (1.11) with an accuracy of a correction term depending on the energy V. The correction which takes into account the multiple minima of the ferroelectric

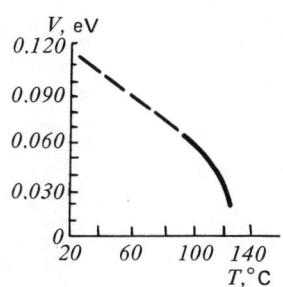

Figure 1.6. Experimental dependence $V(T)$ for crystals of $BaTiO_3$ [26].

semiconductor makes a significant contribution to F_2 only for $V \gg kT$. Actually, for $V < kT$, we have

$$N \frac{V}{2\exp\left(-\frac{V}{kT}\right)+1} \simeq N \frac{V}{3}\left[1 + \frac{2}{3}\frac{V}{kT} + \frac{2}{9}\left(\frac{V}{kT}\right)^2\right], \quad (1.82)$$

and since $V \ll \tilde{E}$, the correction is insignificant. For $V \gg kT$, the correction term is approximately NV and consideration of the multiple minima can be significant. Figure 1.6 presents the temperature dependence $V = V(T)$ for $BaTiO_3$ obtained in [26] from the data of measurements of the electrical conductivity. These data show that $V \simeq 0.03$ to 0.05 eV for $BaTiO_3$ near the phase transition, and the correction term is insignificant. However, it should be kept in mind that the effect of electrons on the phase transition is determined by derivatives of the energy with respect to the polarization, for example, $\partial^2 V/\partial P^2$, which can be comparable to the corresponding derivatives of the energy $\tilde{E}$ and, consequently, consideration of the multiminimum structure can be significant. However, just as significant a contribution may be made by the trapping levels [cf. (1.11)] if one takes into account the dependence of their energies u_1 and u_2 on polarization (1.75). If one also keeps in mind that several levels or the entire system of levels can contribute to the free energy in high-resistance semiconductor ferroelectrics, then consideration of the multiple minima is one of the many corrections which are difficult to isolate experimentally.

2

Microscopic Theory of Ferroelectric Semiconductors

2.1. Ferroelectric Phase Transition and the "Soft" Mode of Vibrations

It is well known that the mechanism of polymorphic transformations in crystals is related to the instability of atomic vibrations. In particular, the mechanism of ferroelectric phase transitions in the current dynamic theory is caused by the instability of one of the transverse optical vibrations at the center of the Brillouin band, where anharmonicity of the corresponding branch of vibrations is considered responsible for the instability. This mechanism was first put forward in the works of V. L. Ginzburg [29] and Cochran [41], who showed the correspondence between this microscopic approach and the thermodynamic theory of ferroelectrics. Before turning to the specifics of the dynamic theory of ferroelectric semiconductors, we will briefly discuss the properties of the optical vibrations in a ferroelectric at k = 0 and their relation to the ferroelectric phase transition.

We consider the equation of motion of an anharmonic oscillator under the effect of an external field $\mathcal{E}$:

$$m\ddot{\xi} + r\dot{\xi} + k\xi + s\xi^3 = q\mathcal{E}. \tag{2.1}$$

Here, m and q are the effective mass and charge, respectively, r, k, and s are coefficients, and ξ is the generalized coordinate of the atom. By considering a set of such oscillators with a concentration N and by defining the polarization

$$\tilde{P} = q\xi N,$$

we obtain, for a variable field $\mathcal{E} = \mathcal{E}_0 \exp(i\omega t)$,

$$\mu\ddot{\tilde{P}} + \nu\dot{\tilde{P}} + \alpha\tilde{P} + \beta\tilde{P}^3 = \mathcal{E}_0/2 \exp(i\omega t), \tag{2.2}$$

where

$$\mu = \frac{m}{2q^2N}, \quad \nu = \frac{r}{2q^2N}, \quad \alpha = \frac{k}{2q^2N}, \quad \beta = \frac{s}{2q^4N^3}.$$

If the external field is constant and $\omega = 0$, then (2.2) becomes the equation of state (1.16) already known from the thermodynamics of ferroelectrics:

$$\alpha \tilde{P} + \beta \tilde{P}^3 = \mathscr{E}_0/2. \qquad (2.3)$$

Thus, the polarization $\tilde{P}$ defined for the system of anharmonic oscillators (2.1) can be identified with the spontaneous polarization P of ferroelectrics. Hence, it follows in turn that the dielectric constant of a ferroelectric can be determined from (2.2) by analogy with (1.26) and (1.27):

$$\varepsilon = \varepsilon_\infty + \frac{2\pi}{\alpha - \mu\omega^2 + i\nu\omega}, \quad T > T_0,$$

$$\varepsilon = \varepsilon_\infty + \frac{\pi}{-\alpha - \mu\omega^2 + i\nu\omega}, \quad T < T_0, \qquad (2.4)$$

where α satisfies the temperature dependence (1.22).

By comparing (2.4) with equation (2.5) describing the frequency dispersion of the dielectric constant,

$$\varepsilon = \varepsilon_\infty + \frac{\varepsilon_0 - \varepsilon_\infty}{1 - \left(\frac{\omega}{\omega_{TO}}\right)^2 + i\gamma\left(\frac{\omega}{\omega_{TO}}\right)}, \qquad (2.5)$$

we arrive at the following temperature dependence for the frequency of the transverse optical vibration ω_{TO}:

$$\omega_{TO}^2 = \frac{\alpha}{\mu} = \frac{\alpha'_T}{\mu}(T - T_0), \quad T > T_0. \qquad (2.6)$$

Thus, consideration of the Ginzburg model of an anharmonic oscillator leads to the conclusion of the existence of an optical vibration near the Curie temperature, whose frequency depends anomalously on temperature and goes to zero at the Curie point. The corresponding branch of the vibrations has been called in the literature the *ferroactive* or *"soft"* mode of vibration. Since the use of the dispersion relations (2.4) and (2.5) are valid for long wavelengths $\lambda \gg a$ (a is the lattice constant), relation (2.6) corresponds to the center of the Brilluoin zone $\mathbf{k} = 0$. We note that the shift in the Curie temperature with a change in the concentration of electrons can be obtained from (2.6) and corresponds in this model to a change in the elastic constant of the oscillator α.

The "soft" mode of vibrations and the corresponding ferroelectric phase transition of the displacement type can be obtained from the general theory of atomic vibrations of a crystal lattice with consideration of anharmonicity (Cochran [41]).

As is well known, in the harmonic approximation the frequency dispersion of lattice vibrations, i.e., the dependence $\omega_i = \omega_i(\mathbf{k})$, where $\mathbf{k}$ is the wave vector ($|\mathbf{k}| = 2\pi/\lambda$), can be obtained from the following condition:

$$\left| B_{\alpha\beta}^{kk'} - m_k \cdot \omega^2 \delta_{kk'} \delta_{\alpha\beta} \right| = 0, \tag{2.7}$$

where $B_{\alpha\beta}^{kk'}$ are coefficients depending on $\mathbf{k}$ and the force constants, and $m_{k'}$ are the masses of the atoms. If a unit cell of the lattice contains s different atoms, then the characteristic equation (2.7) of degree $3s$ defines in the general case $3s$ solutions $\omega_i = \omega_i(\mathbf{k})$ ($i = 1, 2, \ldots, 3s$) or branches of vibrations. The characteristic equation (2.7) is simplified if one considers a diatomic lattice and takes into account all the elements of its symmetry. In this case, the problem reduces to determining the frequency dispersion of vibrations of a linear chain of atoms characterized by three branches of vibrations: one longitudinal and two transverse.

To obtain the "soft" transverse optical mode in the long-wave length approximation ($\mathbf{k} = 0$), Cochran considered a special model of atomic interaction in the diatomic crystal. The interaction forces between neighboring atoms in this model can be divided into Coulomb (long-range) depending on the polarization P and non-Coulomb (short-range). The introduction of the non-Coulomb interaction forces defines at once the anharmonicity of vibrations in this model. We take as a specific model two neighboring interacting ions: a positive ion with charge Zq and mass m_1, and a negative ion having a positive nucleus with charge Xq and mass m_2 and a negative electron cloud with charge Yq, where, from the condition of electrical neutrality, $X + Y + Z = 0$. Denoting the displacement from the equilibrium position of the positive ion, the nucleus of the negative ion, and the cloud of the negative ion by u_1, u_2, and v_2, respectively, we write the equations of motion in the form

$$m_1 \ddot{u}_1 = R_0(v_2 - u_1) + \frac{4\pi}{3} PZq,$$

$$m_2 \ddot{u}_2 = k_1(v_2 - u_2) + \frac{4\pi}{3} PXq, \tag{2.8}$$

$$k_1(u_2 - v_2) + R_0(u_1 - v_2) + \frac{4\pi}{3} PYq = 0.$$

Here, R_0 and k_1 are the force constants of short-range non-Coulomb interaction between the cloud and positive ion and between the cloud and nucleus of the negative ion, respectively, and $(4\pi/3)P$ is the Lorentz field. Substituting

$$u_1 = U_1 \exp(-i\omega_T_0 t), \quad u_2 = U_2 \exp(-i\omega_T_0 t),$$
$$v_2 = V_2 \exp(-i\omega_T_0 t) \tag{2.9}$$

into (2.8) and eliminating $(V_2 - U_2)$, we arrive at the system of equations

$$m_1\omega_{TO}^2 U_1 = R'_0(U_1 - U_2) - \frac{4\pi}{3} PZ'q,$$

$$m_2\omega_{TO}^2 U_2 = R'_0(U_2 - U_1) + \frac{4\pi}{3} PZ'q, \qquad (2.10)$$

where

$$R'_0 = \frac{k_1 R_0}{k_1 + R_0} < R_0, \quad Z' = Z + \frac{YR_0}{k_1 + R_0} < Z. \qquad (2.11)$$

We define the polarization P appearing in (2.10) in the following manner:

$$P = \frac{q}{v}(ZU_1 + XU_2 + YV_2), \qquad (2.12)$$

where v is the volume of the unit cell. By using the third equation of system (2.8), we write (2.12) in the form

$$P\left\{1 - \frac{4\pi(Yq)^2}{3v(k_1 + R_0)}\right\} = \frac{Z'q}{v}(U_1 - U_2). \qquad (2.13)$$

We now consider the behavior of the crystal in an external high-frequency electric field $\mathscr{E}$ under the effect of which the electron cloud is displaced, while the ions are fixed, which corresponds in turn to the high-frequency electron polarizability $\alpha = \alpha_\infty$. The equilibrium condition in this field [the third equation of system (2.8)], the expression for the polarization (2.12), and the expression for the electron polarizability $\alpha = \alpha_\infty$ have, respectively, the forms,

$$\mathscr{E}Y = (k_1 + R_0)V_2, \qquad (2.14)$$

$$P = \frac{q}{v}YV_2, \qquad (2.15)$$

$$\alpha_\infty = \frac{Pv}{\mathscr{E}} = \frac{(Yq)^2}{k_1 + R_0}. \qquad (2.16)$$

By using the Clausius–Mossoti equation

$$\frac{4\pi\alpha_\infty}{3v} = \frac{\varepsilon_\infty - 1}{\varepsilon_\infty + 2}, \qquad (2.17)$$

one can convert (2.13) to the form

$$P = \frac{Z'q(\varepsilon_\infty + 2)(U_1 - U_2)}{3v}. \qquad (2.18)$$

Substituting (2.18) into (2.10) and eliminating $U_1 - U_2$, we obtain the final expression for the frequency of the transverse optical vibration at $k = 0$:

$$\mu\omega_{TO}^2 = R'_0 - \frac{4\pi(\varepsilon_\infty + 2)(Z'q)^2}{9v}, \qquad (2.19)$$

where the reduced mass $\mu = m_1 m_2/(m_1 + m_2)$.

The frequency of the longitudinal optical vibration ω_{LO} at k = 0 can be obtained in an analogous manner if, in addition to the Lorentz field in the first part of equations (2.8), one takes into account the macroscopic field $4\pi P$:

$$\mu \omega_{LO}^2 = R_0' + \frac{8\pi (\varepsilon_\infty + 2)(Z'q)^2}{9v\varepsilon_\infty}. \tag{2.20}$$

Relations (2.19) and (2.20) must be supplemented by the Lidden-Sachs-Teller relation

$$\frac{\omega_{LO}^2}{\omega_{TO}^2} = \frac{\varepsilon_0}{\varepsilon_\infty} \tag{2.21}$$

and the relation between the force constant and the compressibility:

$$R_0 = 6 \frac{r_0}{\chi}, \tag{2.22}$$

where r_0 is the distance between the nearest atoms in the simplest structure of the NaCl type.

Comparison of (2.19) and (2.21) shows that Curie-Weiss law (1.28) is satisfied under the condition

$$R_0' - \frac{4\pi (\varepsilon_\infty + 2)(Z'e)^2}{9v} \sim (T - T_0). \tag{2.23}$$

This in turn shows that the frequency of the transverse optical vibration goes to

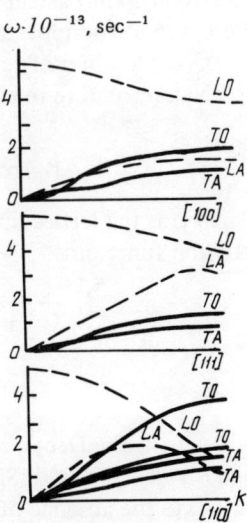

Figure 2.1. Dispersion of the modes of vibrations in a lattice of the NaCl type [47].

zero at the Curie point in the Cochran model and, consequently, the phase transition and dielectric anomalies are related to the existence of the "soft" mode of vibrations. Analysis of the left side of relation (2.23) shows that the temperature dependence of v and ϵ_∞ can be neglected and that, consequently, the temperature dependence of ω_{TO}^2 is caused by the strong dependence of the force constant R_0 on the amplitude of displacement of the ions. The latter indicates the significant anharmonicity of the vibrations, since the constant R_0 according to (2.22) and (1.68) is proportional to the Grüneisen constant. According to (2.20), the frequency of the longitudinal optical vibration does not go to zero at the Curie point. Thus, instability of one of the modes of vibration is sufficient for the phase transition, in particular, the ferroelectric. Figure 2.1 presents the optical and acoustical branches of vibration for $\chi = 7.16 \cdot 10^{-12}$ cgs esu, $\alpha_\infty = 0.01v$, $Z' = Z = 1$, and other parameters (for example, r_0) corresponding to the NaCl lattice.

2.2. Effect of Screening on the "Soft" Mode of Vibrations

With the presence in a ferroelectric semiconductor of a sufficiently high concentration of free carriers, the latter screen the Coulomb interaction of the lattice ions. In terms of the Cochran theory discussed above, this leads to a decrease in the Lorentz field (and also the macroscopic field) on the right side of equation (2.8), and this correspondingly shifts the temperature at which the frequency of the transverse optical mode of vibration goes to zero (shift of the Curie point). The temperature dependence of ω_{TO} with consideration of screening was calculated in [42]. It is significant to note that both in [42] and in other works devoted to the effect of screening on the Curie point [43-45], the electrons do not affect the anharmonicity of the vibrations, but only screen the Coulomb ion-ion interaction.

Consideration of screening in the Cochran model for an ionic lattice with the NaCl structure reduces relation (2.19) to the form

$$\mu \omega_{TO}^2 = R_0' - \frac{4\pi (\epsilon_\infty + 2)(Z'q)^2}{9v} G\left(\lambda \frac{a}{2}\right), \qquad (2.24)$$

where a is the lattice constant, λ is the screening constant, and $G(x)$ is the correlation function:

$$G(x) = \frac{1}{2\pi} \sum_{n,m,p} \frac{x^2}{(n^2 + m^2 + p^2)^{1/2}} \exp[-x(n^2 + m^2 + p^2)^{1/2}], \qquad (2.25)$$

$$l_D = \lambda^{-1} = \left(\frac{\epsilon kT}{8\pi q^2 n}\right)^{1/2}, \qquad (2.26)$$

where l_D is the Debye screening length, n is the concentration of free electrons, and $n + m + p$ is odd. Since $G(x) = 1$ for $x = 0$, relation (2.24) transforms to (2.19) in the absence of electrons ($\lambda = 0$).

At the Curie temperature, $\omega_{TO}^2 = 0$ (the "soft" mode). The shift of the Curie point caused by the electrons can be obtained from the relations

$$R_0' - \frac{4\pi(\varepsilon_\infty + 2)(Z'q)^2}{9v} = 0 \quad \text{for} \quad T = T_0, \tag{2.27}$$

$$R_0' - \frac{4\pi(\varepsilon_\infty + 2)(Z'q)^2}{9v} G\left(\lambda \frac{a}{2}\right) = 0 \quad \text{for} \quad T = T_0', \tag{2.28}$$

where $G[\lambda(a/2)]$ depends on the temperature, since λ and a are functions of the temperature.

To calculate T_0' from (2.28) and (2.27), it is necessary to determine the force constant R_0' as a function of temperature. Assuming, as has been emphasized above, that the electrons do not affect R_0', we determine the temperature dependence $R_0' = R_0'(T)$ from relations (2.19)-(2.21) and the Curie-Weiss law in the form

$$\varepsilon_0 = \varepsilon_\infty + \frac{C}{T - T_0}. \tag{2.29}$$

For $\epsilon_\infty \simeq 1$, these calculations lead to the following dependence for $R_0' = R_0'(T)$:

$$R_0' = \frac{4\pi(Z'q)^2}{3v}\left[\frac{3}{C}(T - T_0) + 1\right]. \tag{2.30}$$

If (2.30) is then substituted into (2.28), then, by neglecting the temperature dependence of the volume of the unit cell v, we obtain

$$T_0' = T_0 - \frac{C}{3}\left[1 - G\left(\lambda \frac{a}{2}\right)\Big|_{T=T_0'}\right]. \tag{2.31}$$

It is seen from (2.31) that since $dG/d\lambda < 0$, an increase in the concentration of free electrons n shifts the Curie point toward lower temperatures. This result was obtained above (Section 1.3) from thermodynamics [cf. (1.36)].

The principal difference between (2.31) and (1.36) involves the fact that N in (1.36) should be understood as the total concentration of carriers, both free as well as localized in deep traps, whereas in (2.31) the screening constant λ, according to (2.26), depends only on the concentration of free carriers n. For small traps in thermal equilibrium with the corresponding band, this difference is insignificant and λ depends on N [46]. However, for $n/kT \gg 1$, i.e., for sufficiently deep traps, $\lambda = \lambda(n)$ and, consequently, the physical meaning of (1.36) and (2.31) is different. This difference becomes particularly apparent with consideration of photoferroelectric phenomena, for example, the shift of the Curie point under the effect of nonequilibrium carriers.

The shift of the Curie point described by (2.31) is a consequence of screening of the internal field in the ferroelectric by free nonequilibrium carriers. The physical meaning of (1.36) involves the effect on the Curie temperature of the optical recharging of levels in the ferroelectric semiconductor. We will discuss

TABLE 1. Numerical Values of Function $G(x)$

x	$G(x)$	x	$G(x)$
0.1	0.99913	0.6	0.96937
0.2	0.99654	0.7	0.95859
0.3	0.99223	0.8	0.94633
0.4	0.98624	0.9	0.93268
0.5	0.97860	1.0	0.91770

one of the possible mechanisms for the effect of the optical recharging of levels on the phase transition in a ferroelectric in the following section. We will present some quantitative evaluations here. Table 1 presents numerical values of the function $G(x)$ for $0.1 \leqslant x \leqslant 1$. By using these values and equations (2.31) and (2.26), we find that for a concentration of free carriers $n = 10^{17}$-10^{18} cm^{-3} and $C \simeq 10^4$-10^5, the shift of the Curie point $\Delta T_0 = T_0 - T_0' = 1$-$10°$. Evaluation from equation (1.36) with consideration of (1.47) or (1.48) shows that the shift of the Curie point $\Delta T_0 = 1$-$10°$ is achieved for a concentration N of the same order (cf. Chapter 5). It is clear that such a high concentration of free nonequilibrium carriers is impossible in a ferroelectric, and n or N in the general cases should be taken as the concentration of the recharged levels (traps) in the crystal. As has already been noted above, the sign of the effect (the direction of the shift of the Curie point) is the same with screening and with optical recharging of the levels. Thus for experimental separation of these two mechanisms, conditions are necessary for which separation of the localized and free carriers is possible (cf. Chapter 5).

An attempt was made in [45] to take into account both free carriers and carriers captured by deep traps within the framework of the screening mechanism and to separate as much as possible their effect on the temperature of the phase transition. The meaning of the approximation used in [45] involves the fact that the screening length (2.26) depends on the dielectric constant ε, while the dielectric constant of the semiconductor in turn depends on the concentration of free and localized carriers. Actually, close to the center of the Brillouin zone (for small values of $|\mathbf{k}| = 2\pi/\lambda$) [47], we have

$$\varepsilon(k) = \varepsilon_\infty + \delta + \frac{\lambda_0^2}{|\mathbf{k}|^2}, \tag{2.32}$$

where

$$\delta = \frac{4\pi\hbar^2 q^2}{u^2 m^*} N, \quad \lambda_0^2 = \frac{8\pi q^2 n}{kT}. \tag{2.33}$$

Here, ε_∞ is the electron dielectric constant in the absence of carriers ($n = N = 0$), n is the concentration of carriers in the band, N is the concentration of carriers at levels with an activation energy u, and m^* is the effective mass of the electrons (holes) localized at the levels. If one further assumes that screening of the ion–ion interaction in the presence of localized electrons is not related to the anharmonicity mechanism and does not affect it, then consideration of the localized carriers can be made by substituting into the expression (2.24) for the frequency of the transverse optical phonon the screening constant λ with consideration of (2.32) and (2.33):

$$\lambda = \lambda_0/(\varepsilon_\infty + \delta)^{1/2}. \quad (2.34)$$

Substitution of (2.34) into (2.24) leads to the following expression for the shift of the Curie point:

$$T_0' = T_0 - \frac{C}{3\varepsilon_\infty}\left[1 - \left(1 - \frac{\delta}{\varepsilon_\infty}\right)G\left(\lambda\frac{a}{2}\right)\right]\bigg|_{T=T_0'}. \quad (2.35)$$

This result permits one to analyze the effect on the Curie point of the free and localized carriers separately. In the absence of localized carriers, $N = 0$ and $\varepsilon_\infty = 1$, (2.35) transforms to (2.31). We now consider the opposite case where the concentration of free carriers can be neglected as compared to the concentration of localized carriers: $n \ll N$. It is this case that is realized in high-resistance ferroelectric semiconductors. We have in this case, from (2.35),

$$T_0' = T_0 - 4\pi\left(\frac{\hbar q^2}{u\varepsilon_\infty}\right)^2 \frac{C}{3m^*}N. \quad (2.36)$$

We draw attention to the fact that the linear dependence of the shift of the Curie point on the concentration of localized electrons N agrees with the linear dependence (1.36) following from thermodynamics. Substituting $N = 10^{-18}$ cm^{-3}, $u = 0.5$ eV, $C = 10^{5}$ °K, and $\varepsilon_\infty = 10$ into (2.36) (cf. Chapter 5), we obtain $|\Delta T_0| = T_0 - T_0' = 9m/m^*$, where m is the mass of the free electron or polaron. For localized carriers, $m/m^* < 1$. In any case, (2.36) agrees qualitatively with the experimental data, which we will discuss in more detail in Chapter 5. Nonetheless, a quantitative separation of the cases of free and localized carriers is not possible on the basis of (2.35), in particular, because of the difficulties in evaluating the parameter m/m^*.

2.3. Ferroelectric Phase Transition and the Interband Electron–Phonon Interaction

The screening mechanism and the participation of the electron susbsystem of the crystal in the phase transition considered in the preceding section were related to a change in the Coulomb ion–ion interaction and did not affect the short-range forces in the Cochran model as a cause of vibration anharmonicity.

This led to renormalization of the frequency of the transverse optical phonon and the corresponding shift of the Curie point with a constant value of the Curie-Weiss constant. Meanwhile, it is clear that the effect of the electrons may also not reduce to such a simple effect, but touch directly on the anharmonicity mechanism responsible for the phase transition. It is not possible to take into account the effect of the electrons on the anharmonicity of the ferroactive mode of vibrations within the framework of the Cochran theory, since the cause of the anharmonicity remains undisclosed in this area.

It was shown in the works of Kristofel', Konsin, and Bersuker [48-56] that, at least for ferroelectric phase transitions, the microscopic mechanism responsible for the direct effect of electrons on the phase transition is the interband electron-phonon interaction. Since the theory of ferroelectric phase transitions has already been considered in the main features, both on the basis of lattice dynamics as well as on the basis of the phenomenological Landau-Ginzburg theory, we will briefly discuss here only the role of the interband electron-phonon interaction in the mechanism of phase transitions. As a consequence, we will consider several effects related to the effect of electron excitations on phase transitions.

It was shown in Section 2.1 that the anharmonicity of one of the transverse optical oscillations at $k = 0$ and its corresponding instability are the cause of the ferroelectric phase transition. One of the possible mechanisms responsible for the instability of vibrations and the corresponding phase transition is the interband electron-phonon interaction. This mechanism involves the interaction of electrons of two neighboring bands of the crystal with one of the optical vibrations, where one of the bands is empty (or nearly empty), while the other is completely filled with electrons. This interaction or the "mixing" of neighboring energy bands leads, on the one hand, to a change in the frequency of the interacting optical phonon and, on the other, to a change in the electronic spectrum (the width of the forbidden band). This interaction of filled and empty bands leads to instability of the "mixing" optical vibration and a corresponding phase transition from a symmetric to a less symmetric phase. It is known from quantum chemistry that the interactions of degenerate and nondegenerate electronic levels of molecules lead to a transition of the latter from a symmetric to an asymmetric configuration (the Jahn-Teller effect). It is clear from the discussion why the phase transition caused by the interband electron-phonon interaction has been called in [48-56] the pseudo-Jahn-Teller effect.

Let us consider two neighboring energy bands of a crystal denoted by the indices $\sigma = 1, 2$, with boundary energies E_1 and E_2 and the corresponding width of the forbidden band $E_{g0} = E_2 - E_1$. Let some active optical vibration with coordinate u and frequency ω interact with electrons in the bands. Neglecting the dependence of E and u on k, i.e., neglecting the dispersion of the bands, we write the Hamiltonian of the crystal in the form

$$H = \sum_\sigma E_\sigma a_\sigma^+ a_\sigma + \frac{1}{2}\left(-\frac{\hbar^2}{M}\frac{\partial^2}{\partial u^2} + M\omega^2 u^2\right) + \sum_{\sigma\sigma'} \frac{V_{\sigma\sigma'}}{\sqrt{N}} a_\sigma^+ a_{\sigma'} u. \quad (2.37)$$

Here, a_σ^+ and a_σ are the electron creation and annihilation operators, $V_{\sigma\sigma'}$ are the constants of the interband electron–phonon interaction, M is the corresponding mass factor, and N is the number of electrons in the lower band (equal in order of magnitude to the number of unit cells). Analysis of (2.37) shows that consideration of the interband interaction leads to renormalization of the electronic spectrum:

$$E_{1,2}^* = \frac{E_1 + E_2}{2} \mp \sqrt{E_{g0}^2 + \frac{4V^2}{N} u^2}, \quad (2.38)$$

where $V \equiv V_{12}$. It follows directly from (2.38) that the interband electron–phonon interaction changes the width of the forbidden band:

$$E_g = 2\sqrt{\frac{1}{4} E_{g0}^2 + \frac{V^2}{N} u_0^2(T)}. \quad (2.39)$$

Here, E_g is the width of the forbidden band in the presence of the pseudo-Jahn-Teller effect, and $u_0(T)$ is the coordinate of the active vibration corresponding to the minimum of the free energy of the crystal. The free energy of the crystal can be represented in the form of the sum of the vibrational energy F_ω corresponding to the active optical vibration and the energy of the electron subsystem F_e:

$$F = F_\omega + F_e, \quad (2.40)$$
$$F_\omega = 1/2 M\omega^2 u^2, \quad (2.41)$$

$$F_e = \sum_\sigma \left\{ n_\sigma E_F - kT \int \ln\left[1 + \exp\left(\frac{E_F - E_\sigma^*}{kT}\right)\right] g(E_\sigma^*) dE_\sigma^* \right\}, \quad (2.42)$$

$$n_\sigma = kT \frac{\partial}{\partial E_F} \int \ln\left[1 + \exp\left(\frac{E_F - E_\sigma^*}{kT}\right)\right] g(E_\sigma^*) dE_\sigma^*, \quad (2.43)$$

where $g(E_\sigma^*)$ is the function of the density of states, $n_1 + n_2 = N$, and E_F is the energy of the Fermi level. One must substitute in the expression for F_e the value of the renormalized energies $E_{1,2}^*$ according to (2.38) where the interband interaction constant appears.

In the approximation of completely filled and empty bands, the following expression for the free energy F is obtained:

$$F(T, u) = \frac{N E_{g0}}{2} -$$
$$- NkT \ln\left\{2\left[1 + \cosh\left(\frac{1}{kT}\sqrt{\frac{1}{4} E_{g0}^2 + \frac{V^2}{N} u^2}\right)\right]\right\} + \frac{M\omega^2}{2} u^2. \quad (2.44)$$

It is seen from (2.44) that consideration of the interband electron–phonon interaction leads to anharmonicity of the branch of vibrations under consideration, since terms with higher powers of u are present in (2.44) along with the quadratic term u^2.

We denote by $u = u_0$ the coordinate of the vibration corresponding to the minimum of the free energy F:

$$\frac{dF}{du_0} = 0. \tag{2.45}$$

Then the following expression is obtained for u_0 from (2.44):

$$u_0^2 = N\left\{\frac{[f_1(u_0) - f_2(u_0)]^2}{M^2\omega^4}V^2 - \frac{1}{4}\frac{E_{g0}^2}{V^2}\right\}, \tag{2.46}$$

where the electron populations of the bands f_1 and f_2 are the Fermi function

$$f_{1,2}(u_0) = \left\{\exp\left[\mp\frac{1}{kT}\left(\frac{1}{4}E_{g0}^2 + \frac{V^2}{N}u_0^2\right)^{1/2}\right] + 1\right\}^{-1}. \tag{2.47}$$

The solutions of (2.46) and (2.47) determine the temperature dependence $u_0 = u_0(T)$ (cf. Figure 2.2). For $T = 0$, we have $f_1 = 1$ and $f_2 = 0$, and it follows from (2.46) that u_0^2 at $T = 0$ is

$$u_0^2(0) = N\left[\frac{V^2}{(M\omega^2)^2} - \frac{E_{g0}^2}{4V^2}\right]. \tag{2.48}$$

With increasing temperature, u_0^2 decreases monotonically from $u_0^2(0)$ to zero. The function $u_0^2(T)$ vanishes at the temperature $T = T_1$, which plays the role of the temperature of the phase transition:

$$kT_1 = \frac{E_{g0}}{4}[\text{arctanh}\,\tau]^{-1}, \tag{2.49}$$

where

$$\tau = \frac{2V^2}{M\omega^2 E_{g0}}. \tag{2.50}$$

It is seen from (2.48) that in order that the value of u_0 be real and the phase transition occur, it is necessary that

$$\tau > 1. \tag{2.51}$$

Thus, τ is the fundamental parameter of the theory of the interband electron–phonon interaction. Condition (2.51) means that for given values of $M\omega^2$ and width of the forbidden band E_{g0}, the phase transition occurs only for a sufficiently strong interband interaction.

The nature of the phase transition based on the mechanism under consideration is determined by the nature of the active vibration. If dispersion is taken into consideration, then the solution $u_0 = u_0(\mathbf{k})$ corresponding to an arbitrary point of the Brillouin zone can satisfy condition (2.45). If $u_0 \neq 0$ for $\mathbf{k} = 0$, then this corresponds to a change in the structure and symmetry of the crystal due to the relative shift of the sublattice. If the crystal is ionic and the shift of the sublattice leads to the generation of spontaneous polarization, then the phase transition is ferroelectric, and the point $T = T_1$ is the Curie point. If $u_0 \neq 0$ at the boundary of the Brillouin zone, then at $T = T_1$ a transition takes place to the antiferroelectric phase. If the phonon inside the Brillouin zone is the active phonon, then the phase transition may be different and, in particular, not ferroelectric. For ferroelectric transitions, expansion of the free energy F determined by (2.44) in a series in even powers of u near $u = 0$ leads to a series analogous to that used in the phenomenological Landau–Ginzburg theory (cf. Section 1.2). In this case the relation between the spontaneous polarization P and u_0 is given by the simple relation

$$P = \frac{\bar{q}}{v} \frac{u_0}{\sqrt{N}}, \qquad (2.52)$$

where $\bar{q}$ is the effective charge corresponding to the active optical vibration and v is the volume of the unit cell. The temperature dependence shown in Figure 2.2 in the case of the ferroelectric phase transition is thereby the dependence of the square of the spontaneous polarization on temperature (cf. Section 1.2). The coefficients in the expansion of the free energy F in the polarization P (or in u) can be expressed in terms of the parameters of the microscopic theory, in particular, in terms of the interband electron photon interaction constant V.

Renormalization of the coefficients of u^2 in the expression for the free energy (2.44) leads to a change in the frequency of the active mode of vibrations ω. For the high-symmetry phase

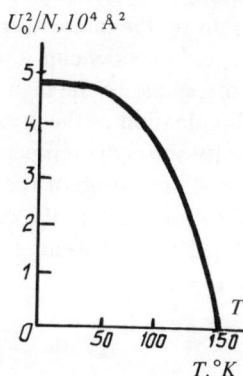

Figure 2.2. Dependence $u_0^2 = u_0^2(T)$ in the low-symmetry phase.

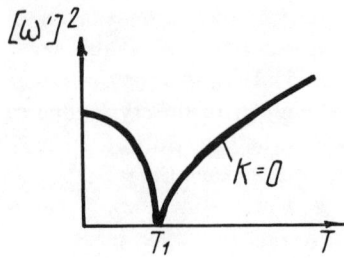

Figure 2.3. Temperature dependence of the "soft" mode.

$$[\omega']^2 = \omega^2 + \frac{2V^2}{ME_{g0}} [f_2(0) - f_1(0)]. \tag{2.53}$$

It follows from (2.53) and (2.49) that the frequency ω' vanishes at the temperature of the phase transition $T = T_1$ (cf. Figure 2.3). The interband electron-phonon interaction with satisfaction of condition (2.51) thereby leads to instability of the vibration interacting with the bands. The "soft" optical mode with frequency ω' at $k = 0$ is responsible for the ferroelectric phase transition (Section 2.1). Thus we see that the presence of the "soft" mode of vibration is caused by an anharmonicity of the vibrations near the phase transition. However, the anharmonicity itself in the model under consideration is a consequence of the electron–phonon interaction.

The mechanism discussed above permits one to investigate, on the basis of (2.44), (2.46), and (2.47), all the basic properties of the crystal, in particular, the ferroelectric properties near the phase transition. Since the latter were considered specially in Section 1.2, we will limit ourselves here to the effects of electrons, which are a direct consequence of the interband electron–phonon interaction.

1. The model under consideration permits one to explain the whole set of photoferroelectric phenomena, for example, the effect of nonequilibrium electrons on the temperature of the phase transition. We consider a photoconducting crystal experiencing a phase transition at $T = T_1$, where the temperatures $T > T_1$ correspond to the high-symmetry phase. It follows from thermodynamics that illumination of the crystal in the spectral region where it displays photoconductivity shifts the temperature of the phase transition downward (cf. Section 1.3). Let illumination of the crystal convert electrons from the lower energy band to the upper so that the concentration of electrons in the upper band increases by Δn. Then, according to (2.49), a shift in the temperature of the phase transition ΔT occurs toward lower temperatures by an amount

$$\Delta T = \frac{E_g}{4k} \left[\left(\operatorname{arctanh} \frac{\tau}{1 + 2\Delta n\tau} \right)^{-1} - (\operatorname{arctanh} \tau)^{-1} \right]. \tag{2.54}$$

The physical meaning of this effect involves the fact that with photoactive absorption of light, the population of the electron bands changes and, consequently, the contribution to the free energy caused by the interband electron–phonon interaction changes. As has already been indicated above, not only does the concentration of free electrons in the band change with photoactive illumination, but the population of all the trapping levels in the forbidden band caused by impurities or defects also changes. Attention has already been drawn to the fact that the shift in temperature of the phase transition for high-resistance ferroelectric semiconductors must be "impurity" in nature, since the concentration of electrons at the levels and its change with illumination can be several orders of magnitude higher than in the band. In contrast to the mechanism based on screening (Section 2.2), there is no fundamental difference between the participation of the free and localized carriers in the mechanism under consideration. Condition (2.51) remains a natural criterion for the applicability of this mechanism to localized electrons, which imposes limitations on the activation energy of the corresponding levels with a given value of the electron–phonon interaction constant.

We also note that equation (2.49) permits one to relate the Curie temperature of the ferroelectric T_1 to the width of the forbidden band E_{g0}. The results agree qualitatively with experiment. For example, it is sufficient to compare the Curie temperature of the narrow-band ferroelectric semiconductor GeTe and the wide-band dielectric $BaTiO_3$.

2. Starting from (2.39) and (2.52), one can expand the width of the forbidden band of the crystal E_g in a series in even powers of u_0 or P. Retaining the quadratic term, we have

$$E_g = E_{g0} + \frac{a}{2} P^2, \qquad (2.55)$$

where the constant $a = 4(v^2/\bar{q}^2)(V^2/E_{g0})$ is proportional to the square of the electron–phonon interaction constant. Thus, the expansion (1.46), considered above as a result of the thermodynamic interpretation of the width of the forbidden band, is a consequence of renormalization of the electronic spectrum caused by the interband electron–phonon interaction. Equation (2.55), as was indicated in Section 1.4, describes the temperature dependence and anomalies of the width of the forbidden band in the region of first- and second-order phase transitions. Actually, for first-order phase transitions, P^2 experiences a discontinuity, and the width of the forbidden band correspondingly experiences a finite discontinuity $\Delta E_g = E_g - E_{g0} \simeq (a/2) P_0^2$. For second-order phase transitions, $P \propto (T - T_0)$ and the temperature coefficient of the width of the forbidden band thus experiences a finite discontinuity. Thus, the change in $E_g = E_g(T)$ can give a direct check of the existence of the pseudo-Jahn–Teller effect in ferroelectrics. From the discontinuity ΔE_g, one can evaluate with the help of equation (2.55) the electron–phonon interaction constant for $BaTiO_3$:

$$V \simeq 0.6 \text{ eV}/\overset{\circ}{A}$$

($\Delta E_g \simeq 0.02$ eV, $P_0 \simeq 20$ $\mu C/cm^2$, $v_0 \simeq 64$ $\overset{\circ}{A}{}^3$, $\bar{q} \simeq 2.4$ q).

2.4. Pseudo-Jahn–Teller Effect in Wide-Band and Impurity Ferroelectrics

In conclusion, we turn to the problem of the applicability of the interband electron–phonon interaction model to narrow-band and wide-band ferroelectrics. Condition (2.51) imposes a limitation on E_{g0}: for fixed values of the electron–phonon interaction constant V and the corresponding anharmonicity of vibrations determined by the frequency ω, the width of the forbidden band E_{g0} must not be too large. The physical meaning of this limitation involves the following. For ferroelectrics with a sufficiently narrow forbidden band, the natural anharmonicity of vibrations does not play a significant role. A phase transition from the high-symmetry paraelectric phase to the low-symmetry ferroelectric phase is related to the anharmonicity of the optical mode of vibrations mixing the bands. In this case, the temperature dependences of the frequency of the "soft" mode, the spontaneous polarization, the dielectric constant, and other macroscopic parameters are determined according to (2.47) by the population of neighboring bands. Let us now imagine a wide-band ferroelectric for which the parameter E_{g0}/kT is so large that the thermal transmissions of electrons of the lower band to the upper can be neglected for all real temperatures including the temperature of the phase transition. In this case, $f_1 = 1$ and $f_2 = 0$, and we have from (2.53)

$$(\omega')^2 = \omega^2 - \frac{2V^2}{ME_{g0}}. \qquad (2.56)$$

According to (2.51) and (2.56), $(\omega')^2 < 0$ for a wide-band ferroelectric over the entire temperature interval including the temperature of the phase transition. Thus, the anharmonicity of the optical vibration at $k = 0$ in this case, although it is also related to the interband electron–phonon interaction, can actually be considered as a natural anharmonicity (by analogy to the natural anharmonicity in the Cochran theory [41] or in the theory of Vaks [6]). In this case, as was shown by Kristofel' and Konsin [54], the temperature of the phase transition and the temperature dependences of all the macroscopic parameters are determined by mixing of the active optical vibration with the acoustical vibrations with consideration of the natural anharmonicity. The temperature dependences thus obtained correspond to a first-order phase transition and satisfactorily describe the phase transition in such a wide-band ferroelectric as, for example, barium titanate.

In this regard, the possible role of local levels in the forbidden band of the ferroelectric should once again be emphasized. It is not excluded that the phase transition in wide-band ferroelectrics is related to the band-level interaction and E_{g0} should be taken as the activation energy of corresponding levels. For such an "impurity" ferroelectric, the Curie temperature and the temperature dependences of the parameters are determined by the population of the level and the nearest band.

Along with "mixing" of neighboring energy bands by active optical vibrations in an impurity ferroelectric, a significant role can be played by the mechanism of the interaction of two (or more than two) local levels (centers) with a phonon corresponding to the local vibrations of the centers. This mechanism was developed in detail in [57] using barium titanate with an iron impurity as an example.

The iron with not too great concentrations enters the lattice of the perovskite ferroelectric so that the charge of two ions of Fe^+ are compensated by one oxygen vacancy. One ion then enters one oxygen octahedron, forming the complex FeO_6, while the other ion replaces titanium in another oxygen octahedron, forming the complex FeO_5 + oxygen vacancy (cf. Figure 2.4). These impurity centers are responsible for the donor and acceptor levels in the forbidden band of $BaTiO_3$. The transition of an electron from the donor level FeO_6 to the acceptor level FeO_5 and the interaction of the recharged centers with one of the local modes of vibrations leads, because of the pseudo-Jahn–Teller effect, to instability of the corresponding modes and to a change in symmetry of the impurity complexes. The symmetry of the FeO_6 and FeO_5 + oxygen vacancy centers changes accordingly ($O_h \to C_{4v}$ and $C_{4v} \to C_s$). This in turn leads to deformation of the cells and makes a contribution to the spontaneous polarization of the crystal. Thus, along with the "intrinsic" pseudo-Jahn–Teller effect related to interband electron transitions, one can observe an "impurity" Jahn–Teller effect in ferroelectrics. The latter can be responsible for the effect of the impurity and

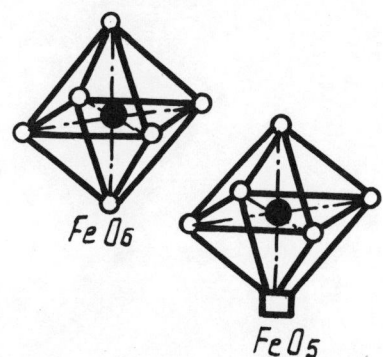

Figure 2.4. Iron impurity centers in $BaTiO_3$.

its optical recharging because of the impurity photoeffect on the phase transitions and the ferroelectric properties near the Curie temperature. Thus, the existence of a number of photoferroelectric phenomena in barium titanate with iron impurity is explained in [57] with the help of this mechanism, for example, the effect of photoactive illumination on the domain structure, broadening of the phase transition, etc. We will discuss this in more detail in Chapters 5 and 6.

3
Screening of Spontaneous Polarization

The subject of the first two chapters was the role of the electron subsystem in the mechanism of ferroelectric phase transitions and the generation of spontaneous polarization. The ferroelectric was considered as an infinite homogeneous crystal. The boundary conditions must be determined for the real ferroelectric, i.e., the values of the polarization P (or the induction D) must be specified at the surface. The boundary conditions in turn determine the screening of the spontaneous polarization at the surface of the ferroelectric. Actually, we imagine a crystal of a ferroelectric in the form of an infinite homogeneous sheet of thickness of L placed in vacuum. Further, let the spontaneous polarization equaling P_0 be homogeneous everywhere, including the surface of the crystal. Then there exists a homogeneous field $4\pi P_0$ outside the crystal, and the energy of the system as a whole is infinite. Thus, it is clear from energy considerations that the normal component of the spontaneous polarization (or the spontaneous induction) must be zero at the surface of the ferroelectric; in other words, there must be screening.

In the absence of free charges inside the crystal (an ideal dielectric), screening of the spontaneous polarization can occur due to the external medium (adsorption of ions or other charged particles). The screening must be "internal" for ferroelectric semiconductors and accomplished accordingly by electrons and holes. Then, as will be shown below, the division of ferroelectrics from the point of view of the screening mechanism into dielectrics and semiconductors is meaningless in a majority of cases. The surface curvature of the bands related to screening must be large even in wide-band ferroelectrics, which leads to a sharp change in the surface electron conductivity as compared to the volume conductivity ("intrinsic" field effect). Thus, even in wide-band ferroelectrics, which are traditionally considered in the literature as ideal ferroelectrics, the screening can be "internal" and be purely electronic in nature. We add that even the presence of surface electronic levels can introduce a significant or even decisive contribution to the screening. For a ferroelectric with electrodes applied to its surface and short-circuited, it would appear that screening is provided by charges on the electrodes flowing from the external circuit. However, the presence of a layer with peculiar dielectric properties (dielectric "gap") on the surface of the ferroelectric makes the effect of screening due to electrodes weak and we thus return to the electron mechanism of "internal" screening.

The subject until now has been a single-domain ferroelectric. Screening, which decreases the free energy of the single-domain crystal, makes energetically possible the existence of single-domain crystals. The solution of the problem of screening of spontaneous polarization of a single-domain crystal determines the coordinate dependence of the curvature of the bands, the induction, and the internal field in the ferroelectric. These functions in turn determine, on the one hand, the physical properties of the ferroelectric as an n- or p-type semiconductor and, on the other hand, such important "classical" properties of the ferroelectric as the polarization reversal mechanism, the distribution and magnitude of the depolarization field, and the critical size of the ferroelectric crystal. If screening of the single-domain crystal does not correspond to the minimum of its free energy, it splits up into domains. The screening in the number of other factors thereby determines the domain structure of the ferroelectric.

Thus, the participation of the electrons of the ferroelectric in the screening of spontaneous polarization determines the basic properties of the crystal as a ferroelectric and semiconductor. This chapter is devoted to an analysis of screening of spontaneous polarization.

3.1. Single-Domain Crystal in the Absence of Surface Levels

The problem of screening of spontaneous polarization of a single-domain ferroelectric in the absence of surface levels was considered successively in the works of Ivanchik [58], Guro [59, 60], and Chenskii [61, 62]. As applied to a ferroelectric with a first-order phase transition, this problem in the one-dimensional case is formulated with the help of the following equations [60][†]:

$$\frac{1}{4\pi}\mathscr{E} = \frac{dF}{dD}, \qquad (3.1)$$

$$\frac{\partial \sigma_h}{\partial z} = 0, \qquad (3.2)$$

$$F = F_0 + \frac{g}{4}D_0^4 + \frac{\varkappa}{2}\left(\frac{dD}{dz}\right)^2 + \frac{a}{2}D^2 + \frac{c}{4}D^4 + \frac{f}{6}D^6, \qquad (3.3)$$

[†]In contrast to the first chapter, here and in a number of cases subsequently, the expansion of the free energy of the crystal F is carried out not in the polarization P, but in the induction D. Actually, we find from the relation $D = \mathscr{E} + 4\pi P$ that $D \approx 4\pi P$ for a ferroelectric, and thus both expansions coincide with an accuracy to coefficients differing by a factor of 4π in the corresponding powers. The derivative with respect to the induction in (3.1) is, generally speaking, variational [63].

$$\frac{dD}{dz} = 4\pi\rho, \qquad (3.4)$$

$$\rho = q\{p_0(\varphi) - n_0(\varphi)\}, \qquad (3.5)$$

$$D|_{z=\pm L/2} = 0, \quad D|_{z=0} = D_0, \quad \varphi|_{z=0} = 0. \qquad (3.6)$$

Here, F is the free energy, D is the spontaneous induction, $\mathscr{E}$ is the field, σ_k are the components of the stress tensor, D_0 is the spontaneous induction at the center of the crystal where there are no stresses, ρ is the charge density, φ is the potential, p_0 and n_0 are the density of holes and electrons, respectively, and $(\varkappa/2)(dD/dz)^2$ is the correlation term describing the energy gradient of the crystal. The coefficient g is determined by the elastic properties of the ferroelectric. The crystal is an infinite homogeneous sheet of thickness L, on whose surface the boundary conditions (3.6) are satisfied. For $\sigma_k = 0$, the expression for the free energy (3.3) coincides, with an accuracy to the energy gradient, with the expansion (1.14), while the equation of state (3.1) coincides in accuracy with equation (1.16). Substitution of (3.3) into (3.1) gives an equation of state of the form

$$\frac{1}{4\pi}\mathscr{E} = -\varkappa\frac{d^2D}{dz^2} + aD + cD^3 + fD^5. \qquad (3.7)$$

We neglected the correlation term in the first chapter with consideration of the thermodynamics of an infinite crystal. It will be shown below that this term can also be neglected in the problem of screening for not too high concentrations of carriers. The equation of state (3.1), the Poisson equation (3.4), the expression for the charge density (3.5), and the boundary conditions (3.6) thus determine the distribution inside the crystal of the spontaneous induction $D = D(z)$, the depolarization field $\mathscr{E} = \mathscr{E}(z)$, the screening charge $\rho = \rho(z)$, and the curvature of the bands $\varphi = \varphi(z)$. Following [60], we indicate some characteristic features in these distributions, both for intrinsic as well as for impurity ferroelectric semiconductors.

We first consider the case in which the ferroelectric is an intrinsic semiconductor. This case corresponds to the band diagram presented in Figure 3.1a. In accordance with the Poisson equation (3.4) and the boundary conditions (3.6), screening of the spontaneous induction leads to the generation of a space charge $\rho = \rho(z) \neq 0$ and the corresponding field $\mathscr{E} = \mathscr{E}(z) = -d\varphi/dz \neq 0$. If the curvature of the bands $\varphi = \varphi(z)$, at least in the volume of the crystal, is not too large and satisfies the condition

$$E_c - E_F - q\varphi \gg kT, \quad E_F - E_v + q\varphi \gg kT, \qquad (3.8)$$

where E_F is the energy of the Fermi level, E_c is the energy of the bottom of the conduction band, E_v is the energy of the top of the valence band, then the

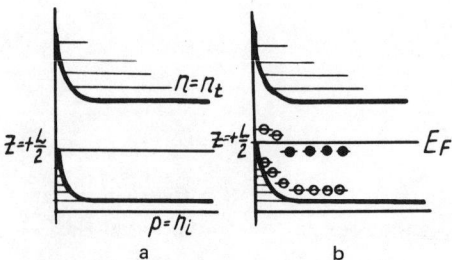

Figure 3.1. Curvature of the bands in the intrinsic and impurity ferroelectric semiconductor.

density of the space charge ρ for the intrinsic semiconductor, according to (3.5), can be represented in the following manner:

$$\rho = qn_i \exp\left(\frac{q\varphi}{kT}\right) - qn_i \exp\left(-\frac{q\varphi}{kT}\right) = -2qn_i \operatorname{sh}\frac{q\varphi}{kT}, \quad (3.9)$$

where n_i is the equilibrium concentration of carriers in the quasi-neutral region of the intrinsic semiconductor. According to [60], the degeneracy of the order of kT in the surface region still permits one to use relation (3.9). By combining (3.7), (3.4), and (3.9), Gurs et al. [60] arrive at the equation describing the distribution of the spontaneous induction:

$$\left\{\varkappa + \frac{kT}{32\pi^2 q^2 n_i}\left[1 + \left(\frac{1}{8\pi q n_i}\frac{dD}{dz}\right)^2\right]^{-1/2}\right\}\frac{d^2D}{dz^2} = aD + cD^3 + fD^5. \quad (3.10)$$

Successive integration of (3.10) gives the relation between the density of the space charge ρ and the spontaneous induction D,

$$\rho = \frac{D_0}{4\pi}\frac{1}{\sqrt{\varkappa\zeta}}\sqrt{1 + 2\xi + \zeta Q\left(\frac{D}{D_0}\right)} - \sqrt{(1 + 2\xi)^2 + 2\zeta Q\left(\frac{D}{D_0}\right)}, \quad (3.11)$$

and the dependence of the spontaneous induction on the coordinate $D = D(z)$,

$$z = \sqrt{\varkappa\zeta}\int_0^{D/D_0}\frac{dt}{\sqrt{1 + 2\xi + \zeta Q(t)} - \sqrt{(1 + 2\xi)^2 + 2\zeta Q(t)}}. \quad (3.12)$$

The following notation is assumed in (3.10)–(3.12):

$$\xi = \frac{\varkappa}{l_D^2}, \quad \zeta = \frac{8\pi^2 D_0^2 \varkappa q^2}{(kT)^2}, \quad t = \frac{D}{D_0},$$

$$Q(t) = (1 - t^2)\left\{\frac{1}{4\pi\varepsilon_0} + (1 - t^2)\left(\frac{cD_0^2}{2} + \frac{fD_0^4}{3}\right) + (1 - t^4)\frac{fD_0^4}{3}\right\}. \quad (3.13)$$

The parameters $\frac{1}{4}\pi\varepsilon_0$ and D_0 in (3.13) are determined from the system of equations

$$\frac{1}{4\pi\varepsilon_0} = -(a + cD_0^2 + fD_0^4), \quad \frac{L}{2} = z(1). \tag{3.14}$$

Here, $z(D/D_0)$ is the function defined by (3.12) and $l_D = (1/4\pi)(kT/q^2 n_i)^{1/2}$ is the Debye screening radius.

If it is assumed that the Debye radius is many times greater than the thickness of the crystal $l_D \gg L$, then the correlation length is the parameter of the distribution of the induction and field in the ferroelectric. It was in this approximation that the solutions of (3.11) and (3.12) obtained by the authors were analyzed in [60]. The correlation length can be introduced in the following manner. We determine from (3.11) the surface charge ρ_s corresponding to $D \simeq 0$ and $t \simeq 0$. Keeping in mind that $\zeta Q(0) \gg 1$, we find from (3.11)

$$\rho_S \simeq \frac{D_0}{4\pi}\left[\frac{Q(0)}{\varkappa}\right]^{1/2} \tag{3.15}$$

Defining the correlation length

$$l_\varkappa = \left[\frac{\varkappa}{Q(0)}\right]^{1/2}, \tag{3.16}$$

we rewrite (3.15) in the form

$$n_S = \frac{1}{q}\rho_S \simeq \frac{D_0}{4\pi q l_\varkappa}. \tag{3.17}$$

The correlation length and surface concentration of free carriers $l_\varkappa \simeq 2 \cdot 10^{-7}$ cm and $n_s \simeq 5 \cdot 10^{20}$ cm^{-3} were evaluated for BaTiO$_3$ ($\varkappa \simeq 10^{-15}$ cm^2, $D_0 \simeq 10^6$ cgs esu) in [60]. Thus, according to [60], in the surface layer of a single-domain crystal of BaTiO$_3$ of thickness equaling the thickness of the 180°-domain wall, the concentration of free carrier reaches $5 \cdot 10^{20}$ cm^{-3}. Substituting in (3.9) the values

$$n_S = \rho_S/q = 5 \cdot 10^{20} \text{ cm}^{-3},$$
$$n_i = N_c \exp\left[-\frac{E_c - E_v}{kT}\right], \quad N_c \simeq 5 \cdot 10^{19} \text{ cm}^{-3},$$

we determine the curvature of the bands $q\Delta\varphi \simeq E_c - E_v$. Thus, screening of the spontaneous induction in an intrinsic ferroelectric semiconductor leads to a symmetric curvature of the bands equal to the order of the width of the forbidden band and the formation of surface layers with electron and hole conductivity, respectively. This phenomenon can be called the "intrinsic" field effect in ferroelectrics in contrast to the field effect known in semiconductor physics in which the surface curvature of the bands is caused by the applied external field.

For an impurity ferroelectric semiconductor, the "intrinsic" field effect inevitably leads to degeneracy as, for example, occurs for "pure" barium titanate. This case was also analyzed in [60]. We consider the band diagram of a ferroelectric with one donor and one acceptor level (Figure 3.1b) where, to be specific, we will assume that the donor level lies below the Fermi level in the paraelectric region. In this case, expression (3.5) for the density of the space charge has the form

$$\rho(\varphi) = q\{p_0(\varphi) + p_d(\varphi) - n_a(\varphi) - n_0(\varphi)\}, \qquad (3.18)$$

where p_0 and p_d are, respectively, the concentration of holes in the valence band and at the donor levels; n_0 and n_a are, respectively, the concentration of electrons in the conduction band and at the acceptor levels. If conditions (3.8) are satisfied, then we have, for p_0 and n_0,

$$p_0(\varphi) = p_0 \exp(-q\varphi/kT), \quad n_0(\varphi) = n_0 \exp(q\varphi/kT), \qquad (3.19)$$

where p_0 and n_0, being the concentrations of holes and electrons in the quasi-neutral regions of the crystal, must satisfy the relation

$$n_0 p_0 = n_i^2. \qquad (3.20)$$

Denoting the energy and concentration of the donor and acceptor levels by E_d and E_a, N_d and N_a, respectively, we have, for p_d and n_a,

$$p_d(\varphi) = N_d \frac{1}{\exp\left(\frac{E_F + q\varphi - E_d}{kT}\right) + 1},$$

$$n_a(\varphi) = N_a \frac{1}{\exp\left(\frac{E_a - E_F - q\varphi}{kT}\right) + 1}. \qquad (3.21)$$

The solution of (3.7) and (3.4) along with (3.18) leads to distributions $D = D(z)$ and $\rho = \rho(z)$ analogous to (3.11) and (3.12). In particular, the surface charge ρ_s, as before, satisfies (3.15) and (3.17). However, the distribution of the potential $\varphi = \varphi(z)$ and the surface curvature of the bands are asymmetric with respect to the center of the crystal and their nature depends significantly on the concentrations of the donor and acceptor impurities.

Two limiting cases are of interest here. Let the concentrations N_d and N_a be so large that the ionized donors completely screen the negative end of the spontaneous induction, while the ionized acceptors completely screen its positive end. According to (3.17), this case is realized when the conditions

$$N_d \gg \frac{D_0}{4\pi q l_x}, \quad N_a \gg \frac{D_0}{4\pi q l_x}, \qquad (3.22)$$

are satisfied. The bands are curved at each of the surfaces only by the magnitude of the energy of donors and acceptors $q\Delta\varphi_1 = E_d$ and $q\Delta\varphi_2 = E_a$. Consequently, the "intrinsic" field effect is weakly expressed and a high surface electrical conductivity is absent accordingly. (Figure 3.1b corresponds to this case.) In the other limiting case, the concentrations N_d and N_a are so small that a large concentration of free electrons and holes is necessary to screen the spontaneous induction, which is provided in turn by a sharp surface curvature of the bands. This limiting case corresponds to the condition

$$N_d \ll \frac{D_0}{4\pi q l_\varkappa}, \quad N_a \ll \frac{D_0}{4\pi q l_\varkappa}. \tag{3.23}$$

The magnitude and asymmetry of the curvature of the bands is determined by the type of conductivity. For example, the condition $n_0 \gg p_0$ is satisfied for an n-type semiconductor. We find from (3.19) expressions for the potential near both surfaces:

$$q\varphi_p = -kT \ln \frac{n_S}{p_0}, \quad q\varphi_n = kT \ln \frac{n_S}{n_0}. \tag{3.24}$$

We find from the condition $n_0 \gg p_0$ that $|\varphi_n| \ll |\varphi_p|$. Thus, for a ferroelectric with n-type conductivity, the curvature of the bands near the hole surface must be many times greater than the curvature of the bands near the electron surface. Moreover, the total curvature of the bands $q\Delta\varphi = q(\varphi_n - \varphi_p)$, as follows from (3.24) and (3.20), is

$$q\Delta\varphi = q(\varphi_n - \varphi_p) = 2kT \ln \frac{n_S}{n_i}. \tag{3.25}$$

By substituting the values of n_s and n_i for BaTiO$_3$ into (3.25), one can confirm that for a weakly alloyed barium titanate, the total curvature of the bands equals the width of the forbidden band as before. Of course, the asymmetry of the "intrinsic" effect has the opposite sign for a p-type crystal.

The "intrinsic" field effect is the basis for a number of electronic phenomena at the surface of ferroelectrics. We refer here to the effect of the spontaneous polarization on the surface conductivity and photoconductivity, electroluminescence with polarization reversal of the ferroelectric, and a number of other phenomena which we will discuss below.

3.2. The Debye Length as a Parameter of Screening Length in a Ferroelectric

The exact solution of the problem of screening of the spontaneous induction of a single-domain ferroelectric [60] was presented in the preceding section. However, numerical evaluation of such parameters as the charge at the surface, the surface curvature of the bands, and the critical size of the crystal for which the single-domain state is still possible depends on the choice of the parameter of screening length. The correlation length l_x equal to the thickness of the 180°-domain wall was used above as such a parameter. Meanwhile, a second characteristic length parameter, the Debye screening length l_D, is always many times greater in real ferroelectrics than the correlation length, which is related to the large values of the dielectric constant ϵ. For example, from the data of the experimental work of [64], the concentration of free carriers $n \simeq 10^{17}$ cm^{-3} is sufficient for screening the single-domain state in BaTiO$_3$ (cf. below for more details). Meanwhile, for $n \simeq 10^{17}$ cm^{-3} and $\epsilon \simeq 10^3$, we have $l_D \simeq 10^{-5}$, i.e., the Debye length is two orders of magnitude greater than the correlation length. It was shown rigorously in [65] that under the condition $l_x \ll l_D$, the correlation energy can be neglected as compared to the screening energy, which in turn permits neglecting the correlation energy in the expression for the free energy (3.3). This approximation is obviously unsuitable for large concentrations, when $l_D < l_x$. However, the condition $l_D < l_x$ corresponds to the limiting concentration of carriers compatible with the existence of ferroelectricity, since the Debye screening length is then comparable with the length of the unit cell.

Thus, following [62], we consider the problem of the distribution of the field and screening charge in a single-domain crystal under the conditions where the length parameter is the Debye screening length. As in the preceding section, we will assume that there are no surface levels. We consider the free energy

$$F(D) = \frac{\alpha}{2} D^2 + \frac{\beta}{4} D^4 + \frac{\gamma}{6} D^6 + \frac{g}{4}(D_0^2 - D^2)^2 \quad (3.26)$$

along with the system of equations and boundary conditions (3.1), (3.4), (3.6), and (3.9):

$$\frac{1}{4\pi}\mathscr{E} = \frac{dF}{dD} \equiv F'(D), \quad \frac{dD}{dz} = 4\pi\rho, \quad \rho = -2qn_i \sinh\frac{q\varphi}{kT},$$

$$D|_{z=\pm L/2} = 0, \quad D|_{z=0} = D_0, \quad \varphi|_{z=0} = 0.$$

The relation between the coefficients in the expansions of the free energy (3.3) and (3.26) have the form $d = \alpha - gD_0^2$, $c = \beta + g$, $\gamma = f$. Moreover, for a first-order phase transition,

$$\alpha = \alpha_T'(T - T_0) = \alpha(T_1)\left(1 + \frac{\Delta T}{\theta}\right), \quad (3.27)$$

where $\alpha(T_1) = \alpha_T'(T_1 - T_0)$, $\Delta T = T - T_1$, $\theta = T_1 - T_0$, T_1 is the temperature of the phase transition, T_0 is the Curie–Weiss temperature, $\alpha(T_1) > 0$, $\beta < 0$, $\gamma > 0$, and $g > 0$.

Differentiating the right and left sides of the Poisson equation (3.4), we have

$$\mathscr{E} = -\frac{1}{4\pi\frac{d\rho}{d\varphi}} \frac{d^2D}{dz^2}. \quad (3.28)$$

SCREENING OF SPONTANEOUS POLARIZATION

With consideration of (3.28), the equation of state (3.21) acquires the form

$$-\frac{1}{(4\pi)^2 \frac{d\rho}{d\varphi}} \frac{d^2 D}{dz^2} = F'(D). \tag{3.29}$$

By integrating (3.29) over the limits from D_0 to D and using the boundary conditions (3.6), we arrive at the expression for the screening energy:

$$-\int_0^\varphi \rho \, d\varphi = F(D) - F(D_0). \tag{3.30}$$

Substituting (3.9) into (3.30), we obtain

$$2n_i kT \left[\operatorname{ch} \frac{q\varphi}{kT} - 1\right] = F(D) - F(D_0). \tag{3.31}$$

By combining (2.31), (3.4), and (3.9), we arrive at the equation describing the distribution of the spontaneous induction:

$$\frac{dD}{dz} = \pm \frac{4\pi q}{kT} [F(D) - F(D_0)] \sqrt{1 + \frac{4n_i kT}{F(D) - F(D_0)}}. \tag{3.32}$$

It is suitable to consider two cases here.

In the first case, $F(D) - F(D_0) \gg 4n_i kT$. The distribution of the space-charge density and the curvature of the conduction band in this case are given by the relations

$$\rho = \pm q \frac{F(D) - F(D_0)}{kT},$$

$$E_c - E_F - q\varphi = -kT \ln \frac{F(D) - F(D_0)}{kT N_c}. \tag{3.33}$$

The corresponding numerical evaluations can be made, for example, for $BaTiO_3$. For barium titanate, $[F(D) - F(D_0)] \simeq 10^6$ ergs/cm^3. Thus, the approximation (3.33) is valid for carrier concentrations $n_i \ll 10^{19}$ cm^{-3}. The solution (3.33) corresponds to a concentration of the carriers in the surface layer:

$$n_S = \frac{\rho_S}{q} \simeq \frac{F(0) - F(D_0)}{kT} \simeq 10^{19} \text{ cm}^{-3}.$$

In the second case, $F(D) - F(D_0) \ll 4n_i kT$ and the distribution of the spontaneous induction is found from the equation

$$\frac{dD}{dz} = \pm 8\pi q \sqrt{\frac{n_i [F(D) - F(D_0)]}{kT}}. \tag{3.34}$$

The approximation (3.34) corresponds to the case of high concentrations $n_i \gtrsim 10^{19}$ cm^{-3}. However, one should keep in mind that such high concentrations can be provided not only by free carriers in the band, but also by carriers captured by shallow traps in thermal exchange with the band. We present here the solution of (3.34), obtained and analyzed in [62]. The integral of (3.34) has the form

$$\int_t^1 \frac{dt}{\sqrt{(1-t^2)\left\{[2(1+\zeta)-a^2]a^2 - \left(1+\frac{\Delta T}{\theta}\right) + [2(1-\zeta)-a^2]a^2 t^2 - a^4 t^4\right\}}} = l_D^{-1}(z). \quad (3.35)$$

The following notation is taken here: $t = D/D_0$, $a = D_0/D(T_1)$, $\zeta = -g/\beta > 0$, and $l_D = (\epsilon_C kT/16\pi q^2 n_0)^{1/2}$ is the Debye length; ϵ_C is the dielectric constant at the Curie point. The value of the spontaneous induction at the Curie point $D(T_1)$, according to (1.32), satisfies the relation

$$D(T_1) = 2\left(-\frac{\alpha_C}{\beta}\right)^{1/2}.$$

The square root in (3.35) is positive since the value of the induction at the center of the crystal D_0 satisfies the following inequality:

$$D(T_1)\sqrt{\frac{2-\sqrt{1-3\Delta T/\theta}}{3}} \leq D_0 \leq D(T_1)\sqrt{\frac{2+\sqrt{1-3\Delta T/\theta}}{3}}. \quad (3.36)$$

The lower limit of D in (3.36) corresponds to the maximum of the free energy (cf. Figure 1.3), while the upper corresponds to the value (1.30) for spontaneous induction in an infinite crystal at any temperature.

The solution of (3.35) gives the following dependence $D = D(z)$:

$$D = D_0 \frac{\sqrt{\tau_1}\,\mathrm{Cn}\,y}{\sqrt{\tau_1 - \mathrm{Sn}^2 y}}, \quad (3.37)$$

where $\mathrm{Cn}\,y$ and $\mathrm{Sn}\,y$ are the elliptical cosine and sine, respectively:

$$y = a^2 \sqrt{\tau_1(1-\tau_2)}\, l_D^{-1} z, \quad (3.38)$$

$$\tau_{1,2} = \frac{2(1-\zeta)-a^2}{2a^2} \pm \sqrt{\frac{\zeta(\zeta-2)}{a^4} + \frac{3\zeta+1}{a^2} - \frac{\Delta T}{\theta a^4} - \frac{3}{4}}. \quad (3.39)$$

The coordinate dependences corresponding to (3.37) of the spontaneous induction, field, and space-charge density are presented in Figure 3.2. We note first of all that the parameter for the distribution of the induction, field, and space charge is the Debye screening length l_D. As is seen in Figure 3.2, the internal field in the single-domain ferroelectric changes sign. Within the volume of the crystal, the direction of the field is directed opposite the spontaneous polarization, i.e., it is depolarizing in sign. The maximum value of the depolarizing field

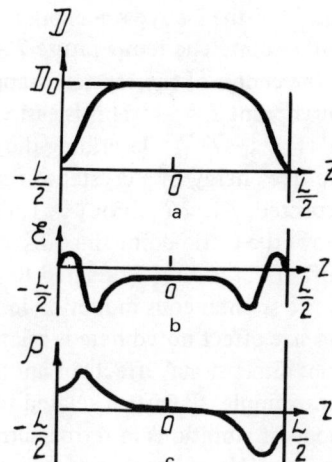

Figure 3.2. Distribution of the induction D, electric field ℰ, and free-charge density ρ in a sheet with single-domain polarization at $T \simeq T_1$ [62].

is equal to the coercive field. The maximum of the screening charge density is located within the volume of the crystal, and the temperature dependence of its coordinate is determined by the temperature dependence of the screening length.

The solution of (3.37) determines the critical size of the single-domain crystal with respect to the screening length. Substituting the boundary condition (3.6) into (3.37) and (3.38) gives

$$a^2 = \left(\frac{D_0}{D_S}\right)^2 = 2\frac{F(q)}{\sqrt{\tau_1(1-\tau_2)}}\frac{l_D}{L}, \qquad (3.40)$$

where F is the elliptical integral and $q = [(\tau_1 - \tau_2)/\tau_1(1-\tau_2)]^{1/2}$ is its argument.

Figure 3.3 gives the dependence of the induction at the center of the crystal on the relative thickness of the crystal L/l_D. The essential result of this analysis involves the fact that, as is seen in Figure 3.3, there exists a critical thickness of the crystal L_{cr}, defined by the condition that the single-domain state in the crystal is impossible for $L < L_{cr}$. It is significant that L_{cr} is the order of mag-

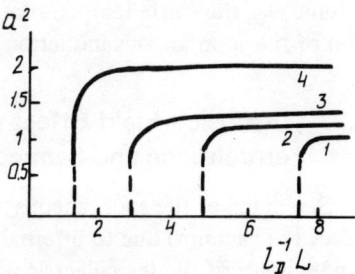

Figure 3.3. Relationship between the induction at the center of the crystal and the thickness of the sheet for various temperatures [62]: (1) $|\Delta T| = 0$; (2) $|\Delta T| = 2.5\theta$; (3) $|\Delta T| = 4\theta$; (4) $|\Delta T| = 9\theta$.

nitude of the Debye screening length and decreases with departure from the Curie point. The temperature $T = T_1^*$, for which the spontaneous polarization at the center of the crystal disappears, then does not generally coincide with the Curie point $T = T_1$. This is easily understood if one keeps in mind that the term $(g/4)(D_0^2 - D^2)^2$ describing the elastic energy is present in the expression for the free energy of a crystal of finite size (3.26). For $\zeta = -g/\beta \ll 1$, as should be expected, $T_1^* \simeq T_1$. For $\zeta > 1$, the crystal of finite size can remain ferroelectric above the Curie point since $T_1^* = T_1 + \frac{1}{3}\theta$. It is clear from the discussion that this shift effect is directly related to elastic stresses generated as a result of screening of the spontaneous induction in the crystal of finite thickness. We add to this that the size effect noted here is related only to the screening of spontaneous induction; dimensional effects of another kind can be superimposed on it in principle, for example, the effect related to a change in the dispersion of the ferroactive mode of vibrations in ferroelectric films [7].

Analysis of the case of low carrier concentrations $n_i \ll 10^{19}$ cm^{-3} leads to analogous conclusions. Integration of (3.32) and (3.33) leads to the following expressions for L_{cr}:

$$L_{cr} = 2l_D, \quad l_D = \left(\frac{kT\varepsilon_C}{16\pi q^2 n^*}\right)^{1/2},$$
$$n^* = \frac{F(0) - F(D_0)}{kT} \cdot \frac{1}{\ln^2\left[\frac{F(0) - F(D_0)}{n_i kT}\right]}. \tag{3.41}$$

It is seen from (3.41) that the critical thickness L_{cr} coincides in order of magnitude with the Debye screening length in this case. The screening length l_D itself is determined in this case not by the concentration n_i corresponding to the quasi-neutral region of the crystal, but to the concentration n^* in the region of strong surface curvature of the bands. The strong curvature of the bands is also caused precisely by the screening of spontaneous induction for small values of n_i.

Thus, consideration of the problem of screening of spontaneous induction by intrinsic carriers of a ferroelectric of finite size significantly refines the conclusions of the thermodynamic theory obtained in Chapter 1 for the infinite crystal. Screening in the state of thermodynamic equilibrium leads to a change in the equilibrium value of such parameters as the spontaneous induction in the volume D_0, the Curie temperature T_1, and others, and determines the distribution of the spontaneous induction, depolarization field, and space charge.

3.3. "Intrinsic" Field Effect with the Contact of a Single-Domain Ferroelectric and Semiconductor

Screening of the spontaneous induction in a ferroelectric with a free surface (placed in vacuum) due to internal carriers caused by the intrinsic or impurity conductivity of the ferroelectric was considered in the preceding two sections.

SCREENING OF SPONTANEOUS POLARIZATION

As another boundary condition, it is interesting to consider the contact of a single-domain ferroelectric with a semiconductor. Screening of the spontaneous induction in this case can be accomplished by carriers both in the ferroelectric and in the surface region of the semiconductor. Surface curvature of the bands related to the formation of screening surface and space charges in the semiconductor is thereby generated in the semiconductor due to the spontaneous induction of the ferroelectric. An "intrinsic" field effect of this kind with contact of a ferroelectric and semiconductor is an analogy of the ordinary field effect in semiconductors [66], but a number of interesting peculiarities are observed. We will consider these peculiarities below, following mainly the works of Vul, Guro, and Ivanchik [67, 68] in which this effect was first investigated theoretically.

The main peculiarity of the field effect with contact of a ferroelectric and semiconductor is the possibility of producing a large surface field causing a sharp curvature of the bands and degeneracy of the free carriers near the surface of the semiconductor. Actually, in the absence of levels at the surface of the ferroelectric and charges within its volume, a field the order of D_0 must be screened in the semiconductor. In perovskite ferroelectrics, $D_0 \simeq 10^8$ V/cm. As is well known, such large fields are not realized in the ordinary field effect.

We consider qualitatively the problem of screening of spontaneous induction at the boundary of a single-domain ferroelectric and semiconductor in the absence of surface levels in the ferroelectric. The direction of the spontaneous induction is assumed perpendicular to the interface. The distribution of the induction D and concentration of carriers n in the semiconductor in equilibrium satisfy the equations

$$\frac{dD}{dz} = 4\pi q n, \qquad (3.42)$$

$$j = \mu n q \mathscr{E} - q \frac{dn}{dz} D^* = 0. \qquad (3.43)$$

Here, (3.42) is the Poisson equation analogous to (3.4), while (3.43) is the condition where the current j through the semiconductor is equal to zero. With consideration of the Einstein relation $D^* = \mu(kT/q)$ between the diffusion coefficient D^* and mobility μ, equation (3.43) takes the form

$$kT \frac{dn}{dz} = q n \mathscr{E}, \qquad (3.43')$$

where $\mathscr{E}$ is the field in the semiconductor.

Equations (3.42) and (3.43') must be supplemented by the boundary conditions. It is natural to use as boundary conditions the continuity of the normal component of the induction at the interface of the semiconductor and ferroelectric (3.44) and the expression for the potential discontinuity at the interface (3.45):

$$\varepsilon \mathscr{E} = D, \qquad (3.44)$$

$$\Delta A = q\varphi_1(D) - q\varphi_2(D). \qquad (3.45)$$

Here, ΔA is the difference in the work functions between the semiconductor and ferroelectric, φ_1 and φ_2 are the potentials at the interface in the semiconductor and ferroelectric, respectively, and D is the value of the induction at the interface. Since the semiconductor is assumed linear, the relation between the induction and field in the semiconductor is given by (3.44), where ϵ is the dielectric constant of the semiconductor. Equations analogous to (3.42) and (3.43) can be written for the ferroelectric.

The system (3.42)–(3.45) determines the distribution of the concentration of carriers of the field and also the curvature of the bands both in the semiconductor and in the ferroelectric. Actually, by solving equation (3.45), we obtain the value of the induction at the interface $D = D_1$, which, according to (3.44), serves as the boundary condition for the solution of equations (3.42) and (3.43'). It depends on the value of induction at the interface $D = D_1$ whether the spontaneous induction D_0 in the semiconductor is screened, leading to the field effect in the semiconductor, or whether the screening will occur due to carriers of the ferroelectric, leading correspondingly to curvature of the bands in the ferroelectric. Thus, if $D_1 \lesssim D_0$, the change in spontaneous induction in the terroelectric can be neglected and the screening occurs practically completely in the semiconductor. In this case, surface curvature of the bands occurs in the semiconductor and the corresponding surface conductivity of the semiconductor differs significantly in magnitude from the volume conductivity and, possibly, in nature (electron or hole). If $D_1 \ll D_0$, then conversely the spontaneous induction is screened practically completely by the carriers in the volume of the ferroelectric, and the field does not penetrate the semiconductor. As was shown by Vul et al. [67], with contact of an intrinsic semiconductor and a single-domain ferroelectric, the solution $D_1 \lesssim D_0$ corresponds to the condition $E_{g1} \ll E_g$ (E_g, E_{g1} are the width of the forbidden bands of the ferroelectric and semiconductor, respectively), while the solution $D_1 \ll D_0$ corresponds to the condition $E_{g1} \gg E_g$ (it was assumed in [67] that $\Delta A = 0$). Thus, the field effect must be observed in the semiconductor with the contact of a narrow-band intrinsic semiconductor and a wide-band ferroelectric.

We evaluate the screening charge and curvature of the bands in the semiconductor in the two limiting cases indicated above. Solving (3.42) and (3.43') together by integrating over the limits from $D = 0$ to $D = D_1$ and setting $n = 0$ at $D = 0$, we have

$$nkT = \frac{D_1^2}{8\pi\varepsilon}. \qquad (3.46)$$

By substituting the value $D = D_1$ at the interface of the semiconductor and ferroelectric, one can evaluate the carrier concentration n, the curvature of the bands φ_1, and the field $\mathscr{E}$ in the semiconductor. To be specific, we consider the contact of barium titanate with intrinsic germanium ($D_0 \simeq 10^6$ cgs esu, $\epsilon \simeq 10$). In the first case, substituting $D_1 \simeq D_0$ into (3.46), we have $n \simeq 10^{23}$ cm^{-3}. Thus, the boundary condition $D_1 \simeq D_0$ corresponds to such a strong degeneracy of the electron gas in the semiconductor that relation (3.46) ceases to be valid and must be replaced by an analogous relation corresponding to a degenerate Fermi gas. It should be kept in mind that for this the physical meaning of (3.46) involves the fact that the pressure of the electron gas nkT is balanced by the pressure of the electric field $D_1^2/8\pi\epsilon$. By using this method [67] and the equation of state of a Fermi gas [28], we write the expression for the pressure of Fermi particles in the form

$$p = \frac{1}{5}\left(\frac{6\pi^2}{g}\right)^{2/3}\frac{\hbar^2}{2m} n^{5/3}. \qquad (3.47)$$

By using (3.47) as the left side of (3.46), and substituting $g = 2$ for electrons, we have

$$n = \left(\frac{5m}{4\pi\epsilon\hbar^2}\right)^{3/5}\left(\frac{1}{3\pi^2}\right)^{2/5} D_1^{6/5}. \qquad (3.48)$$

From the Fermi distribution function and expression (3.48) for the concentration n, we obtain the dependence of the difference in the energies of the Fermi level E_F and the bottom of the conduction band E_c on the induction D_1:

$$E_F - E_c = \frac{\hbar^2}{2m}(3\pi^2 n)^{2/3} = \left(\frac{15}{16}\frac{\pi}{\epsilon}\right)^{2/5}\left(\frac{\hbar^2}{2m}\right)^{3/5} D_1^{4/5}. \qquad (3.49)$$

Thus, evaluation of the concentration of carriers and the curvature of the bands in the semiconductor can be carried out with consideration of degeneracy from (3.48) and (3.49). Substituting $D \simeq D_0$ into (3.48) and (3.49), we have $n \simeq 10^{22}$ cm^{-3} and $E_F - E_c \simeq 1$ eV. Of course, in the second limiting case $D_1 \ll D_0$, degeneracy of the electrons in the semiconductor cannot occur and the concentration of carriers in the semiconductor satisfies (3.46). Thus, for example, substituting the value $n \simeq N_c = 3 \cdot 10^{19}$ cm^{-3} corresponding to weak degeneracy into (3.46), we obtain $D_1 \simeq 2 \cdot 10^4$ cgs esu. Thus, for all boundary values of $D_1 < 2 \cdot 10^4$ cgs esu, the surface curvature of the bands in the semiconductor is so small as not to lead to degeneracy of the electrons.

The "intrinsic" field effect with contact of a ferroelectric and semiconductor has a number of other peculiarities. Thus, polarization reversal of the ferroelectric must lead to a change in sign of the surface curvature of the bands and a corresponding change in sign of the carriers in the surface region of the semiconductor. Curvature of the bands and concentration of the carriers in the semiconductor must display the characteristic temperature dependence caused by the temperature dependence $D_1 = D_1(T)$ and the phase transition in the ferroelectric. One should keep in mind that the calculation presented above for this effect, as well as the solution of the problem on the internal screening of spontaneous induction in a ferroelectric, does not take into account the possible role of surface levels in the ferroelectric. We will show below that the role of surface levels in screening is actually very significant and that the analysis of screening of spontaneous induction in Sections 3.1-3.4 is thus qualitative and incomplete.

3.4. Screening in a Ferroelectric Capacitor. Opposing Domains

The most widely used experimental case of screening of spontaneous induction occurs with the contact of a ferroelectric with metal electrodes (ferroelectric capacitor). It is customarily assumed that if the electrodes are short-circuited, then all the screening charge is located on the electrodes, while $D \equiv D_0$ in the volume and on the surface of the ferroelectric and the corresponding internal field equals zero, $\mathcal{E} \equiv 0$. [By $D = D_0$ we mean the equilibrium value of the induction satisfying (1.30).] Nonetheless, experiments—above all, photoelectric investigations (cf. below)—show that an internal depolarization field differing from zero exists in the volume of a short-circuited ferroelectric capacitor. Thus, even in the presence of short-circuited metal electrodes, internal screening of spontaneous induction occurs in the general case in a ferroelectric.

The internal screening mechanism in a short-circuited ferroelectric capacitor can be twofold. First, this can be the same mechanism which occurs with contact of a semiconductor and ferroelectric (cf. preceding section) and is related to the electron work function from the metal to the ferroelectric and in the opposite direction. Second, this mechanism can be related to the existence of peculiar surface layers on the surface of the ferroelectric (dielectric "gaps"); these dielectric properties differ from the properties of the crystal in the volume. We will discuss the second possible mechanism in the following section. Here, following [69-71], we consider the first mechanism of internal screening in the short-circuited ferroelectric capacitor.

The condition of internal screening in the short-circuited ferroelectric capacitor can be obtained from the boundary conditions (3.44) and (3.45), which now refer to the metal-ferroelectric interface. In the metal, $q\varphi_1(D) \simeq qRD$, where R is the Debye screening radius in the metal. If the spontaneous induction D is mainly screened by charges on the electrodes, then $D \equiv D_0$, $\mathcal{E} \equiv 0$ inside the ferroelectric and the corresponding surface with curvature of the bands in the ferroelectric is much less than the half-width of the forbidden band $E_g/2$:

$$q\varphi_2 \ll E_g/2, \quad q\varphi_1 \simeq qRD_0. \qquad (3.50)$$

Substituting (3.45) into (3.50), we are led to the following condition:

SCREENING OF SPONTANEOUS POLARIZATION

$$qRD_0 - \Delta A \ll E_g/2. \tag{3.51}$$

For example, if the condition

$$qRD_0 - \Delta A \gtrsim E_g/2 \tag{3.52}$$

is satisfied, then, in spite of the presence of the short-circuited metal electrodes, the spontaneous induction is screened by carriers inside the ferroelectric. Thus, condition (3.52) can be considered as a criterion of internal screening in the short-circuited ferroelectric capacitor. If one assumes, as was done in [67], that, to be specific, the difference of the work functions $\Delta A = 0$, then the criterion (3.52) relates the screening mechanism in the ferroelectric capacitor to the forbidden band of the ferroelectric. For a ferroelectric with a not too wide forbidden band, which satisfies the condition $E_g \lesssim 2qRD_0$, the screening must be internal and thus the electrodes do not have a significant effect on the nature of the distribution of the spontaneous induction and the internal field in the ferroelectric. Thus, when condition (3.52) is satisfied, the ferroelectric with short-circuited electrodes can be considered as a ferroelectric with a free surface and the same solution applies to it as was presented in Sections 3.1 and 3.2. We stipulate once again that the role of levels at the surface of the ferroelectric is not taken into account in the internal screening mechanism.

The electron processes at the metal–ferroelectric interface are the basis of the phenomenon called opposing domains [69-71]. Below, following [70], we will consider this phenomena, since the analysis of screening at the boundary of opposing domains can be performed by analogy with the case of contact of the ferroelectric and semiconductor. We consider a ferroelectric located between plane-parallel and short-circuited metal electrodes.

Further, let the spontaneous induction D_0 be directed perpendicular to the ferroelectric-electrode interface. We assume that the work function from the metal to the ferroelectric is less than in the opposite direction. Then a negative space charge caused by electrons passing from the electrode into the volume of the ferroelectric is generated near each of the electrodes. Corresponding to this, an electric field directed from the positive electrode to the negatively charged volume of the crystal arises in the ferroelectric near each of the electrodes. If the intensity of this field is close to the coercive field, then it leads to such a ferroelectric polarization where the spontaneous inductions of the two electrode regions are in the opposite directions. As a result, two opposing domains with opposite direction of the polarization are formed in the crystal (Figure 3.4a). If the work function from the ferroelectric to the metal is lower than in the opposite direction, then the electrodes will be charge negatively and the volume of the crystal near each of the electrodes is charged positively. This case corresponds to the opposing domains presented in Figure 3.4b.

Screening of the spontaneous induction occurs at the boundary between the

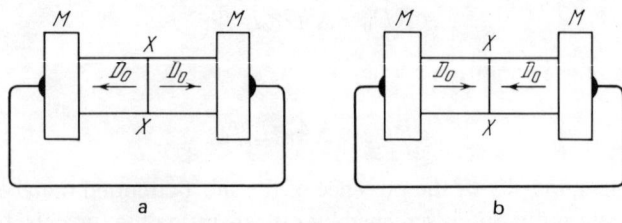

Figure 3.4. Opposing domains with electron (a) and hole (b) boundaries [70].

opposing domains (it is indicated by the line X-X in Figure 3.4), because of which a screening layer is formed with a high concentration of electrons (Figure 3.4a) or holes (Figure 3.4b). Accordingly, this boundary in Figure 3.4a can be called the electron boundary, and that in Figure 3.4b the hole boundary. With screening on the spontaneous induction D_0 in BaTiO$_3$, a surface charge density $2D_0/4\pi q \simeq 10^{14}$ cm^{-2} and space charge density $n = 2D_0/4\pi q l_D$ are formed at the boundary of the opposing domains, where l_D is the effective thickness of the screening layer. According to the evaluation made in [70], $n \simeq 10^{19}$ cm^{-3}, i.e., the layer at the boundary of the opposing domains has a high electron or hole conductivity.

Let us consider quantitatively the screening at the boundary of n-type opposing domains. Denoting the volume density of electrons near the boundary by n, we use the Poisson equation (3.42) and the condition (3.43) expressing zero current in the screening layer. Eliminating n from (3.42) and (3.43), we arrive at the equation

$$kT\,dn = \frac{1}{4\pi}\mathscr{E}\,dD. \qquad (3.53)$$

Combining (3.53) and the equation of state (3.1), we arrive at the equation

$$dF/dn = kT. \qquad (3.54)$$

Integrating (3.54) and setting $n = 0$ at $D = D_0$, we have

$$nkT = F(D) - F(D_0). \qquad (3.55)$$

Expression (3.55) coincides in essence with the expression obtained above for the density of the screening charge (3.33). The expression for the free energy F in relation (3.55) is given by equation (3.26). For BaTiO$_3$, as has already been indicated above, $[F(D) - F(D_0)] \simeq 10^6$ ergs/cm^3, and it then follows from (3.55) that $n \simeq 10^{19}$ cm^{-3}. This value corresponds to a degenerate electron gas in the screening layer. Thus, a more correct calculation of the concentration n should be made using the method given in Section 3.3 by substituting, on the left side of (3.55), expression (3.47) for the pressure of a degenerate electron gas ($g = 2$):

SCREENING OF SPONTANEOUS POLARIZATION

$$n = \left(\frac{10m}{\hbar^2}\right)^{3/5} \left(\frac{1}{3\pi^2}\right)^{2/5} [F(0) - F(D_0)]^{3/5}. \tag{3.56}$$

The curvature of the bonds at the boundary between the opposing domains, which provides the concentration n in accordance with (3.56), is given by equation (3.49):

$$E = E_F - E_c = q\varphi - \frac{1}{2}E_g = \frac{\hbar^2}{2m}(3\pi^2 n)^{2/3}. \tag{3.49'}$$

Substituting the expression for n (3.56) into (3.49'), we have

$$q\varphi \simeq \frac{1}{2}E_g + \left(\frac{\hbar^2}{2m}\right)^{3/5} \left(\frac{15\pi^2}{2}\right)^{2/5} [F(0) - F(D_0)]^{2/5}. \tag{3.57}$$

Expression (3.57) permits one to evaluate the difference in work functions between the metal and ferroelectric ΔA, for which opposing domains can be observed. Since the curvature of the bands is caused by the contact difference, then

$$\Delta A \gtrsim q\varphi, \tag{3.58}$$

where $q\varphi$ satisfies (3.57). A numerical evaluation of the right side of (3.57) for BaTiO$_3$ gives $\Delta A \gtrsim 1.0$ eV. Since the work function for pure barium titanate is $A_2 = 2.8$ eV, the work function of the metal according to (3.58) has two possible values: $A_1 \gtrsim 4.7$ eV or $A_1 \lesssim 0.9$ eV. The first value corresponds to the formation of a hole boundary of the opposing domains in barium titanate, and the second, an electron boundary. Since metals with a work function $A_1 \lesssim 0.9$ eV are not encountered in practice, opposing domains with a hole boundary must be observed for pure barium titanate and electrodes corresponding to $A_1 \gtrsim 4.7$ eV.

The experimental methods for observing opposing domains in ferroelectrics may differ from the simple diagram presented in Figure 3.4. We add that the processes of screening spontaneous induction at the interphase boundaries in ferroelectrics lead to an analogous increase in the conductivity. We will discuss this problem specially in Section 6.3.

3.5. Screening of Spontaneous Polarization in the Presence of a Surface Layer

We will discuss here another possible mechanism of internal screening in a ferroelectric capacitor, which is related to the existence of surface layers with peculiar dielectric properties on the surface of the ferroelectrics.

It is well known that a whole series of experimental investigations have indicated unambiguously the existence of such layers. We refer, for example, to data on the effect of the thickness of the crystal on the dielectric constant, the coercive field, and the polarization reversal mechanism. The data of electrographic and x-ray analysis can serve as unambiguous evidence. Thus, according to these data, there exists the surface layer of thickness 10^{-4}-10^{-6} cm with a low dielectric constant ϵ_l of the order of several units at the (001) surface in BaTiO$_3$. There is a strong electric field 10^4-10^6 V/cm in the volume of this layer. In the opinion of Merz and Fatuzzo [15], this layer is not a foreign film (arising, for example, with the growth of the crystal), but is a surface region of the crystal in which dielectric saturation and piezoelectric compression occur because of the strong electric field. The latter two mechanisms are responsible for the decrease in the dielectric constant in the surface layer. As for the physical nature of the surface layer, it is possibly related to screening of spontaneous polarization. For example, in the opinion of Triebwasser [72], the surface layer in BaTiO$_3$ is a Schottky barrier formed with contact of the semiconductor with the metal electrode. The presence of Schottky-type barriers with the contact of BaTiO$_3$ with the electrodes was confirmed by independent electric and photoelectric measurements. The model of the surface layer as a Schottky barrier permits one to explain the asymmetry of the surface layers at the opposite ends, the dependence of the capacitive reactance and the resistance of the layer on the applied external field, and the dependence of the dielectric constant of BaTiO$_3$ in the paraelectric region on the constant electric field bias. Much work has been devoted to investigation of the properties and nature of surface layers in ferroelectrics; a detailed survey of this problem can be found in [15].

It is significant here only that the presence on the ferroelectric surface of the surface layer with dielectric properties differing from the properties in the volume independently of the nature of the layer introduces the specific properties in the nature of the distribution of the internal field with screening of the spontaneous induction. In particular, as has already been indicated above, the surface layer can lead to the formation of an internal depolarizing field in a ferroelectric capacitor with short-circuited electrodes. This field is responsible for depolarization of the single-domain ferroelectric and the effect of nonequilibrium carriers (photodomain effect), short-circuit photocurrents in the ferroelectrics, optical distortion effects, and many other phenomena of electronic origin, which we will discuss specially in the following chapters.

We consider the model of a short-circuited ferroelectric with dielectric "gaps." As everywhere above, the single-domain ferroelectric is an infinite plane-parallel sheet, whose surface is perpendicular to the direction of spontaneous induction (Figure 3.5). There is a surface layer (gap) at each of the surfaces of the sheet.

We further assume that the screening charge consists of a surface charge Q_p localized on the surface between the gap and the crystal and a space charge with a constant density $p = Q_f/l_D$, where Q_f is the screening space charge in the crystal per unit surface area and l_D is the screening length in the ferroelectric.

SCREENING OF SPONTANEOUS POLARIZATION

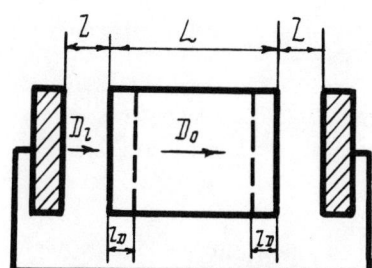

Figure 3.5. Model of a short-circuited ferroelectric with gaps.

We will assume here that the surface charge is localized in surface levels, whose concentration is $N_p \lesssim D_0/4\pi q$ (D_0 is the spontaneous polarization). For $N_p = D_0/4\pi q$, the spontaneous polarization is completely screened by the charge of the surface levels. It was indicated above that for $BaTiO_3$ $D_0/4\pi q \simeq 10^{14}$ cm^{-2}. If the screening charge Q_f is localized at sufficiently deep levels of the crystal, Q_p can be identified with the pyroelectric charge released with the phase transition from the ferroelectric region to the paraelectric. Of course, such a division is impossible in the general case and the pyroelectric charge can be localized both at surface levels and in the volume in shallow trapping levels.

We replace the origin of the z coordinate at the boundary between the gap and the crystal. The distribution of the induction D, the field $\mathscr{E}$, and the charge density ρ in the ferroelectric for the model under consideration can be obtained from the system of equations (3.4), (3.1), and (1.18), the boundary conditions

$$\frac{dD}{dz} = 4\pi\rho, \quad \frac{1}{4\pi}\mathscr{E} = \frac{dF}{dD}, \quad F = F_0 + \frac{\alpha}{2}D^2 + \frac{\beta}{4}D^4 + \cdots,$$

$$D_0 - 4\pi Q_f - 4\pi Q_p = D_l,$$
$$D|_{z=l_D} = D_0, \quad D|_{z=0} = D_0 - 4\pi Q_f, \qquad (3.60)$$

and the short-circuit condition

$$\frac{D_l}{\varepsilon_l}l + \int_0^{l_D} \mathscr{E}\,dz = 0. \qquad (3.61)$$

For simplicity, expression (1.18) for the free energy is written neglecting the elastic energy for a crystal with a second-order phase transition. Here, F is the free energy, D_0 is the spontaneous induction, D_l is the induction in the surface layer, l is the thickness of the surface layer, l_D is the screening length in the crystal, ϵ_l is the dielectric constant of the surface layer, and $\epsilon_s = \frac{1}{4}\pi|\alpha|$ is the dielectric constant of the ferroelectric.

Equation (3.1) determines the depolarization field $\mathscr{E}$ as a function of the induction:

$$\frac{1}{4\pi}\mathscr{E} = \alpha D + \beta D^3 \qquad (3.62)$$

It follows from (3.62) that for

$$D = \tilde{D} = \frac{1}{3^{1/2}}\left(-\frac{\alpha}{\beta}\right)^{1/2} = \frac{1}{3^{1/2}}D_0 \qquad (3.63)$$

the depolarization field has a maximum of the order of the coercive field:

$$\mathscr{E} = \tilde{\mathscr{E}} = -\frac{8\pi}{3^{3/2}}\frac{|\alpha|^{3/2}}{\beta^{1/2}}. \qquad (3.64)$$

Integration of the Poisson equation (3.4) under the assumption that $\rho = Q_f/l_D$ = const for the boundary conditions (3.60) gives

$$D = 4\pi\frac{Q_f}{l_D}z + D_0 - 4\pi Q_f. \qquad (3.65)$$

The solution of (3.62) and (3.65) in parametric form determines the distribution of the depolarization field in the volume of the crystal. In particular, by substituting (3.65) into (3.63), we determine the position of the maximum of the depolarization field $\mathscr{E}(\tilde{z}) = \tilde{\mathscr{E}}$

$$\tilde{z} = l_D\left(1 - \frac{1}{q^*}\frac{\sqrt{3}-1}{\sqrt{3}}\right), \qquad (3.66)$$

where

$$q^* = 4\pi\frac{Q_f}{D_0}.$$

The nature of the distribution of the internal depolarization field $\mathscr{E}$ (and the corresponding curvature of the bands) is determined by the magnitude of the screening parameter q^*, which can vary over the range $0 \leqslant q^* \leqslant 1$. The value of q is determined in turn by the screening length l_D and the parameters of the surface layer l and ϵ_l. In explicit form, this dependence follows from relations (3.61) and (3.59):

$$\frac{D_l}{\epsilon_l}l - D_0\frac{l_D}{4\epsilon_s}(2-q^*)[1-(1-q^*)^2] = 0. \qquad (3.67)$$

It is suitable to consider two regions of values of the parameter q^*. The first region is characterized by the following limits:

$$1 - \frac{1}{\sqrt{3}} \leqslant q^* \leqslant 1. \tag{3.68}$$

It follows from (3.66) that for these values of q^* the depolarization field in the crystal reaches a value of the order of the coercive field. Thus, the distribution of the internal field has the following nature. The field equals zero at the center of the crystal and, increasing in the direction toward each of the surfaces, the field reaches a maximum equal to the coercive force at $z = \tilde{z}$. The direction of the internal field everywhere in the crystal is directed opposite to the spontaneous polarization, which corresponds to the depolarizing nature of the field. Thus, for values of q^* from the region (3.68), the distribution of the internal field in the ferroelectric is similar to that presented in Figure 3.2b, except that the field inside the crystal does not change sign. The field in the gap has the opposite sign in respect to the polarizing field and can be determined from (3.67). The fields in the gap and crystal can be called, according to sign, the direct and reverse, respectively. In the particular case for $\tilde{z} = l_D(1/\sqrt{3})$, $q^* = 1$. The case $q^* = 1$ corresponds to complete internal screening of the spontaneous induction in the crystal, where the induction, according to (3.60), vanishes at the boundary between the crystal and the gap, while the induction in the gap $D_l = -4\pi Q_p$. Substituting $q^* = 1$ into (3.67), we have

$$D_l = -\frac{1}{4} \frac{\varepsilon_l}{\varepsilon_s} \frac{l_D}{l} D_0, \tag{3.69}$$

where D_l is the induction in the gap. Relation (3.69) can be used to determine the voltage drop at the surface layer $\mathscr{E}_l l$ if one first evaluates the screening length l_D. For BaTiO$_3$ and SbSI, the concentration of equilibrium carriers $n \simeq 10^{10}$ cm^{-3}. Substituting $n \simeq 10^{10}$ cm^{-3} and $\varepsilon_s \simeq 10^3$ into (3.41), we have $l_D \simeq 10^{-2}$ cm. For these values of l_D, ε_s, and $D_0 \simeq 20 \cdot 10^{-6}$ C/cm^2, evaluation from (3.69) leads to $\mathscr{E}_l l \simeq 1$ V. Thus, the voltage drop in the gap $\mathscr{E}_l l$ calculated from (3.69) satisfactorily agrees with the data on the surface layer in BaTiO$_3$ obtained from a number of independent measurements and presented above.

For the values of q^* from the second characteristic region,

$$0 \leqslant q^* < 1 - \frac{1}{\sqrt{3}}, \tag{3.70}$$

the depolarization field everywhere in the crystal is less than the coercive field $\mathscr{E} < \tilde{\mathscr{E}}$, where for $q^* \ll 1$, the field $\mathscr{E} \ll \tilde{\mathscr{E}}$. When $q \simeq 0$, the induction in the crystal is constant, $D \equiv D_0$, the depolarization field equals zero, and the spontaneous depolarization D_0 is screened by the charge of the surface levels. The field in the gap also becomes zero in this case, as follows from (3.59) and (3.61). For values of $q^* \ll 1$, we have, from relation (3.67),

$$4\pi Q_f = \frac{l/\varepsilon_l}{l/\varepsilon_l + l_D/\varepsilon_s} (D_0 - 4\pi Q_p). \tag{3.71}$$

The case $q^* \simeq 0$ corresponds to a crystal with a free surface, at which the spontaneous induction is screened by the charge at the surface levels. The internal field of the short-circuited crystal increases with an increase in the screening parameter, and the charge at the surface decreases simultaneously with the increase in screening space charge. The change in the relation between the screening charge at the surface and in the volume can be accomplished both by field emission through the gap as well as by photoelectrically active illumination of the short-circuited ferroelectric. The presence of the internal field in the short-circuited ferroelectric capacitor causes at once many photo-ferroelectric phenomena related to the screening.

3.6. Role of Surface Levels in the Screening of Spontaneous Polarization

It was emphasized above that the role of surface levels of the ferroelectric was not taken into account in analyzing internal screening in the ferroelectric and also screening at the ferroelectric–semiconductor and ferroelectric–metal interfaces. Meanwhile, as a number of experimental data have indicated, the surface states must play a significant role both in phenomena related to screening as well as in the thermodynamics of a crystal of finite size.

As an example, we indicate here two experimental facts. Analysis of the internal screening in a single-domain crystal indicates that the single-domain state is impossible in ferroelectric films thicker than the Debye screening length (cf. Section 3.2). This conclusion contradicts the experimental data in investigations of finely dispersed barium titanate by the method of electron paramagnetic resonance [73]. These experiments have shown that the spontaneous polarization does not change with fragmentation of $BaTiO_3$ crystals to dimensions of the order of 10^{-5} cm. A long series of experiments performed on triglycine sulfate and other ferroelectrics [74, 75] showed that, conversely, the predominating single-domain structure is observed with thinning of thin films of ferroelectrics. As another experimental fact, one can indicate investigations of the "intrinsic" field effect with contact of a ferroelectric and semiconductor. These experiments indicate that, although the effect is observed in principle, its magnitude is much lower than that predicted by the theory. Thus, the theory, which does not take into account screening due to surface levels, predicts for the contact of barium titanate and intrinsic germanium that screening of the spontaneous polarization leads to a degeneracy of the electrons in the germanium (cf. Section 3.3). At the same time, the experiment showed a change in the conductivity of germanium by 10% at most, and this change was even less in other experiments. Analogous results have been obtained for triglycine sulfate and triglycine selenate. Attempts to detect the "intrinsic" field effect at the free surface of a ferroelectric, so far as we know, have not given definite results at all.

SCREENING OF SPONTANEOUS POLARIZATION

The nature of the surface states in elementary semiconductors is comparatively well studied at present [66]. The surface states can be caused by disruption of the periodicity of the crystal lattice (the Tamm levels), by unsaturated valences of atoms at the surface (the Schockley levels), by surface defects, and by chemisorbed particles. The concentration of surface levels in semiconductors can vary over the range 10^{10}-10^{14} cm^{-2}. With a high concentration of surface levels, the latter form surface bands, in which the energy varies quasi-continuously. In contrast to semiconductors, the surface states in ferroelectrics have not been completely investigated experimentally, and this is related not only with the specific properties of ferroelectrics, for example, with the presence of surface layers with peculiar dielectric properties (cf. Section 3.5). A majority of the known ferroelectrics are high-resistance semiconductors with a wide forbidden band and a complex system of local levels in the volume. This makes it difficult to investigate the surface levels from experiments on the field effect, photoconductivity kinetics, etc. Nonetheless, the application of a definite model of the surface levels to the analysis of screening of spontaneous induction in ferroelectrics not only removes a number of difficulties in explaining the experimental phenomena related to screening, but also designates possible ways of investigating the surface states themselves. This approach to investigating screening effects was developed in the works of Selyuk [76-79], following whom we will consider below the contribution of surface electron states to the free energy of the short-circuited ferroelectric. The maximum concentration of surface levels in the ferroelectrics N_P can be evaluated starting from the concept of complete screening of spontaneous induction by the charge at the surface $N_P \simeq D_0/4\pi q$. For the maximum values of spontaneous induction D_0 observed for all presently known ferroelectrics, $N_P \simeq 10^{14}$ cm^{-2}. This value N_P is simultaneously the largest known concentration of surface states in semiconductors (for example, slow states on a natural surface or states on a surface subjected to processing in ultrahigh vacuum). It is possible that this coincidence is not by chance, which in turn indicates the role of surface states in screening spontaneous polarization. We turn now to the model of surface levels used in [76-79]. Let there be, at the surface of a single-domain ferroelectric, levels forming a quasicontinuous band with density N per unit energy. We denote the electrochemical potential of the uncharged surface by ξ_{i0} and the electron affinity by χ_i. The index i can have two values (+ and -) corresponding to the positive and negative end of the spontaneous induction vector, where $\xi_{+0} \neq \xi_{-0}$ and $\chi_+ \neq \chi_-$. The nonequivalence of the opposite surfaces of the crystal perpendicular to the spontaneous induction is caused by the polarity of the crystal and by the difference in the physicochemical properties of the opposite ends of the domain. By applying the well-known Bardeen model to the band of surface levels, one can represent the surface charge Q_{pi} localized at the levels of the ith surface in the following manner:

$$Q_{pi} = -qN(\xi_i - \xi_{i0}). \tag{3.72}$$

Here, the electrochemical potential of the surface in the presence of the charge Q_{pi} is denoted by ξ_i. According to the Bardeen model, the sign of the charge on the surface is determined by the direction of the shift of the electrochemical potential.

In order to take into account the specific contributions which can be made by the surface electronic states to the free energy of the ferroelectric, it was advisable to consider the screening in a sufficiently thin crystal for which the thickness of the direction of the spontaneous induction is much less than the Debye screening length, i.e., $L \ll l_D$. In this case, one can neglect the screening space charge in the crystal by assuming that the spontaneous induction is effectively screened by the charge on the surface, as a result of which the internal field does not lead to degeneracy. We consider a plane-parallel sheet of a single-domain ferroelectric located between shoft-circuited electrodes. There is a "dielectric" gap of thickness l with a dielectric constant ϵ_l, which is an air gap or surface layer, on the surfaces between the crystal and the electrodes. The band diagram of the crystal is presented in Figure 3.6. The following notation is taken everywhere below: μ_i is the distance from the level ξ_{i0} to the bottom of the conduction band, ξ_F is the Fermi level for the system in electronic equilibrium, and φ_i is the distance from the Fermi level at the ith surface to the bottom of the conduction band. With consideration of the assumed notation, the charge on the ith surface satisfies the following relation, according to (3.72):

$$Q_{pi} = -qN(\xi_i - \xi_{i0}) = -qN(\xi_F - \xi_{i0}) = qN(\varphi_i - \mu_i). \tag{3.73}$$

Since the screening space charge is absent and $Q_f = 0$, accordingly, the boundary conditions (3.59) for each of the surfaces of the boundary between the crystal and gap has the form

$$\varepsilon_l \mathscr{E}_{l-} = D - 4\pi Q_{p-}, \quad \varepsilon_l \mathscr{E}_{l+} = D + 4\pi Q_{p+}. \tag{3.74}$$

Here, $\mathscr{E}_{li}$ is the field in the gap. Denoting the field in the volume of the ferro-

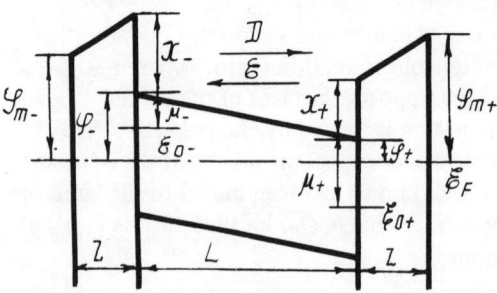

Figure 3.6. Band diagram of a short-circuited single-c-domain crystal.

SCREENING OF SPONTANEOUS POLARIZATION

electric by $\mathscr{E}$, one can represent the difference in potentials between the electrodes by the sum of the potential discontinuities on the band diagram of Figure 3.6:

$$q(l\mathscr{E}_{l+} + l\mathscr{E}_{+} + \mathscr{E}L) = \varphi_{m+} - \varphi_{m-} + \chi_{-} - \chi_{+}. \quad (3.75)$$

Substituting (3.73) and (3.74) into (3.75) and taking into account the relation $\varphi_{+} - \varphi_{-} = qL\mathscr{E}$, one obtains the following expression:

$$-q\mathscr{E}L = (1 + \alpha^*)^{-1} \left[\frac{D}{2\pi qN} - (\mu_{+} - \mu_{-}) - \right.$$
$$\left. - \alpha^* (\chi_{-} - \chi_{+} + \varphi_{m+} - \varphi_{m-}) \right], \quad (3.76)$$

where

$$\alpha^* = \frac{\varepsilon_l}{4\pi q^2 N l}. \quad (3.77)$$

The equation of state (3.1) relates the field in the ferroelectric to the spontaneous induction D:

$$\frac{1}{4\pi} \mathscr{E} = \alpha D + \beta D^3 + \gamma D^5. \quad (3.78)$$

The expansion of $\mu_{+} - \mu_{-}$ and $\chi_{+} - \chi_{-}$ in a series in the spontaneous induction D in the linear approximation was also used in [76] (it is assumed that the contribution of the terms in D^3 and D^5 is insignificant):

$$\mu_{+} - \mu_{-} = \Delta\mu + \chi_{\mu}D, \quad (3.79)$$

$$\chi_{-} - \chi_{+} = \Delta\chi + \chi_{\chi}D, \quad (3.80)$$

where $\Delta\mu$, $\Delta\chi$, χ_{μ}, and χ_{χ} do not depend on D. Substitution of (3.78)–(3.80) into (3.76) leads to the final equation relating the spontaneous induction with the parameters of the ferroelectric capacitor:

$$-(1+\alpha^*)^{-1}(qL)^{-1}\left[\Delta\mu + \alpha^*(\Delta\chi + \varphi_{m+} - \varphi_{m-})\right] +$$
$$+ \left[\chi + (1+\alpha^*)^{-1}L^{-1}\left(\frac{2}{q^2N} - \chi_{\mu} - \alpha^*\chi_{\chi}\right)\right]D + \beta D^3 + \gamma D^5 = 0. \quad (3.81)$$

By comparing (3.81) with the equation of state of an infinite homogeneous crystal, one can represent the free energy of the ferroelectric capacitor in the form

$$F = F_0 - \mathscr{E}_1 D + \frac{1}{2}\alpha_L D^2 + \frac{1}{4}\beta_L D^4 + \frac{1}{6}\gamma_L D^6, \quad (3.82)$$

where

$$\mathscr{E}_1 = (1+\alpha^*)^{-1}(qL)^{-1}[\Delta\mu + \alpha^*(\Delta\chi + \varphi_{m+} - \varphi_{m-})],$$
$$\alpha_L = \alpha + (1+\alpha^*)^{-1} L^{-1}\left(\frac{2}{q^2N} - \chi_\mu - \alpha^*\chi_\chi\right), \quad \beta_L = \beta, \quad \gamma_L = \gamma. \tag{3.83}$$

It follows from (3.82) and the equation of state (3.1) that $\mathscr{E}_1$ plays the role of an effective internal field, whose magnitude is determined both by the structure of the surface levels of the opposite ends of the domain as well as the properties of the surface layer and electrodes. Moreover, it is seen from (3.83) that screening of the spontaneous induction by the charge of the surface levels and the field in the surface layer is equivalent to renormalization of the Devonshire coefficient α in the expression for the free energy. Thus, the effect of the surface levels on the ferroelectric properties of a thin crystal in the approximation $L \ll l_D$ can be investigated with the help of expression (3.82) for the free energy. This investigation can be performed in the same sequence in which the effect of the local electronic states in the volume of an infinite homogeneous crystal on its ferroelectric properties and phase transition was analyzed in Section 1.3.

For this, we analyze (3.83) in two limiting cases. The first limiting case corresponds to a high concentration of surface states and the corresponding complete internal screening of the spontaneous induction by the charge of the surface levels. Actually, by setting $N \simeq 10^{14}$ cm$^{-2} \cdot$ eV^{-1} and $\epsilon_i^{-1}l \simeq 10^{-7}$ cm, we find from (3.77) $\alpha^* \simeq 10^{-1}$. In this case, expressions (3.83) have the form

$$\mathscr{E}_1 \simeq (qL)^{-1}\Delta\mu, \quad \alpha_L \simeq \alpha + L^{-1}\left(\frac{2}{q^2N} - \chi_\mu\right). \tag{3.84}$$

As is seen from (3.84), neither the electrodes nor the properties of the surface layer affect the free energy (3.82) in the limiting case under consideration. The ferroelectric capacitor in this case is equivalent to a crystal with a free surface, and the internal screening of the spontaneous induction is provided by the high concentration of surface levels. Corresponding to this, the free energy (3.82) depends only on the concentration and structure of the surface levels. In particular, the field $\mathscr{E}_1$ causing the unipolarity of the crystal depends on the difference in the electrochemical potentials $\Delta\mu$ of the opposite ends of the domain. Let the high concentration of surface levels satisfy the condition

$$2/q^2N < \chi_\mu. \tag{3.85}$$

If the contribution of the field $\mathscr{E}_1$ to the free energy (3.82) is insignificant, then when condition (3.85) is satisfied the free energy of a single-domain crystal in the presence of surface levels is less than the free energy of the same volume isolated in an infinite homogeneous crystal. Thus, the electronic surface states make a negative contribution to the free energy of the ferroelectric, where, as is seen from (3.84), the thinner the crystal, the greater this contribution in absolute value. This explains why the tendency toward the single-domain structure is observed in sufficiently thin layers of ferroelectrics. At the same time, if the surface levels are not taken into account, then, as was shown in Section 3.2, thinning of the ferroelectric layer for $L < l_D$ leads to the opposite tendency—splitting up of the crystal into domains. The expression for the free energy (3.82) with consideration of (3.84) permits one to determine all the phenomenological parameters and evaluate the effect of the surface levels on them. Thus, renormalization of the coefficient α according to (3.84) leads to a shift of the Curie point toward higher temperatures:

$$\Delta T = T_0' - T_0 = \frac{C}{4\pi L}\left(\chi_\mu - \frac{2}{q^2 N}\right), \tag{3.86}$$

where C is the Curie-Weiss constant. To evaluate the magnitude of the shift ΔT, we set $q\chi_\mu D \simeq 1$ eV and $L \simeq 10^{-4}$ cm. The shift ΔT with these values is of the order of a degree for triglycine sulfate, and of the order of several degrees for BaTiO$_3$. A decrease in L simultaneously leads to an increase of the spontaneous induction. These effects were actually observed for thin crystals of triglycine sulfate.

The second limiting case corresponds to low concentration of surface levels. For example, for $N < 5 \times 10^{12}$ cm$^{-2}\cdot$eV^{-1} and $\epsilon_l^{-1} l \simeq 10^{-7}$ cm, we obtain $\alpha^* \gg 1$. Substitution of $\alpha^* \gg 1$ into (3.83) leads to the following expressions:

$$\begin{aligned}\mathscr{E}_1 &\simeq (qL)^{-1}(\Delta\chi + \varphi_{m+} - \varphi_{m-}), \\ \alpha_L &\simeq \alpha + L^{-1}\left(\frac{8\pi l}{\varepsilon_l} - \chi_\chi\right).\end{aligned} \tag{3.87}$$

It is seen from (3.87) that for a sufficiently low concentration of surface levels, the free energy of the ferroelectric capacitor ceases to depend on the parameters of these levels and is completely determined by the properties of the surface layer and the difference in potential discontinuities at the opposite surfaces $\chi_- - \chi_+$ and $\varphi_{m+} - \varphi_{m-}$. This is caused by the absence of internal screening in the thin ferroelectric because of the low density of space and surface charges. All the phenomenological parameters and the temperature of the phase transition can be obtained from (3.82) and (3.87). Equations (3.87) illustrate well the role of surface levels and the difference in values of the electron affinity χ for the opposite ends of the domain. Actually, if the surface levels are neglected ($N < 5 \cdot 10^{12}$ cm$^{-2}\cdot$eV^{-1}) and $\chi_\chi \simeq 0$ is assumed in (3.87), then renormalization of the coefficient α leads to a shift of the Curie point toward lower temperatures by an amount $\Delta T = 2lC/\epsilon_l L$, which is hundreds of degrees for BaTiO$_3$ with $\epsilon_l^{-1} l \simeq 10^{-7}$ cm and $L \simeq 10^{-4}$ cm.

Thus, the surface electron states in the ferroelectric, like the local electronic levels in the volume, give a significant contribution to the free energy of the crystal, and it is necessary to take them into account in analyzing screening of the spontaneous polarization.

3.7. Effect of Surface Levels on the Schottky Barrier in Ferroelectrics

It is well known that the surface levels in semiconductors play a significant role in the formation of a barrier with contact of the semiconductor with a metal electrode.

Actually, let a metal with a work function A_1 contact a semiconductor having a work function A_2, and let $A_1 > A_2$. In this case, balancing of the chemical potentials of the metal and semiconductor will be accompanied by a transfer of electrons from the semiconductor to the metal. As a result, a space charge and a corresponding surface curvature of the bands are formed in the semiconductor. The thickness d of this barrier (a depletion layer in the simplest case) is determined by the concentration of donors (or acceptors) N_d in the semiconductor:

$$d = \left(\frac{2\varphi_0 \varepsilon}{4\pi q N_d}\right)^{1/2}, \qquad (3.88)$$

where φ_0 is the contact potential difference or the barrier height. The magnitude of $q\varphi_0$ equals in turn the difference of the work functions $A_1 - A_2$. The theory of the semiconductor–metal contact, proposed by Schottky [80], ignored the surface levels in the semiconductor. Bardeen [81] showed that when the surface levels in the Schottky barrier theory are considered, a screening effect occurs in the presence of surface states, where the barrier height ceases to depend on the work function of the metal. Numerous experiments with semiconductors have confirmed the conclusion about the essential role of the surface levels.

A Schottky barrier is also formed with the contact of a ferroelectric and a metal, which has been experimentally investigated in a number of works, predominantly for $BaTiO_3$ [82-85]. The physical nature of the Schottky barrier for ferroelectrics must have interesting peculiarities. The formation of the barrier at the metal-ferroelectric boundary must be caused by internal screening of the spontaneous induction in the ferroelectric. Then, as was shown in the preceding section, the surface level must play an essential role. It follows from relation (3.88) that the parameters of the Schottky barrier for ferroelectrics must depend significantly on the temperature, particularly near the phase transition. However, the temperature dependence cannot be obtained directly from (3.88) and $\epsilon = \epsilon(T)$, since this formula does not take into account the ferroelectric nonlinearity. We will discuss below, following the work of Selyuk [77], the mechanism of the formation of the Schottky barrier in a ferroelectric and, in particular, the role of surface levels in this mechanism.

We will consider the contact of a single-domain ferroelectric with a metal in the presence of a gap or surface layer between them, not taking into account, as everywhere above, the nature of this gap or layer. The band diagram of the ferroelectric in equilibrium with the metal is presented in Figure 3.7. The Fermi level in Figure 3.7 is denoted by the dashed line. The following notation will be used everywhere below: φ_{i0} is the height of the surface barrier, χ_i is the electron affinity, μ_i is the distance of the electrochemical potential of the uncharged surface from the bottom of the conduction band, E_F is the distance from the Fermi level to the bottom of the conduction band in the quasi-neutral region of the crystal, Q_{pi} is the charge on the ferroelectric-gap interface, ϵ_l and l are the dielectric constant and gap thickness, respectively, and $\mathscr{E}_{li}$ is the field in the gap. The index i takes two values, plus and minus, for the positive and negative ends of the domain, respectively. Screening of the spontaneous induction D is provided both by the surface charge Q_{pi} as well as by the space charge with density ρ. For the simplest model of a depletion layer, the density of the space charge $\rho = qN_d$ is constant within the barrier, where N_d is the concentration of donors or acceptors in the ferroelectric. We denote the effective thickness of the barrier by d_i. The origin of the coordinates is placed at the ferroelectric–gap boundary.

For the model described above, the problem of the screening of the spontaneous induction is formulated with the help of the equation of state (3.1), the Poisson equation (3.4), the expression for the surface charge Q_{pi} (3.73), and the following boundary conditions:

$$\epsilon_l \mathscr{E}_{li} - D_i^* = 4\pi Q_{pi}, \tag{3.89}$$

$$\mathscr{E}_{li} = \frac{\varphi_m - (\varphi_{i0} + \chi_i)}{l}, \tag{3.90}$$

$$D_{z=0} = D_i^*, \quad D_{z=d_i} = D_0. \tag{3.91}$$

The problem under consideration was solved in [77] in the linear approximation:

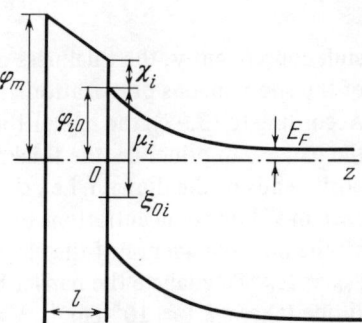

Figure 3.7. Curvature of the bands near the metal–ferroelectric contact; ξ_{0i} is the electrochemical potential of the neutral surface.

$$D_i = 4\pi P_i + \varepsilon \mathscr{E}(z), \tag{3.92}$$

where D_i and P_i are the projections onto the z axis of the spontaneous induction and polarization, $\mathscr{E}(z)$ is the field in the ferroelectric, and ϵ is the dielectric constant of the ferroelectric. By definition $P_- = P_0$, $P_+ = -P_0$, where P_0 is the spontaneous polarization. We present here the expressions obtained in the linear approximation for the height and thickness of the Schottky barrier φ_{i0} and d_i:

$$\varphi_{i0} = \frac{f_i}{1+\alpha^*} + \frac{\psi_{1i}}{(1+\alpha^*)^2} - \left\{ \frac{\psi_{1i}^2}{(1+\alpha^*)^4} + \frac{2\psi_{1i}}{(1+\alpha^*)^2} \left[\frac{f_i}{1+\alpha^*} - E_F \right] \right\}^{1/2}, \tag{3.93}$$

$$d_i = \frac{D_0}{qN_d} - \frac{\mathscr{E}_{li}}{qN_d}\varepsilon_l - 4\pi \frac{N(\varphi_{i0} - \mu_i)}{N_d}, \tag{3.94}$$

where

$$\psi_{1i} = (4\pi q^2 N^2)^{-1}\varepsilon N_d, \tag{3.95}$$

$$f_i = \frac{P_i}{qN} + \alpha^*(\varphi_m - \chi_i) + \mu_i, \tag{3.96}$$

while α^* satisfies (3.77).

We first analyze expression (3.94) for the thickness of the barrier d_i. Since $\mathscr{E}_{li} \ll D_0$, the second term in (3.94) can be neglected. If one then keeps in mind (3.73), then the final expression for d_i has the form

$$d_i = \frac{D_0}{qN_d} - \frac{4\pi Q_{pi}}{qN_d}. \tag{3.97}$$

If the spontaneous induction of the ferroelectric is completely screened by the charge of the surface levels, then $D_0 = 4\pi Q_{pi}$ and the barrier is not formed in accordance with (3.97). The field in the gap $\mathscr{E}_{li}$ equals zero in this case. If, conversely, the charge of the surface levels can be neglected as compared to the spontaneous polarization (if, for example, for $BaTiO_3$, $N \ll 10^{14}$ cm$^{-2} \cdot$ eV^{-1}), then

$$d_i \simeq D_0/qN_d \tag{3.98}$$

and, consequently, the thickness of the barrier is determined by the magnitude of the spontaneous polarization and the concentration of donors (acceptors). According to (3.97), the higher the concentration of surface levels, the thinner the barrier. In principle, the thickness of the barrier is not the same at the opposite ends of the domain, i.e., $d_+ \neq d_-$. This can be caused both by the difference in volume concentrations of the donors and acceptors $N_d \neq N_a$ as well as by the noncoincidence of the electrochemical potentials of the opposite surfaces $\xi_{+0} \neq \xi_{-0}$. To analyze the barrier height φ_{i0}, we make use of the following values. If one takes $\epsilon_l l^{-1} \simeq 10^8$ cm^{-1}, $N \simeq 4 \cdot 10^{14}$ cm$^{-2} \cdot$ eV^{-1}, $N_d \simeq 10^{18}$ cm^{-3}, $\epsilon \simeq$

200, and $P_0 \simeq 8 \cdot 10^4$ cgs esu, then according to (3.77) and (3.95), $\alpha^* \simeq 10^{-1}$, $\psi_{li} \simeq 10^{-3}$ eV, and $(qN)^{-1}P_0 \simeq 0.4$ eV.

The second term in (3.93) can be neglected in this case, and the expression for the barrier height φ_{i0} has the form

$$\varphi_{i0} \simeq (1 + \alpha^*)^{-1} \left[\frac{P_i}{qN} + \alpha^* (\varphi_m - \chi_i) + \right.$$

$$\left. + \mu_i - \sqrt{2\psi_{1i} \left(\frac{f_i}{1 + \alpha^*} - E_F \right)} \right]. \quad (3.99)$$

As is seen from (3.99), the barrier height φ_{i0} depends on the concentration of surface levels. One should set $P_i = 0$ in (3.99) to determine φ_{i0} in the paraelectric region. Expression (3.99) then coincides with the expression found in [86] for the height of the Schottky barrier in the presence of surface levels in the semiconductor, and, as was shown in [86], with the proper choice of α^*, the theoretical values of φ_{i0} agree well with the experimental values. According to (3.99), the height of the Schottky barrier, in principle, is not the same at the opposite ends of the domain. If one also neglects the difference of μ_i and χ_i for the opposite ends, then, according to (3.99):

$$\varphi_{0-} - \varphi_{0+} \simeq 2 \frac{P_0}{qN}. \quad (3.100)$$

Substituting $P_0 \simeq 20 \cdot 10^{-6}$ C/cm^2 and $N \simeq 4 \cdot 10^{14}$ cm$^{-2} \cdot$ eV^{-1} into (3.100), we have $\varphi_{0-} - \varphi_{0+} \simeq 0.6$ eV. This agrees with the measurements of Murakami [84, 85] for barium titanate. In SbSI, on the contrary, $\varphi_{0+} > \varphi_{0-}$. This result is based on the investigation of SbSI as a single-domain diode, in which the rectification factor reached 100 [87]. Investigation of the photoconductivity in the charge-screening region in SbSI leads to the same conclusion, which we will discuss in Chapter 8. The height of the barrier $|\varphi_{0i}|$ and also the difference $\varphi_{0-} - \varphi_{0+}$ increase with a decrease in concentration of surface levels. Thus, for $N \ll 10^{14}$ cm$^{-2} \cdot$ eV^{-1}, $|\varphi_{0i}| \gg 0.6$ eV. This is due to the fact that in the absence of surface levels, screening of the spontaneous polarization by the space charge in the ferroelectric leads to a sharp surface curvature of the bands (cf. Section 3.1). Thus, the agreement of the experimental and theoretical data in the study of the Schottky barrier in ferroelectrics is based essentially on the assumption of the presence of surface levels.

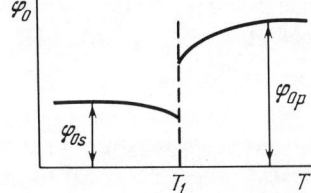

Figure 3.8. Dependence of the height of the potential barrier on temperature.

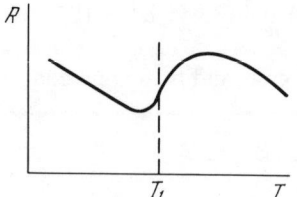

Figure 3.9. Experimental dependence of the resistance of a ferroelectric diode on temperature.

According to (3.99) and (3.95), the height of the Schottky barrier depends on the temperature, since $\epsilon = \epsilon(T)$. An anomaly of φ_{i0} related to the Curie-Weiss law must be observed near the phase transition. This anomaly is more notable, the higher the concentration of donors (acceptors) N_d. Figure 3.8 presents the temperature dependence of φ_0, calculated by Selyuk under the assumption that the level of the electrochemical potential drops in the paraelectric phase ($\mu_p > \mu_s$). Figure 3.9 presents the experimental temperature dependence of the resistance of a single-domain Schottky diode in $BaTiO_3$. Comparison of the experimental and theoretical curves indicates the possible role of the actual domain structure, particularly near the phase transition, and also the deviation of the experimental dependence $\epsilon = \epsilon(T)$ from the theoretical, related possibly to broadening of the phase transition.

3.8 Effect of Screening and Surface Levels on the Domain Structure

Until now, we have considered screening of the spontaneous polarization in a single-domain crystal. However, as was shown in Sections 3.2 and 3.6, the possibility of the existence of a single-domain crystal of finite size is determined both by the screening length and by the concentration of surface levels. Thus, by changing the concentration of carriers in the volume and at the surface levels (for example, due to nonequilibrium carriers), one can in principle change the domain structure of the ferroelectric. The effect of illumination and the corresponding nonequilibrium carriers on the domain structure has been called the photo-domain effect in the literature.

The equilibrium domain structure can be obtained from the condition of the minimum of the free energy of the multidomain crystal F consisting of the electrostatic energy of the domains F_{el} (or the depolarization energy) and the elastic energy of the domain walls F_{ela} (neglecting the internal stresses and other defects in the crystal) [14]:

$$F = F_{el} + F_{ela}. \qquad (3.101)$$

We consider a plane-parallel sheet of ferroelectric of thickness L, whose surface is perpendicular to the direction of spontaneous polarization with plane domain boundaries parallel to the direction of the spontaneous polarization (c-domain structure). The width of the domains $d_0 \ll L$. The expression for the surface electrostatic energy density in this case has the form [14]

$$F_{el} = \frac{3,4}{1 + \sqrt{\epsilon_z \epsilon_x}} P_0^2 d_0, \qquad (3.102)$$

where ϵ_z, ϵ_x are the components of the dielectric tensor and P_0 is the spontaneous polarization.

The surface elastic energy density is given by the expression

$$F_{ela} = \frac{1}{d_0} L\sigma, \qquad (3.103)$$

where σ is the elastic energy per unit area of the domain wall.

The condition of the minimum of F leads to the following expression for the "equilibrium" width of the domains:

$$d_0 = \frac{1}{P_0} \left[\frac{\sigma L \sqrt{\epsilon_z \epsilon_x}}{3,4} \right]^{1/2} \qquad (3.104)$$

Screening of the spontaneous polarization by carriers in the volume of the crystal decreases the electrostatic energy F_{el} and correspondingly increases d_0 as compared to (3.104). Thus, for a sufficiently high concentration of carriers $n_0 = \tilde{n}_0$ in the volume of the crystal, the latter must remain single domain. We evaluate $\tilde{n}_0$ starting from the mechanism of volume screening [62]. The crystal becomes single domain if the screening energy for the two opposite ends of the domain equals the energy of the domain walls. We will assume that the screening energy F_{scr} is proportional to the Debye length l_D, i.e., $F_{scr} = 2l_D k$, while the energy of the domain walls per unit area of the crystal F_{ela} is proportional to the thickness of the 180° domain wall, i.e., $F_{ela} = (L/d_0)(4\pi\epsilon_z \varkappa k)^{1/2}$. Here $(4\pi\epsilon_z \varkappa)^{1/2}$ is the thickness of the 180° domain wall [62] and k is a proportionality coefficient. Substituting the value of the Debye length (3.41) and the expression for d_0 (3.104) into the equality $F_{scr} \simeq F_{ela}$, we arrive at the following value for the concentration $n_0 = \tilde{n}_0$ corresponding to the stable single-domain state:

$$\tilde{n}_0 = \frac{kT\sigma}{3.4\pi q \varkappa P_0^2 L} \sqrt{\epsilon_z \epsilon_x}. \qquad (3.105)$$

Substituting the values of $P_0, \epsilon_z, \epsilon_x, \sigma$, and $\varkappa$ for a sheet of BaTiO$_3$ of thickness $L \simeq 0.1$ cm into (3.105) leads to $\tilde{n}_0 \simeq 10^{16}$ cm^{-3}, according to [62]. Such a high concentration of carriers, as was shown in Section 3.2, can occur at the screening length in BaTiO$_3$. One should also keep in mind that the concentration of carriers captured by shallow traps can also appear in the expression for the screening length l_D, which also makes the value of $\tilde{n}_0$ plausible.

Nonetheless, experimental data in a number of cases do not fit into this simple theory or contradict it. In this respect, we recall once again the tendency toward the c-single-domain structure in thin-sheet crystals of $BaTiO_3$, in which it would seem that the a-single-domain structure would be more favorable from the point of view of the mechanism of volume screening. But data from investigation of the photodomain effect is most interesting in this respect. Thus, for example, an increase in concentration of carriers in the volume with elimination of $BaTiO_3$ accelerates the transition of the c-single-domain crystal to the c,a-multidomain state and affects in several cases the equilibrium domain structure by changing the relation between the volume and configurations of the c- and a-domains. This will be discussed in more detail in Chapter 6. We will discuss here one of the possible mechanisms of these phenomena: the effect of surface electronic states on the domain structure of the ferroelectric.

3.8.1. *Free Energy of a Multidomain Crystal with Consideration of Free Surface Electronic States.* The equilibrium domain structure corresponds to the minimum of the free energy of the multidomain crystal. In order to take into account the effect of the surface electronic states on the domain structure, one should obtain an expression for the free energy F with consideration of both the space and surface charges.

As before, we consider a crystal free of mechanical stresses and defects:

$$F = \int F(\mathbf{r}) \, dv. \qquad (3.106)$$

Here, $F(\mathbf{r})$ is the free energy density at a point whose radius vector is $\mathbf{r}$, and dv is the volume element. Integration is carried out over all space. We represent $F(\mathbf{r})$ in the form

$$F(\mathbf{r}) = F_D(\mathbf{r}) + F_\zeta(\mathbf{r}) + F_\varkappa(\mathbf{r}), \qquad (3.107)$$

where the terms $F_D(\mathbf{r})$ and $F_\zeta(\mathbf{r})$ are caused by the electric induction and the change in number of charge carriers, respectively, and $F_\varkappa(\mathbf{r})$ is the inhomogeneity energy, or the correlation energy proportional to the square of the derivatives of the polarization with respect to the coordinates [cf., for example, (3.3)]. In the particular case, the term $F_\zeta(\mathbf{r})$ is the change in free energy of the crystal caused by optical recharging of the levels, i.e., without the formation of a space charge, and corresponds to the term describing the free energy of the electron subsystem in expression (1.14). Thus, $F_D(\mathbf{r})$ and $F_\zeta(\mathbf{r})$ describe, respectively, the two basic mechanisms for the effect of the electron subsystem on the free energy: the effect of the space charges caused by spontaneous polarization and the effect of the electronic states in the quasi-neutral volume and at the surface.

It is well known [88] that

$$F_D(\mathbf{r}) = F_{\mathscr{E}=0} + \int_{\mathscr{E}=0}^{\mathscr{E}} \frac{\mathscr{E}}{4\pi} d(4\pi\mathbf{P} + \mathscr{E}). \tag{3.108}$$

Starting from the definition of the chemical potential, one can write

$$F_\zeta(\mathbf{r}) = \int_0^{n(\mathbf{r})} \zeta(\mathbf{r}) \, dn(\mathbf{r}), \tag{3.109}$$

where $\zeta(\mathbf{r})$ and $n(\mathbf{r})$ are the chemical potential and concentration of charges at the point $\mathbf{r}$, i.e., $n(\mathbf{r}) = -q^{-1}\rho(\mathbf{r})$, where $\rho(\mathbf{r})$ is the charge density. The integration in (3.109) is carried out for $\mathbf{D}$ = const, i.e., under the condition that the change in concentration of the carriers is not accompanied by the formation of charges. This condition is most simply taken into account in calculating $F_\zeta(\mathbf{r})$ by considering that the carriers are not charged ($q = 0$).

Expression (3.108) can be transformed to the form

$$F_D(\mathbf{r}) = F_0 + \int_0^P \mathscr{E} \, d\mathbf{P} + \frac{\mathscr{E}^2}{8\pi}, \tag{3.110}$$

where F_0 corresponds to $P = 0$, or to the form

$$F_D(\mathbf{r}) = F_{\mathscr{E}=0} + \int_{P_0}^P \mathscr{E} \, d\mathbf{P} + \frac{\mathscr{E}^2}{8\pi}, \tag{3.111}$$

where $\mathbf{P}_0$ is the spontaneous polarization. It is seen from (3.110) that the first two terms are the well-known expression for the free energy density of the ferroelectric, since $\mathscr{E} = dP(\mathbf{P})/d\mathbf{P}$. Consequently, one can write

$$F_D(\mathbf{r}) = F(\mathbf{P}) + \frac{\mathscr{E}^2}{8\pi}. \tag{3.112}$$

We represent (3.111) in an analogous manner:

$$F_D(\mathbf{r}) = F(\mathbf{P}_0) + \int_{P_0}^P \mathscr{E} \, d\mathbf{P} + \frac{\mathscr{E}^2}{8\pi}. \tag{3.113}$$

In the case of a linear relation of the induction and field [cf. (3.92)],

$$D_k = D_{0k} + \varepsilon_{hh}\mathscr{E}_k \quad (k = x, y, z), \tag{3.114}$$

expression (3.113) takes the form

$$F_D(\mathbf{r}) = F(\mathbf{P}_0) + \frac{\mathscr{E}\mathbf{D}}{8\pi} - \frac{1}{2}\mathbf{P}_0\mathscr{E}. \tag{3.115}$$

It is assumed in (3.114) that the dielectric tensor ϵ_{kl} is reduced to the principal axes x, y, z. Since

$$\int \frac{\mathscr{E} \mathbf{D}}{8\pi} dv = \frac{1}{2} \int \rho\varphi \, dv, \qquad (3.116)$$

where ρ and φ are the density of free charges and the potential of the point $\mathbf{r}$ and the integration is carried out over all space, then it is natural to call the second and third terms in (3.115) the energy density of the free and bound charges, respectively. The term $\frac{1}{2}\mathbf{P}_0\mathscr{E}$ is the depolarization energy density if the source of the field $\mathscr{E}$ is the discontinuity of the vector $\mathbf{P}$ at the surface of the crystal. This term was denoted by F_{el} above.

Strictly speaking, relations (3.114) and (3.115) can be used for a ferroelectric only for

$$|\mathbf{P} - \mathbf{P}_0| \ll |\mathbf{P}_0|. \qquad (3.117)$$

Nonetheless, the approximations (3.114) and (3.115) have to be used in the theory of domain structure, since the obtained equations cannot be integrated otherwise. The source of the error thus committed and the method for decreasing it are clearly seen in Figure 3.10. The curve represents the dependence of P on $\mathscr{E}$ typical for ferroelectrics in the one-dimensional case. The area of the shaded figure is $\int_{P_0}^{P} \mathscr{E} \, dP$. The curve is replaced by the tangent in the linear approximation, and we obtain a right triangle with hypotenuse 1 in place of this figure. The error thus committed can be decreased by taking a larger value of ϵ, i.e., by replacing the tangent with the line 2. Thus, in using the linear approximation (3.114), one must remember this produces a necessarily larger value of the free energy of the crystal broken into domains. The result can be improved if one takes ϵ exceeding the experimental value found in small fields for the single-domain crystal.

The inhomogeneity energy $F_\varkappa(\mathbf{r})$ appearing in (3.107) is significant in the boundary region of the neighboring domains. We have then neglected the correlation energy, as everywhere above (cf. Section 3.2). Thus, as at the beginning of this section, $F_\varkappa(\mathbf{r})$ means the elastic energy of the domain walls.

We find $F_\zeta(\mathbf{r})$ [cf. (3.109)] for carriers localized in the volume of the crystal

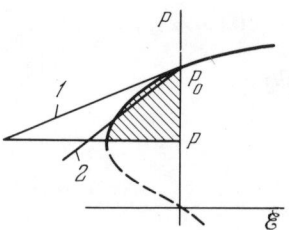

Figure 3.10. *Linear approximation of the equation of state of a ferroelectric.*

SCREENING OF SPONTANEOUS POLARIZATION

and at its surface. The condition $q = 0$ in the first case permits one to replace the chemical potential $\zeta(\mathbf{r})$ in the integral by the Fermi level ξ_F, constant along the entire crystal. In fact, the dependence $\zeta(\mathbf{r})$, generally speaking, can be caused by (1) the difference in chemical binding energy of the charge carriers with the lattice at different points of space because of inhomogeneity of the crystal structure or distribution of the impurities, and (2) the macroscopic electric field. The condition $q = 0$ eliminates the second cause, while the assumption of homogeneity of the crystal eliminates the first. Thus, $\zeta(\mathbf{r}) = $ const. The chemical potential equals the Fermi level ξ_F in the neutral region of the crystal. Thus, $\zeta(\mathbf{r}) = \xi_F = $ const within the homogeneous crystal. If the change in concentration of carriers is small as compared to the total concentration in the crystal, one can neglect the dependence of ξ_F on n. Then we obtain, for the part $F_\zeta(\mathbf{r})$ related to the volume,

$$F_\zeta(\mathbf{r}) = \int_0^n \zeta \, dn \simeq -\xi_F q^{-1} \rho(\mathbf{r}). \tag{3.118}$$

For carriers localized at the surface, the chemical potential depends on the concentration of carriers $n_i = -q^{-1} Q_i$ (Q_i is the surface charge density) in accordance with (3.72). The difference in the electrochemical potentials $\xi_i - \xi_{0i}$ appearing in (3.73) can be replaced by the difference in the corresponding chemical potentials $\zeta_i - \zeta_{0i}$. Then we obtain from (3.109)

$$F_{\zeta i} = -q^{-1} \int_0^{Q_i} \left(\zeta_{0i} - \frac{Q_i}{qN} \right) dQ_i = \frac{Q_i^2}{2q^2 N} - \int_0^{Q_i} \zeta_{0i} \frac{dQ_i}{q}, \tag{3.119}$$

where ζ_{0i} can depend on Q_i. Just this case was considered in Section 3.6, since the dependence of μ_i on P (or on D) determined by equation (3.79) is equivalent to the dependence of ζ_{0i} on Q_i. In fact, for $\alpha^* = 0$,

$$Q_i = P_i \ (P_- = P_0, \ P_+ = -P_0), \text{ while } \zeta_{0i} = -\mu_i \text{ and } \xi_F = -E_F,$$

if one measures the chemical potential of the electrons from the bottom of the conduction band. Then, μ_i is the distance between the bottom of the conduction band and the level of the chemical potential for the neutral surface, and E_F is the energy of the Fermi level for the quasi-neutral region of the crystal. By using this notation, one can easily find the contribution of the surface states to the free energy of the finite crystal in this case. For this, it is sufficient to add expressions (3.119) for the opposite ends:

$$\sum_i F_{\zeta i} = -\int_0^P \frac{\mu_+ - \mu_-}{q} dP + \frac{P^2}{q^2 N}.$$

If a linear relation between $\mu_+ - \mu_-$ and P is assumed in (3.79), we obtain

$$\sum_i F_{\zeta i} = -\frac{1}{2}\Delta\mu\frac{P}{q} - \frac{P}{q}\left(\frac{\mu_+ - \mu_-}{2} - \frac{P}{qN}\right). \tag{3.120}$$

The same correction to the free energy is obtained from (3.82) after substitution of (3.83) and $\alpha^* = 0$. Thus, the contribution (3.120) to the free energy is the result of the effect of the surface states on the ferroelectric properties. This effect as applied to the single-domain crystal was the subject of Section 3.6.

The expressions found here permit one to calculate the free energy of the multidomain crystal with consideration of the space and surface charges. Knowing it, one can establish which of the domain structures under consideration is energetically more favorable, and one can also trace the possible effect of non-equilibrium electrons on the domain structure.

3.8.2. *Comparison of the Energy Advantage of the a- and c-Single-Domain Structures.* We now turn to an investigation of the effect of the space charges and surface levels on the tendency in orientation of the polarization **P** with respect to the sheet crystal of the $BaTiO_3$ type. We compare the energy of single-domain crystals with the orientation of the polarization vector **P** parallel to the surface (*a*-single-domain crystal) and perpendicular to it (*c*-single-domain crystal). The thickness of the sample L is assumed much greater than the Debye length l_D. The crystal is placed in vacuum.

The free energy per unit surface area of the sheet can be represented on the basis of (3.106), (3.107), (3.115), and (3.116) in the form

$$F = F(P_0) + F_P + F_\rho + F_\zeta, \tag{3.121}$$

where $F(P_0)$ is the volume energy density of an infinite sample and F_P is the depolarization energy:

$$F_P = -\frac{1}{2}\sum_i \int_0^\infty P_{0i}\mathscr{E}_i dz = -\frac{1}{2}\sum_i P_{0i}\varphi_{Si}, \tag{3.122}$$

where P_{0i} and $\mathscr{E}_i$ are the projections of the spontaneous polarization and field onto the z axis perpendicular to the surface of the crystal. The index i, as before, denotes "+," "−," and "*a*" for the corresponding surfaces. At the surface of the crystal, $z = 0$; $z \to \infty$ at its center. Moreover, $P_{0-} = P_0 > 0$, $P_{0+} = -P_0$, $P_{0a} = 0$. The symbol Σ indicates summation over all surfaces of the crystal. The energy of the free charges is given by the expression

$$F_\rho = \frac{1}{2}\sum_i \left\{\int_0^\infty \rho_i\varphi_i dz + Q_i\varphi_{Si}\right\}, \tag{3.123}$$

where ρ_i is the space-charge density, Q_i is the surface charge density, φ_i and φ_{Si}

are the potentials of points of the volume and surface measured from the potential at the middle of the crystal. The surface charge is produced by carriers localized at surface levels forming a quasicontinuous band with density N per unit energy interval. From Figure 3.11, which represent the curvature of the bands near the ith end of the ferroelectric, and equation (3.72), we obtain

$$Q_i = -qN(\mu_i - E_F + q\varphi_{si}). \qquad (3.124)$$

In this case, with consideration of the fact that $\zeta_{0i} = -\mu_i$ and $\xi_F = -E_F$, the electrochemical potential of the uncharged surface is $\xi_{0i} = \zeta_{0i} - q\varphi_{si} = -\mu_i - q\varphi_{si}$, while $\xi_i = \xi_F = -E_F$.

We determine F_ζ, the part of the free energy per unit area caused by the redistribution of the charge carriers over the states. In accordance with equations (3.106), (3.118), and (3.119) we obtain

$$F_\zeta = -q^{-1}\xi_F \sum_i \int_0^\infty \rho_i dz + \frac{Q_i^2}{2Nq^2} - \int_0^{Q_i} \frac{\zeta_{0i} dQ}{q}. \qquad (3.125)$$

Calculation of the integrals in (3.123) and (3.125) is related to the necessity of first solving the problem of the distribution of $\varphi_i(z)$ and $\rho(z)$. We limit ourselves to rough estimates based on the simplest approximation for the distribution of the charge density:

$$\rho_i = \begin{cases} \rho_{0i} & \text{for } 0 < z < L_i, \\ 0 & \text{for } z > L_i, \end{cases} \qquad (3.126)$$

where the effective length L_i is determined by the condition

$$\int_0^\infty \rho_i dz = Q_{bi} = \rho_{0i} L_i. \qquad (3.127)$$

Here, Q_{bi} is the space charge per unit surface area of the crystal and ρ_{0i} is the space-charge density at $z = 0$. The solution of the Poisson equation in the

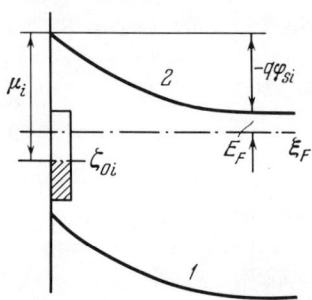

Figure 3.11. Curvature of the bands near the surface of a ferroelectric crystal; 1 and 2 are the boundaries of the forbidden band.

approximation of a linear relation between the induction and field with the distribution (3.126) and boundary conditions

$$\varphi_i = \begin{cases} \varphi_{Si} & \text{for } z = 0, \\ 0 & \text{for } z = L_i \end{cases}$$

gives

$$\varphi_i = \varphi_{Si}\left(1 - \frac{z}{L_i}\right)^2, \qquad (3.128)$$

where $L_i = (2\pi\rho_{0i})^{-1/2}(-\epsilon_i\varphi_{Si})^{1/2}$ and ϵ_i is the dielectric constant of the crystal ($\epsilon_+ = \epsilon_- = \epsilon_c$).

We have from (3.127)

$$Q_{Si} = -\frac{\varphi_{Si}}{|\varphi_{Si}|}\sqrt{\frac{-\epsilon_i\rho_{0i}\varphi_{Si}}{2\pi}}. \qquad (3.129)$$

The potential φ_{Si} is found from the boundary condition for the polarization,

$$P_{0i} = Q_i + Q_{Si}, \qquad (3.130)$$

by substituting (3.124) and (3.129):

$$-q\varphi_{Si} = \frac{P_{0i}}{qN} + \mu_i - E_F + f_i + \frac{\varphi_{Si}}{|\varphi_{Si}|}\left[f_i^2 + 2f_i\left(\frac{P_{0i}}{qN} + \mu_i - E_F\right)\right]^{1/2}, \qquad (3.131)$$

where

$$f_i = (4\pi q^3 N^2)^{-1}\rho_{0i}\epsilon_i. \qquad (3.132)$$

The distributions (3.126) and (3.128) permit one to calculate the integrals of (3.123) and (3.125) and to obtain

$$F_\rho = \sum_i \left(\frac{1}{2}P_{0i}\varphi_{Si} - \frac{1}{3}\varphi_{Si}Q_{bi}\right), \qquad (3.133)$$

$$F_\zeta = \sum_i \left(\frac{Q_i^2}{2q^2 N} - q^{-1}v_i\mu_i Q_{bi} + q^{-1}v_i\mu_i P_{0i}\right), \qquad (3.134)$$

where $v_i \lesssim 1$ is determined by the equality

$$\int_0^{Q_i} \zeta_{0i}dQ_i = v_i\zeta_{0i}Q_i = -v_i\mu_i Q_i. \qquad (3.135)$$

The boundary condition for the polarization (3.130) is taken into account in writing (3.134).

Finally, we find, from (3.121), (3.122), (3.133), (3.134), (3.124), and (3.129),

$$F = F(P_0) + \sum_i \left\{ \frac{P_{0i}}{q} v_i \mu_i + \frac{N}{2} (\mu_i - E_F + q\varphi_{si})^2 + \right.$$
$$\left. + \frac{\varphi_{si}}{|\varphi_{si}|} \left[\frac{1}{3} \varphi_{si} + q^{-1}(v_i\mu_i - E_F) \right] \sqrt{\frac{-\varepsilon_i \rho_{0i} \varphi_{si}}{2\pi}} \right\}. \quad (3.136)$$

If $N \to 0$, we obtain naturally from (3.131), (3.132), (3.124), (3.129), and (3.136):

$$\varphi_{si} = -\frac{2\pi P_0^2}{\varepsilon_i \rho_{0i}} \quad \text{and} \quad F = F(P_0) + \sum_i \frac{2\pi P_{0i}^3}{3\rho_{0i}\varepsilon_i}.$$

Thus, it is seen that the c-domain structure makes a positive contribution to the free energy of the crystal, since ρ_{0i} and P_{0i} are of the same sign.

We assume that N is sufficiently large. For the estimates we use $N = 4 \cdot 10^{14}$ cm^{-2}·eV^{-1}. For $|q^{-1}\rho_{0i}| = 10^{19}$ cm^{-3}, $P_0 = 7.8 \cdot 10^4$ cgs esu, and $\varepsilon_i = 200$, we obtain $|(qN)^{-1}P_0| \simeq 0.4$ eV, and $|f_i| \simeq 0.6 \cdot 10^{-2}$ eV. Moreover, if $\mu_i - E_F \gg f_i$, we have, from (3.136) for the c sheet

$$F_c \simeq F(P_0) - \frac{P_0}{q}\left[(v_+\mu_+ - v_-\mu_-) - \frac{P_0}{qN}\right]. \quad (3.137)$$

It is seen from a comparison with (3.120) that (3.137) coincides with the value of the contribution of the surface levels found in Section 3.6 if $v_+ = \frac{1}{2} + \Delta\mu/4\mu_+$ and $v_- = \frac{1}{2} - \Delta\mu/4\mu_-$. In contrast to (3.120), expression (3.137) is free of the assumption about the linear relation of $\mu_+ - \mu_-$ and P.

For the a sheet, assuming as before that $f_i \simeq f_a$ is small, we obtain, from (3.131), $-q\varphi_{sa} \simeq \mu_a - E_F$, and from (3.136),

$$F_a \simeq F(P_0) - \frac{4}{3}\sqrt{\frac{\varepsilon_a \rho_{0a}(\mu_a - E_F)^3}{2\pi q^3}}, \quad (3.138)$$

where $F_a - F(P_0) < 0$. Thus, the necessary condition for the c-domain structure to be favored is $F_c - F(P_0) < 0$, i.e.,

$$v_+\mu_+ - v_-\mu_- > P_0/qN. \quad (3.139)$$

If $F_a < F_c$ then the a-domain structure must be realized. This condition can be written in the form

$$\frac{4}{3}\sqrt{\frac{\varepsilon_a \rho_{0a}(\mu_a - E_F)}{2\pi q P_0^2}} > \frac{(v_+\mu_+ - v_-\mu_-) - P_0/qN}{\mu_a - E_F}. \quad (3.140)$$

For $|q^{-1}\rho_{0a}| \simeq 10^{19}$ cm^3 relation (3.140) is satisfied for $\varepsilon_a = 4 \cdot 10^3$. For smaller $|\rho_{0a}|$, the inequality (3.140) can be realized only near the transition point where ε_a increases sharply.

Thus, the assumption of the presence of surface levels of sufficiently high density, whose position differs for the opposite ends of the domain, permits one to explain the existence of the c-domain structure in sheet crystals of BaTiO$_3$ and its transformation to the a structure near the transition point. This model also permits one to explain in principle the effect of nonequilibrium carriers on the domain structure and, in particular, on the transition of the c-domain crystal to a c-,a-multidomain state (the photodomain state). This explanation can be based on the assumption of the optical recharging of the surface levels. Actually, we assume that a filling of the surface levels occurs in darkness, for which the relationship between $\nu_+\mu_+ - \nu_-\mu_-$ and $\mu_a - E_F$ provides the advantage to the c-domain structure [i.e., condition (3.140) is violated]. Recharging of the surface levels providing satisfaction of condition (3.140) occurs with illumination of the crystal, which leads to the appearance of a domains. We note also that according to (3.140) the effect of illumination and the corresponding recharging of the surface levels must be more effective near the phase transition where ϵ_a has a maximum.

3.8.3. *The Effect of Screening on the c-Domain Structure of Ferroelectrics.* We now investigate the effect of the screening charges, both space and surface, on the size of the c domains. It is usually assumed that free charges play a secondary role, only fixing the domain structure arising at the phase transition. But the accumulation of compensating charges can make this structure energetically unfavorable and cause its restructuring, as a result of which the equilibrium domain structure corresponding to the minimum of the free energy must ultimately be established. The establishment process can be slow, appearing as aging. Finding the equilibrium structure is advisable even in this case, since the tendency of the domain restructuring is thereby revealed. This process can be accelerated by external effects, for example, by a variable electric field, ultrasonics, or illumination. The resulting equilibrium domain structure must have the greatest stability under the given conditions.

We consider the c-domain structure of a plane ferroelectric sheet of thickness L without electrodes. The crystal is free of internal stresses and is placed in vacuum. Domains of width $d_0 \ll L$ form a periodic banded structure. The screening charge can be localized both in the surface region as well as at the surface itself, if there are surface states of sufficient concentration. We assume that the thickness of the region in which the compensating space charge can be localized is $l_\rho \ll L$. The free energy per unit area of the sheet in the case under consideration can be represented by analogy to (3.121) on the basis of (3.106), (3.107), (3.115), and (3.116) in the form

$$F = F(P_0) + F_P + F_\rho + F_\sigma + F_\zeta + F_\varkappa. \qquad (3.141)$$

We assume here the linear relation (3.114) of the induction and field:

$$D_x = \varepsilon_a \mathscr{E}_x, \qquad D_z = 4\pi P_{0z} + \varepsilon_c \mathscr{E}_z; \qquad (3.142)$$

P_{0z} is the projection of the spontaneous polarization onto the ferroelectric z axis; the domain walls are perpendicular to the x axis.

The depolarization energy F_P is

$$F_P = -\frac{1}{2} \sum \frac{1}{2d_0} \int_0^{2d_0} dx \, P_{0z} \int_0^\infty E_z dz. \qquad (3.143)$$

The summation symbol indicates summation over both surfaces of the crystal. Because $d_0 \ll L$ and $l_\rho \ll L$, the point $z = \infty$ corresponds to the center of the sheet.

Calculating F_ρ in (3.141), we obtain

$$F_\rho = \frac{1}{2} \sum \frac{1}{2d_0} \int_0^{2d_0} dx \left(\varphi_S Q + \int_0^\infty \rho \varphi \, dz \right), \qquad (3.144)$$

where F_ρ is the energy of the free charges whose volume density is ρ and whose surface density is Q, and φ and ϕ_s are the potentials of points of the volume and surface.

We obtain from (3.109) and (3.106) an expression for the change in coupling energy of the charge carriers with the crystal lattice with their redistribution over the states:

$$F_\zeta = \sum \frac{1}{2d_0} \int_0^{2d_0} dx \left[\int_0^\infty dz \int_0^n \zeta \, dn + \int_0^{n_s} \zeta_S \, dn_S \right]. \qquad (3.145)$$

Here $n = -q^{-1}\rho$ and $n_s = -q^{-1}Q$ are the increase in densities of the carrier in the crystal and on the surface with the redistribution; ζ and ζ_s are the chemical potential of the corresponding localized electrons. The elastic energy of the domain walls is

$$F_\sigma = \frac{\sigma L}{d_0}. \qquad (3.103)$$

The correlation energy, not taken into account in the energy density of the domain boundaries σ, is

$$F_\varkappa = \frac{\varkappa}{2} \sum \frac{1}{2d_0} \int_0^{2d_0} dx \int_0^\infty [(\nabla D_x)^2 + (\nabla D_y)^2 + (\nabla D_z)^2] \, dz. \qquad (3.146)$$

If the screening charge is localized in the space region, whose effective thickness l_ρ is much less than the size of the domains,

$$l_\rho \ll d_0, \qquad (3.147)$$

then simple estimates permit one to reach an important conclusion on the nature of the equilibrium domain structure. In fact, because of (3.147), equations (3.143)–(3.146) give, in order of magnitude,

$$F_P \simeq -\frac{1}{2}\sum \overline{P_{0z}\mathscr{E}_z}|_{z=0}\, l_\rho, \quad F_\varkappa \simeq \frac{\varkappa}{2}\rho^2|_{z=0}\, l_\rho,$$

$$F_\rho \simeq \frac{1}{2}\sum \left[\overline{(\rho\varphi)}|_{z=0}\, l_\rho + Q\overline{\varphi}_s\right],$$

$$F_\zeta \simeq -\sum \left[q^{-1}\overline{(\zeta\rho)}|_{z=0}\, l_\rho - \int_0^{n_s} \zeta_s\, dn_s\right].$$

Here the bars over the letters designate averaging over x. It is seen that F_ρ, F_P, $F_\varkappa$, and F_ζ do not depend on x. Thus, (3.141) takes the form

$$F = \frac{\sigma L}{d_0} + \text{const}.$$

The growth of d_0 is energetically favored. Consequently, under the condition (3.147), the single-domain structure corresponds to the equilibrium state. The equilibrium domain structure of a real crystal in this case must be determined by internal stresses caused, for example, by crystal defects.

Now let

$$l_\rho \gg d_0, \tag{3.148}$$

which is valid for dielectric crystals if, moreover, the change in potential at the surface is sufficiently small (in order that degeneracy not begin). Condition (3.148) indicates that the magnitude of the screening space charge in each domain per unit surface area of the crystal is much less than the total screening charge (the order of the spontaneous polarization) localized at the length l_ρ. Because of this, the screening space charge can be neglected. Then we are justified in setting $\rho = \rho(z) + \rho(x, z) \simeq \rho(z)$, even if $\rho(z)$ and $\rho(x, z)$ are of the same order of magnitude at several points, since $\rho(x, z)$ represents here the space-charge density screening the spontaneous polarization, while $\rho(z)$ is the space-charge density due to nonferroelectric causes, for example, surface states. Since $L \gg L_\rho$, the total surface charge can be comparable to the bound charge. Thus, it is impossible, generally speaking, to completely neglect the space charge, but this does not interfere with finding the distribution of the potential and calculating the free energy as a function of d_0.

We represent the potential in the form

$$\varphi = \varphi(z) + \varphi(x, z), \tag{3.149}$$

where $\varphi(z)$ satisfies the Poisson equation

$$\nabla^2 \varphi(z) = -\frac{4\pi}{\varepsilon_c}\rho(z). \tag{3.150}$$

SCREENING OF SPONTANEOUS POLARIZATION

The alternating part of the potential $\varphi(x, z)$ is then found from the Laplace equation

$$\frac{\varepsilon_c}{\varepsilon_a}\frac{\partial^2 \varphi(x, z)}{\partial z^2} + \frac{\partial^2 \varphi(x, z)}{\partial x^2} = 0 \quad \text{for} \quad z \geqslant 0,$$
$$\frac{\partial^2 \varphi(x, z)}{\partial z^2} + \frac{\partial^2 \varphi(x, z)}{\partial x^2} = 0 \quad \text{for} \quad z < 0. \quad (3.151)$$

The solution of equations (3.151) must satisfy the following boundary conditions:

$$\varphi(x, z) = 0, \quad \frac{\partial \varphi(x, z)}{\partial x} = \frac{\partial \varphi(x, z)}{\partial z} = 0 \quad \text{for} \quad z = \pm \infty, \quad (3.152)$$

$$\varphi(x, z) = \varphi(x + 2md_0, z), \quad m = \pm 1, \pm 2, \ldots, \quad (3.153)$$

$$\left.\frac{\partial \varphi(x, z)}{\partial x}\right|_{z=+0} = \left.\frac{\partial \varphi(x, z)}{\partial x}\right|_{z=-0}, \quad (3.154)$$

$$\varepsilon_c \left.\frac{\partial \varphi(x, z)}{\partial z}\right|_{z=+0} = \left.\frac{\partial \varphi(x, z)}{\partial z}\right|_{z=-0} + 4\pi [P_{0z} - Q(x)]. \quad (3.155)$$

Here, $Q(x)$ is the variable part of the surface-charge density $Q = Q^0 + Q(x)$. The term Q^0, not depending on x, satisfies the boundary condition

$$\varepsilon_c \left.\frac{\partial \varphi(z)}{\partial z}\right|_{z=+0} = \left.\frac{\partial \varphi(z)}{\partial z}\right|_{z=-0} - 4\pi Q^0. \quad (3.156)$$

The equalities (3.155) and (3.156) express the boundary conditions for the induction vector at the surface of the crystal. The density of the surface charge Q is determined by the energy structure of the surface states and the position of the Fermi level at the surface. The equilibrium charge Q from equation (3.72) is

$$Q = -qN(\xi_F - \xi_0), \quad (3.157)$$

where ξ_F is the Fermi level and ξ_0 is the level of the electrochemical potential of the neutral surface. It is clear from Figure 3.12 that

$$\xi_F - \xi_0 = \mu - E_F + q\varphi(0) + q\varphi(x, 0). \quad (3.158)$$

The magnitude of μ determines the position of the surface levels in the forbidden band of the crystal. Since, generally speaking, they can differ for the opposite ends of the domains, we set

$$\mu = \begin{cases} \mu_+ & \text{for} \quad P_{0z} < 0 \text{ (positive end)} \\ \mu_- & \text{for} \quad P_{0z} > 0 \text{ (negative end)} \end{cases}$$

It is convenient to expand the functions $P_{0z}(x)$ and μ in a Fourier series:

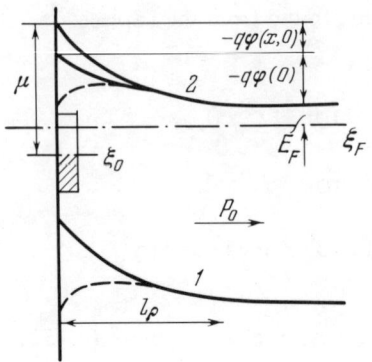

Figure 3.12. Curvature of the band near one of the c-domains. The dashed line shows the curvature of the bands in the neighboring domain; 1 and 2 are the boundaries of the forbidden band.

$$P_{0z} = \frac{4}{\pi} P_0 \sum_{m=0}^{\infty} \frac{\sin(2m+1)\frac{\pi}{d_0}x}{2m+1}, \qquad (3.159)$$

$$\mu = \frac{\mu_+ - \mu_-}{2} - \frac{2(\mu_+ - \mu_-)}{\pi} \sum_{m=0}^{\infty} \frac{\sin(2m+1)\frac{\pi}{d_0}x}{2m+1}. \qquad (3.160)$$

Substituting (3.160) and (3.158) into (3.157), we obtain $Q = Q^0 + Q(x)$, where

$$Q^0 = qN\left[E_F - \frac{\mu_+ + \mu_-}{2} - q\varphi(0)\right], \qquad (3.161)$$

$$Q(x) = -qN\left[q\varphi(x,0) - \frac{2(\mu_+ - \mu_-)}{\pi} \sum_{m=0}^{\infty} \frac{\sin(2m+1)\frac{\pi}{d_0}x}{2m+1}\right]. \qquad (3.162)$$

The solution of equation (3.151) satisfying the conditions (3.152)–(3.155), with consideration of (3.159) and (3.162), has the form

$$\varphi(x,z) = \begin{cases} \sum_{m=0}^{\infty} \varphi_{m0} \exp\left[-(2m+1)\frac{\pi}{d_0}\sqrt{\frac{\varepsilon_a}{\varepsilon_c}}z\right] \sin(2m+1)\frac{\pi}{d_0}x \\ \qquad\qquad\qquad\qquad\qquad\qquad\qquad \text{for } z \geqslant 0, \\ \sum_{m=0}^{\infty} \varphi_{m0} \exp\left[(2m+1)\frac{\pi}{d_0}z\right]\sin(2m+1)\frac{\pi}{d_0}x \\ \qquad\qquad\qquad\qquad\qquad\qquad\qquad \text{for } z < 0, \end{cases} \qquad (3.163)$$

where

$$\varphi_{m0} = -\frac{16P_0 d_0\left[1 - \frac{qN}{2P_0}(\mu_+ - \mu_-)\right]}{\pi(2m+1)^2\left[1 + \sqrt{\varepsilon_a \varepsilon_c} + (2m+1)^{-1}4Nq^2 d_0\right]}. \qquad (3.164)$$

As for the first term in equation (3.149), it is determined by equation (3.150) and the boundary conditions (3.156) and (3.161). Consequently, it does not depend on the presence of domains and is due only to the fact that the positions of the chemical potential within the crystal and at its surface are not the same [i.e., the difference $E_F - \frac{1}{2}(\mu_+ + \mu_-)$].

As $N \to 0$, equations (3.163) and (3.164) naturally do not differ from the analogous equations obtained in the theory not taking into account the screening charges. It gives such a large curvature of the bands at the surface ($\varphi_{m0} \simeq 10$ V for $d_0 \simeq 10^{-4}$ for BaTiO$_3$) that degeneracy begins and the condition for their applicability is violated. Degeneracy leads in turn to condition (3.147) and, consequently, to the favorability of the single-domain state. Only if

$$q|\varphi_0| = 16 P_0 d_0 q (\pi \sqrt{\varepsilon_a \varepsilon_c})^{-1} < E_F \qquad (3.165)$$

can the structure be formed with the equilibrium width of the domains d_0 equal to

$$d_0 = \sqrt{\frac{\sigma \sqrt{\varepsilon_a \varepsilon_c} L}{3.4 P_0^2}}. \qquad (3.104)$$

This can occur in the direct vicinity of the point of a phase transition, particularly a second-order transition in the region of temperatures where $P_0 \to 0$. Thus, the temperature for which the inequality (3.165) is violated corresponds to a transition from the single-domain to the multidomain equilibrium structure. It is natural to identify it with the experimentally observed point for the disappearance of unipolarity. At lower temperatures, the existing domains caused by internal stresses cannot completely eliminate the tendency toward the single-domain structure.

Let us consider what changes the existence of surface states can lead to. In the case under consideration, when condition (3.148) is satisfied, the correlation energy (3.146) can be neglected. Equations (3.149), (3.163), and (3.164) determine the field in the crystal and permit one to find the free energy (3.141). In calculating the first integral in (3.145), we will assume as above that the change in the number of carriers is negligibly small as compared to their total number. Then, one can make use of equation (3.118). The second integral is found from (3.119) and the conditions $\xi_{0i} = -\mu_i$ and $\xi_F = -E_F$:

$$\int_0^{n_S} \zeta_S \, dn_S = \frac{Q^2}{2q^2 N} + \frac{\nu}{q} \mu Q, \qquad (3.166)$$

where $\nu \lesssim 1$ is determined by the equality (3.135):

$$\int_0^Q \zeta_S \, dQ = -\nu \mu Q.$$

We will further neglect the possible dependence of the value of v on Q, approximating v by some constant parameter of the order of unity. After making use of equations (3.161)-(3.163) and also (3.141), (3.143), (3.144), and the condition of electrical neutrality,

$$\int_0^\infty \rho\, dz + \frac{1}{2d_0}\int_0^{2d_0} Q\, dx = 0,$$

we finally obtain

$$F = F(P_0) + \frac{\sigma}{d_0} L - \frac{2v-1}{4} N(\mu_+ - \mu_-)^2 +$$

$$+ \sum_{m=0}^\infty \frac{32 P_0^2 d_0 \left[1 - \frac{qN}{2P_0}(2v-1)(\mu_+ - \mu_-)\right]\left[1 - \frac{qN}{2P_0}(\mu_+ - \mu_-)\right]}{\pi^2(2m+1)^3\left(1 + \sqrt{\varepsilon_a \varepsilon_c} + \frac{4Nq^2 d_0}{2m+1}\right)} +$$

$$+ F\left(E_F - \frac{\mu_+ + \mu_-}{2}\right), \tag{3.167}$$

where the last term does not depend on d_0 and is due to the difference $E_F - \frac{1}{2}(\mu_+ + \mu_-)$:

$$F\left(E_F - \frac{\mu_+ + \mu_-}{2}\right) = \int_0^\infty \rho(z)\varphi(z)\, dz + Q^0\varphi(0) -$$

$$- 2N\left(E_F - \frac{\mu_+ + \mu_-}{2} - v\right)\left[E_F - \frac{\mu_+ + \mu_-}{2} - q\varphi(0)\right]. \tag{3.168}$$

If $P_{0z} = 0$ (the paraphase or a-domain structure), then equation (3.168) determines the energy of the surface charged layers caused by the surface state, which are located at a distance $\mu = (v/2)(\mu_+ + \mu_-)$ below the edge of the construction band.

It was assumed above that $l_\rho \ll L$. If, on the contrary, $L \ll l_\rho$, the volume distribution of the charge can be neglected. Then the problem is simplified and leads to equation (3.167) with $F[E_F - (\mu_+ + \mu_-/2)] = 0$. We assume for the estimates that the density of surface states $N \simeq 4.2 \cdot 10^{14}$ cm^{-2} · eV^{-1} [89], and $\mu_+ - \mu_- \simeq 1$ eV. Then the quantities in brackets in equations (3.167) and (3.164) are of the order of unity if $P_0 \simeq 10^5$ cgs esu. Thus, for

$$4Nq^2 d_0 \ll \sqrt{\varepsilon_a \varepsilon_c} \tag{3.169}$$

the equilibrium width of the domain is determined by a formula differing from (3.104) only by a factor of the order of unity for P_0^2. When the inequality (3.169) changes to the inequality in the opposite sense, F decreases with increasing d_0, i.e., the single-domain structure becomes more favored.

SCREENING OF SPONTANEOUS POLARIZATION

Thus, for crystals with surface states, the condition of applicability of relation (3.104) is the simultaneous fulfillment of inequalities (3.169) and (3.165). The temperature at which at least one of them is violated is the point for the disappearance of unipolarity. For temperatures sufficiently low compared to the Curie temperature, the single-domain structure is more favored only if internal stresses and defects do not lead to the multidomain structure.

The model of screening with consideration of surface levels also permits one to explain the effect of illumination on the c-domain structure of a ferroelectric semiconductor. As the analysis carried out above indicates, it makes sense to consider separately the effect of nonequilibrium carriers on the c-domain structure in the ferroelectric region and on the c-domain structure arising with the transition of the crystal from the paraelectric region to the ferroelectric (i.e., at unipolarity or natural polarization).

In the first case, we imagine a single-domain ferroelectric at a temperature below the Curie temperature. Let conditions (3.165) and (3.169) be satisfied, and let condition (3.139) be also simultaneously satisfied, i.e.,

$$1 - \frac{qN}{2P_0}(\mu_+ - \mu_-) \ll 1. \qquad (3.170)$$

In this case, the depolarization energy represented by the sum in (3.167) is small and, correspondingly, $d_0 \simeq L$, i.e., the ferroelectric is in the c-single-domain state. If it is assumed that illumination in the ferroelectric region leads to the recharging of the surface levels for which $\mu_+ - \mu_-$ decreases and inequality (3.170) accordingly changes to the condition

$$1 - \frac{qN}{2P_0}(\mu_+ - \mu_-) \simeq 1, \qquad (3.171)$$

then the single-domain state changes to the multidomain, with the equilibrium width of the domain of the order of (3.104). This effect can display a temperature dependence due to the temperature dependence of the spontaneous polarization P_0. Thus, with separation from the Curie point and increase of P_0, the effect of recharging the surface levels and the corresponding photodomain effect decreases.

In the second case, we imagine a ferroelectric near the phase-transition temperature or in the paraelectric region. Neglecting the space charge, i.e., the term $F[E_F - (\mu_+ + \mu_-)/2]$, the free energy of the crystal has the form

$$F \simeq F(P_0) - \frac{2v-1}{4}(\mu_+ - \mu_-)^2 \simeq$$
$$\simeq F(P_0) - \frac{P_0}{q}\left[v(\mu_+ - \mu_-) - \frac{P_0}{qN}\right], \qquad (3.172)$$

i.e., it coincides in essence with (3.137). As was shown in Section 3.6, (3.172) corresponds to the presence in the crystal of an effective internal field $\mathcal{E}_1$, which according to (3.83) is proportional to $\Delta\mu$. In the paraelectric region, according to (3.79), $\Delta\mu = \mu_+ - \mu_-$; this causes unipolarity of the crystal not related to spontaneous polarization. This unipolarity, i.e., the condition $\Delta\mu \neq 0$ and the presence of the corresponding field $\mathcal{E}_1$, causes natural polarization of the crystal which arises with the transition to the ferroelectric region in the absence of an external field. We assume further that illumination of the crystal in the paraelectric region increases the unipolarity $\Delta\mu$ due to recharging of the surface levels. Thus, illumination in the paraelectric region increases the natural polarization and consequently leads to enlargement of the domains. The photodomain effect in the ferroelectric region near the Curie point is a superposition of these two phenomena.

3.9. Screening and Periodic Structure of Interphase Boundaries in a Ferroelectric

In ferroelectrics displaying a first-order phase transition, the coexistence of both phases, ferroelectric and paraelectric, can occur near the Curie temperature, which leads to the formation of boundaries between the phases. Near the interphase boundaries, the spontaneous polarization is screened by carriers, which leads to curvature of the bands and the corresponding "intrinsic" field effect. The problem of the screening near the interphase boundary can be considered by analogy to the "intrinsic" field effect at the contact of a ferroelectric and semiconductor (Section 3.3). It is essential here that the screening energy be not too large compared with the elastic energy of the interphase boundaries; in the opposite case, the interphase boundary is not formed and corresponding coexistence of both phases in a finite temperature interval becomes energetically unfavored.

Diffusion of carriers through the crystal can lead to a volume redistribution where a periodic or layered structure consisting of alternating sections of para- and ferroelectric phases arises. The regions of the paraelectric phase have a higher concentration of carriers, while the regions of the ferroelectric phase have a corresponding lower concentration, which is due to the shift in the Curie point (cf. Chapter 1). The period of such a structure is determined by the diffusion length, since the carriers cannot diffuse over a greater distance over their lifetime. Thus, the generation of an equidistant system of interphase boundaries in the ferroelectric is the result, on the one hand, of diffusion-drift equilibrium in the ferroelectric semiconductor and, on the other hand, the interaction of the electron subsystem with the crystal lattice. This phenomenon was first observed experimentally by the author (cf. Chapter 6), and was considered theoretically by Larkin and Khmel'nitskii, whose work [90] we discuss here.

In accordance with (1.7), we write the free energy of the ferroelectric near the Curie point in the form of a sum of the free energy of the lattice F_1 and the free energy of the electron subsystem F_2, where F_1 satisfies, for example, (3.3), while F_2 satisfies the relation analogous to (1.10):

SCREENING OF SPONTANEOUS POLARIZATION

$$F_2 = E_n n_n + E_p n_p, \tag{3.173}$$

where E_n and E_p are the energies of the electrons and holes, respectively, n_n and n_p are their concentration, and $E_g = E_n + E_p$ is the width of the forbidden band. Here, as in Chapter 1, we neglect the change in effective masses of the carriers near the phase transition and, moreover, consider only the free carriers in the bands, neglecting the localized states. The equation of state (3.1) with consideration of (3.3) and (3.173) has the form[†]

$$\alpha'_T (T - T_0) P - |c| P^3 + f P^5 - \varkappa \frac{d^2 P}{dz^2} + n_n \frac{dE_n}{dP} + n_p \frac{dE_p}{dP} - \mathscr{E} = 0. \tag{3.174}$$

The electric field $\mathscr{E}$ satisfies the Poisson equation

$$\text{div}(\mathscr{E} + 4\pi P) = 4\pi q (n_n - n_p). \tag{3.175}$$

We consider the temperature interval near the phase transition $\Delta T \simeq T_1 - T_0 \simeq 3(c^2/4\alpha'_T f)$ in which the coexistence of both phases is possible. We show that this possibility is determined by the concentration of electrons, on which the position and concentration of interphase boundaries also depends. For this, we supplement the system (3.174)–(3.175) by the continuity equation

$$\frac{d\rho}{dt} + \frac{dj}{dz} = 0, \tag{3.176}$$

where ρ is the charge density and j is the current density. Substituting the expressions for the current and charge density for the electrons and holes, respectively, into (3.176), we arrive at the following equation:

$$\frac{d}{dz}\left[D_n \left\{\frac{dn_n}{dz} + n_n \left(\frac{dE_n}{dz} + qE\right)\Big/kT\right\}\right] + G - R = 0,$$
$$\frac{d}{dz}\left[D_p \left\{\frac{dn_p}{dz} + n_p \left(\frac{dE_p}{dz} - qE\right)\Big/kT\right\}\right] + G - R = 0, \tag{3.177}$$

where D_n and D_p are the diffusion coefficients of the electrons and holes, respectively, and G and R are the generation and recombination rates, respectively, of the nonequilibrium electron–hole pairs.

We find the periodic solutions of the system of equations (3.174), (3.175), and (3.177) by assuming that the period is greater than the correlation length $l_\varkappa = [\varkappa/\alpha'_T(T_1 - T_0)]^{1/2}$ and the Debye length l_D, where we will assume that $L_d \gg l_\varkappa, l_D$, where $L_d = (D\tau)^{1/2}$ is the diffusion length (τ is the lifetime). One can neglect the current through the boundary and the last two terms in (3.177) near the interphase boundaries. This leads to the following values of the concentration:

[†]The coefficients in (3.174) differ from the coefficients in (3.3) by a factor of 4π in the corresponding powers.

$$n_n \simeq \exp\left[-\frac{E_n + q\varphi}{kT}\right], \quad n_p \simeq \exp\left[-\frac{E_p + q\varphi}{kT}\right], \quad (3.178)$$

where φ is the potential. For distances from the boundary large compared to l_x and l_D ($q\varphi \ll E_n, E_p$), we obtain from (3.175) and (3.178) the condition of electrical neutrality $n_n = n_p = n/2$. We have the ratio of the concentration of carriers in the ferro- and paraelectric phase (on both sides of the boundary):

$$n^s/n^p = \exp(-\Delta E_g/2kT), \quad (3.179)$$

where $\Delta E_g = E_g^s - E_g^p$ is the discontinuity in the width of the forbidden band at the first-order phase transition. Far from the interphase boundary, one can neglect the derivatives dE_n/dz and dE_p/dz, and the corresponding equation (3.177) with consideration of $n_n = n_p$ leads to the form

$$\frac{d^2n}{dz^2} + \frac{1}{D}(G - R) = 0, \quad (3.180)$$

where the bipolar diffusion coefficient $D = 2(D_n + D_p)^{-1}$. If one substitutes into (3.180) the recombination rate $R = n/\tau$ and the generation rate $G = n_0/\tau$, where n_0 is the concentration of carriers being generated, which is proportional to the light intensity and the quantum yield, then equation (3.180) takes the form

$$\frac{d^2n}{dz^2} + \frac{1}{L_d^2}(n_0 - n) = 0. \quad (3.181)$$

Here, L_d is the diffusion length depending, through D and the lifetime τ, on the phase in which the carriers are excited. The solution of equation (3.181) must correspond to the boundary conditions in the form (3.179) and (3.182):

$$D^s \frac{dn^s}{dz} = D^p \frac{dn^p}{dz}. \quad (3.182)$$

The latter condition is the equality of the diffusion currents through the interphase boundary. The condition of boundary equilibrium in the presence of electrons can be obtained from (3.174). Multiplying (3.174) by dP/dz and integrating over z in the region $l_x, l_D < z < L_d$, we obtain, with consideration of (3.175),

$$F^s(P) + kTn^s + \frac{(\mathscr{E}^s)^2}{8\pi} = F^p(P) + kTn^p + \frac{(\mathscr{E}^p)^2}{8\pi}, \quad (3.183)$$

where $\mathscr{E}^s$ and $\mathscr{E}^p$ are the electric field in the ferro- and paraelectric phase, respectively. Thus, the condition of boundary equilibrium involves the equality of the pressure on it from both phases. Neglecting the pressure of the electric field in (3.183) and expressing $F^s - F^p$ in terms of the heat of transition ΔQ, we have, with consideration of (3.179),

$$\frac{\Delta Q}{T_1^2}(T_1 - T) = n^s \left[\exp\left(\frac{\Delta E_g}{2kT}\right) - 1 \right]. \qquad (3.184)$$

The concentration n^s appearing in (3.184), equal to n_0 in order of magnitude, is determined from the diffusion equation (3.181). Relation (3.184) can be considered as the dependence of the temperature interval $T_1 - T$, within which the interphase boundaries exist, on the concentration of carriers $n^s \simeq n_0$. The temperature interval within which the interphase boundaries exist broadens with an increased concentration of the nonequilibrium carriers. On the contrary, below some critical carrier concentration n_k, interphase boundaries cannot be observed at all in a finite temperature interval. To determine from (3.184) the temperature interval for the existence of an interphase boundary, we find an expression for n^s appearing in (3.184). The solution of equation (3.181) for the boundary conditions indicated above gives

$$n(z) = n_0^s - \frac{n_0^s - n_0^p \exp(-\Delta E_g/2kT)}{1 + \sqrt{D^s \tau^p / D^p \tau^s} \exp(-\Delta E_g/2kT)} \exp\left(\frac{z}{L_d^s}\right), \qquad z < 0,$$

$$n(z) = n_0^p + \frac{n_0^s \exp\left(\frac{\Delta E_g}{2kT}\right) - n_0^p}{1 + \sqrt{D^p \tau^s / D^s \tau^p} \exp(\Delta E_g/2kT)} \exp\left(-\frac{z}{L_d^p}\right), \qquad z > 0. \qquad (3.185)$$

Here, $z < 0$ corresponds to the ferroelectric phase, $z > 0$ corresponds to the paraelectric phase, and $z = 0$ corresponds to the boundary; $n_0^s = n_{z=-\infty}$, and $n_0^p = n_{z=+\infty}$. Using the solution (3.185) and setting $D_s \simeq D^p$, we obtain the following expression for the concentration of carriers near the boundary n_s:

$$n_S = n_0^s \frac{\alpha}{1+\alpha}\left(1 + \sqrt{\frac{\tau^p}{\tau^s}}\right), \qquad (3.186)$$

where $\alpha = (\tau^p/\tau^s)^{1/2} \exp(-\Delta E_g/2kT)$ taking into account that $n_0^p = (\tau^p/\tau^s)n_0^s$. We will omit the upper index in n_0^s below. Substituting (3.186) into (3.184), we determine the temperature of the phase transition:

$$T_1' = T_1 - \frac{T_1^2}{\Delta Q} n_0 \frac{\alpha}{1+\alpha}\left(1 + \sqrt{\frac{\tau^p}{\tau^s}}\right)\left[\exp\left(\frac{\Delta E_g}{2kT}\right) - 1\right]. \qquad (3.187)$$

Here, $T_1' - T_1$ is the shift in temperature of the phase transition caused by the contribution of the electron subsystem. It is seen from (3.187) that this shift is directly proportional to the concentration of carriers n_0. Thus for $\Delta E_g \ll kT$ and $\tau^s \simeq \tau^p$, equation (3.187) coincides with the analogous result obtained in Chapter 1 [cf. (1.36)]. The dashed line represents the line of the phase transition on the (n_0, T) diagram (Figure 3.13).

It follows from (3.187) that for a specified carrier concentration n_0, one interphase boundary can be in equilibrium at one specific temperature T'_1. The system of interphase boundaries must correspond to a finite temperature interval $T_1 - T$. This is easily understood from equations (3.186) and (3.181). The concentration n^s depends not only on the light intensity, but also on the distance between the boundaries. The group of boundaries must then form an equidistant structure. This follows from equation (3.181) and boundary condition (3.179), since only the periodic solution of (3.181) leads to the same value of n^s at each boundary. The condition of boundary equilibrium (3.183) determines the relations between the dimensions x and y of the ferroelectric and paraelectric sections of this structure. We determine n^s from equation (3.181) for the periodic solution and, by substituting this value into the equilibrium condition (3.184), we have

$$\tanh \frac{x}{2L^s_d} \coth \frac{y}{2L^p_d} = \frac{1}{\alpha} \frac{t - \alpha \sqrt{\tau^p/\tau^s}}{1 - t}, \qquad (3.188)$$

where

$$t = \frac{(T_1 - T) \Delta Q}{T_1^2 n_0 \left[\exp\left(\frac{\Delta E_g}{2kT}\right) - 1 \right]}. \qquad (3.189)$$

It is seen from (3.188) that the periodic structure of the interphase boundaries can exist in the temperature interval

$$\frac{T_1^2 n_0 \left[\exp\left(\frac{\Delta E_g}{2kT}\right) - 1 \right]}{\Delta Q} > T_1 - T > \frac{T_1^2 n_0 \left[\exp\left(\frac{\Delta E_g}{2kT}\right) - 1 \right]}{\Delta Q} \alpha \sqrt{\frac{\tau^p}{\tau^s}}. \qquad (3.190)$$

This region is bounded by the dot-dashed lines in Figure 3.13. There can be any period of the structure, but the limitation on the dimensions of the phases exists:

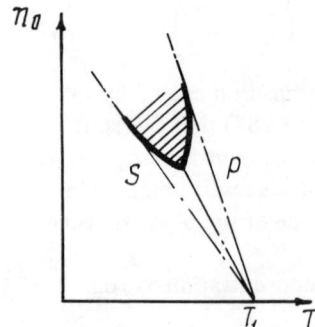

Figure 3.13. (n_0, T) diagram of the state of coexisting phases.

$$\tanh\frac{x}{2L_d^s} \leqslant \frac{1}{\alpha}\frac{t-\alpha\sqrt{\tau^p/\tau^s}}{1-t},$$

$$\alpha\sqrt{\tau^p/\tau^s} < t < \frac{\alpha}{1+\alpha}(1+\sqrt{\tau^p/\tau^s}), \qquad (3.191)$$

$$\tanh\frac{y}{2L_d^p} \leqslant \alpha\frac{1-t}{t-\alpha\sqrt{\tau^p/\tau^s}}, \quad \frac{\alpha}{1+\alpha}(1+\sqrt{\tau^p/\tau^s}) < t < 1.$$

The equality in (3.191) corresponds to equilibrium of one layer of the ferroelectric phase inside the paraelectric and a layer of the paraelectric phase inside the ferroelectric.

The condition of mechanical equilibrium, while imposing significant limitations on the equilibrium dimensions of the layers, does not determine them uniquely. To determine the period, the obtained equilibrium condition (3.188) must be supplemented by the condition of stability with respect to fluctuations in the growth and disappearance of regions. Such a condition at thermodynamic equilibrium is equivalent to the condition of the minimum of the free energy. One can assume that, in the case under consideration, the probability of fluctuations is determined by the minimum work. Then, the probability for the growth of a new layer is proportional to $\exp(-2\sigma S/kT)$, while the probability of the disappearance is proportional to $\exp(-R_{\min}S/kT)$, where σ is the surface tension and $R_{\min}$ is the minimum work which must be accomplished against the forces of the pressure in order to close the boundaries of the layer. Stability corresponds to the equality of these probabilities:

$$R_{\min} = 2\sigma. \qquad (3.192)$$

The minimum work is proportional to the concentration of carriers n_0. Thus, for sufficiently small concentrations, $R_{\min}$ at any temperature is less than the surface energy, and the homogeneous state will be stable while the layered structure is metastable (cf. Figure 3.13). The boundary of the region in the (n_0, T) diagram, within which the layered structure is stable while the homogeneous state is metastable, is determined from condition (3.192) for the case where there is at most one layer in the sample. To calculate the minimum work in this case, we find the pressure at the boundary of the layer, which equals the difference of the right and left sides of equation (3.183). Finding n^s from the diffusion equation, we obtain

$$p = Tn_0\left(e^{\Delta E_g/2kT} - 1\right)\alpha\frac{\sqrt{\tau^p/\tau^s}+\tanh(x/2L_d^s)}{1+\alpha\tanh(x/2L_d^s)} - \frac{\Delta Q}{T_0}(T_0 - T). \qquad (3.193)$$

In order to obtain the analogous expression for the layer of the paraelectric phase, it is necessary to replace $\tanh(x/2L_d^s)$ by $\coth(y/2L_d^p)$ in equation (3.193). The equilibrium size is determined from the condition that the pressure vanish, which corresponds to the equality in expression (3.191). Integrating the pressure over x in the limits from the equilibrium size to zero, we find the minimum work. Condition (3.192) has the form

$$\sigma = L_d^s T_0 n_0 \left(e^{\Delta E_g/2kT} - 1\right) \times$$

$$\times \frac{\alpha}{1-\alpha} \left\{ \left[\frac{1-\alpha}{\alpha} t + 1 - \sqrt{\frac{\tau^p}{\tau^s}}\right] \operatorname{arctanh}\left(\frac{t - \alpha\sqrt{\tau^p/\tau^s}}{\alpha(1-t)}\right) - \right.$$

$$\left. - \frac{1 - \alpha\sqrt{\tau^p/\tau^s}}{1+\alpha} \ln\left[\alpha\frac{1 - 2t + \alpha\sqrt{\tau^p/\tau^s}}{\alpha(1-t) + t - \alpha}\right]\right\}. \qquad (3.194)$$

It determines the relation of n_0 and T at the boundary of the region of stability of the layered structure. The second branch of this boundary corresponding to the loss of stability with respect to conversion to the homogeneous ferroelectric phase is described by an equation analogous to (3.194). The phase diagram in the (n_0, T) plane is represented in Figure 3.13. The critical concentration below which the layered structure is metastable at all temperatures is obtained from (3.194) at a temperature equal to the transition temperature:

$$n_k = \frac{\sigma(1-\alpha^2)}{L_d^s T_1 \left[\exp\left(\frac{\Delta E_g}{2kT}\right) - 1\right] \alpha \left(1 - \alpha\sqrt{\tau^p/\tau^s}\right) \ln[2/(1+\alpha)]}. \qquad (3.195)$$

The period of the layered structure is infinite at the boundary of the region of stability and decreases inside. It can be determined from equation (3.192). Calculating the work $R_{\min}$ in closing the boundaries of one layer of the layered structure with a given period and substituting it into equation (3.192), one can determine the period for any values of n_0 and T in the stability region.

The time for establishing the equilibrium period can actually be much greater (much greater than the time for establishing the diffusion-drift equilibrium). It is this that causes the presence of hysteresis in the first-order phase transition.

4

Optical Absorption and the Band Structure of Ferroelectrics

The investigation of the intrinsic optical absorption of ferroelectrics is of significant interest. First, the determination of the nature of the interband transitions from the form of the intrinsic absorption edge permits one to refine the band structure of the crystal in both phases, determine its possible change at the phase transition, and compare it with the calculated data obtained by various methods. Second, one can determine from measurement of the intrinsic absorption edge the anomalies of the width of the forbidden band of the ferroelectric at first- and second-order phase transitions. As already indicated in the first chapter, these anomalies are directly related to anomalies of the heat capacity and can be obtained from the thermodynamic theory of ferroelectric phase transitions. From the point of view of the microscopic model of the interband electron–phonon interaction, the anomaly of the width of the forbidden band E_g at a first-order phase transition of the displacement type is directly the pseudo-Jahn–Teller effect in the crystal. As has been indicated in the second chapter, observation of this effect permits one to determine the electron–phonon interaction constant and calculate with its help the magnitude of a number of photoferroelectric effects, for example, the shift of the Curie point due to nonequilibrium carriers, etc.

Of course, investigation of the intrinsic optical absorption of ferroelectrics permits verifying experimentally the relation between E_g and the heat capacity of the crystal. This relation in the general case for semiconductors can be investigated experimentally only near absolute zero, since it is with a decrease in temperature beginning at the Debye temperature that the heat capacity of the crystal decreases and tends to zero at $T \to 0$. Accordingly, the temperature dependence of the width of the forbidden band must become weak at $T \to 0$. The experimental material for Ge and Si, published in the work of Keyes [37], confirms relation (1.69) presented in Chapter 1 (Figure 4.1). A second way of verifying the general relation between $(\partial E_g/\partial T)_p$ and the heat capacity was noted and developed in the works of the author with his colleagues [91-102].

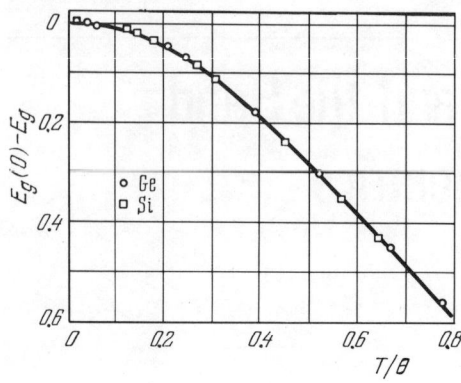

Figure 4.1. Optical width of the forbidden band E_g for Ge and Si as a function of T/θ [37].

This method is based on the change in heat capacity in phase transitions. Phase transitions in ferroelectrics are most convenient in this respect since anomalies in the heat capacity are predicted previously by the Landau–Ginzburg–Devonshire theory in this case. The anomalies of heat capacity of ferroelectrics at first- and second-order phase transitions are different in nature. Since the changes in the temperature dependence of E_g are related to changes in the heat capacity according to the general relation of Keyes, one can expect an experimental verification of the difference in the nature of the anomalies of E_g in the case of first- and second-order phase transitions. This verification was obtained for a number of ferroelectric crystals. The presence of two types of anomalies of the width of the forbidden band is also of practical interest for recording and identifying phase transitions in connection with the difficulty in measuring the heat of transition and the heat capacity, and also in several cases in connection with the difficulty in measuring the dielectric anomalies or with the ambiguity of their nature. This chapter is devoted to a study of both these and other problems related to optical absorption of ferroelectrics.

4.1. Temperature Dependence of the Width of the Forbidden Band near First- and Second-Order Phase Transitions

Determination of the width of the forbidden band E_g in ferroelectrics, for example in [19–102], was carried out over the steepest portion of the dependence of the absorption coefficient on energy, which most likely corresponded to direct optical transitions. As is well known from the physics of semiconductors, determination of E_g is related to a preliminary analysis of the form of the intrinsic optical absorption edge. This analysis for ferroelectrics is a complex independent problem, which we will discuss in Sections 4.2 and 4.3 for several specific examples.

OPTICAL ABSORPTION AND FERROELECTRICS

Investigation of the behavior of the width of the forbidden band in the region of phase transitions was performed first of all for well-studied ferroelectrics with second- and first-order phase transitions.

According to the data from measurements of the heat capacity and dielectric properties, a typical second-order phase transition occurs in triglycine sulfate (abbreviated TGS) and its isomorph triglycine selenate (TGSe). According to [15], the Curie temperature of these crystals are 49 and 22°C, respectively. Above the Curie temperature, the crystals belong to the centrosymmetric point group $2/m$ of the monoclinic system. The ferroelectric properties appear in the direction of the second-order polar axis. Investigation of the intrinsic absorption of TGS and TGSe in unpolarized light was carried out in the range of temperature from -20 to +90°C. The width of the forbidden band for TGS and TGSe at room temperature, determined from the intrinsic absorption edge, is approximately 5.21 eV. The width of the forbidden band E_g of TGS depends linearly on temperature (Figure 4.2) with coefficients $(\partial E_g/\partial T)_p = -(6 \pm 0.25) \cdot 10^{-4}$ eV/deg in the ferroelectric region, and $(\partial E_g/\partial T)_p = -(7.1 \pm 0.35) \cdot 10^{-4}$ eV/deg in the paraelectric region. The magnitude of $(\partial E_g/\partial T)_p$ in the nonpolar phase is greater than in the polar. A discontinuity in the temperature coefficient of the width of the forbidden band $\Delta(\partial E_g/\partial T)_p = +(1.1 \pm 0.6) \cdot 10^{-4}$ eV/deg occurs at the phase transition. Analogous dependences $E_g = E_g(T)$ are obtained for all the investigated samples of TGS of ferroelectric and arbitrary sections. The course of the temperature dependences with heating and cooling coincided.

Figure 4.3 presents the temperature dependence of the width of the forbidden band E_g for TGSe. The magnitude of $(\partial E_g/\partial T)_p$ in the nonpolar phase is

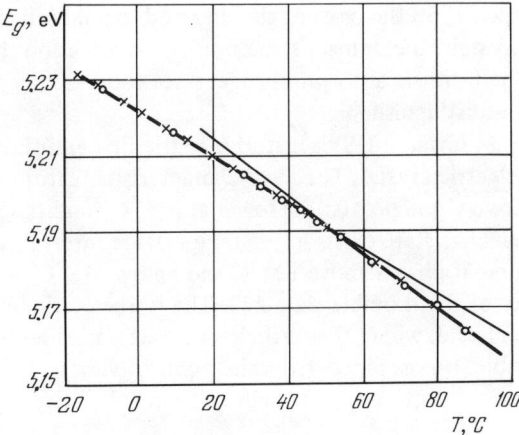

Figure 4.2. Temperature dependence of the width of the forbidden band of TGS in the region of the phase transition. ○, *heating;* ×, *cooling.*

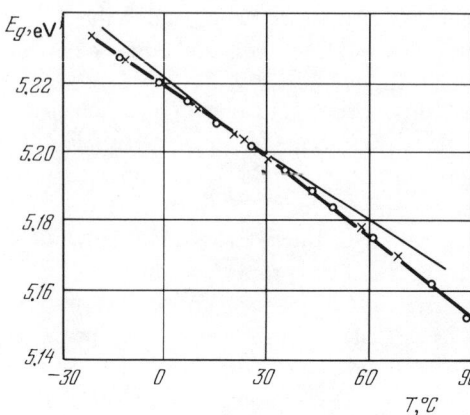

Figure 4.3. Temperature dependence of the width of the forbidden band of TGSe. ○, heating; ×, cooling.

greater than in the polar and is $-(7.8 \pm 0.35) \cdot 10^{-4}$ and $-(6.6 \pm 0.25) \cdot 10^{-4}$ eV/deg in the paraelectric and ferroelectric phases, respectively. A discontinuity in the temperature coefficient of the width of the forbidden band $\Delta(\partial E_g/\partial T)_p = +(1.2 \pm 0.6) \cdot 10^{-4}$ eV/deg occurs at the Curie point and at 22°C for TGSe, analogous to the discontinuity in TGS at 48°C. The Curie temperature for these same crystals of TGS and TGSe was recorded with the help of measurements of the temperature dependence of the spontaneous polarization and dielectric constant. The value of P_0 for TGS at room temperature is about $3 \cdot 10^{-6}$ C/cm², which agrees with the data in the literature. Thus, the second-order phase transition from the ferroelectric phase to the paraelectric in TGS and TGSe at 48 and 22°C, respectively, is accompanied by a discontinuity in the temperature coefficient of the width of the forbidden band $\Delta(\partial E_g/\partial T)_p$, which corresponds to the conclusions reached on the basis of the thermodynamic calculation (Chapter 1). For the TGS crystals, the domain structure has no effect on the results of measurement of the intrinsic absorption edge, since the ferroelectric domains in TGS are optically indistinguishable.

Rochelle salt (abbreviated RS) is historically the first and thus the most investigated ferroelectric crystal. The most characteristic feature of Rochelle salt is the presence of two Curie points: the lower at $-18°$C† and the upper at $+24°$C. The ferroelectric state exists in the temperature interval from -18 to $+24°$C. In the nonpolar phases above $+24°$C and below $-18°$C, the crystals have the symmetry of the rhombic class 222. The ferroelectric phase belongs to the monoclinic class 2, where the ferroelectric axis is parallel to the direction of the initial rhombic X axis. Since, from the point of view of the electrical properties, the phase transitions in Rochelle salt have all the features of a

†Translator's note. The minus sign was inadvertently omitted in the Russian original.

second-order phase transition far from the critical Curie point, it was natural to investigate the behavior of the intrinsic absorption edge of RS in the regions of the upper and lower Curie points. The temperature dependence of the width of the forbidden band E_g obtained by us for RS in the region of the Curie point for a sample of the ferroelectric section (X section) is presented in Figure 4.4. The width of the forbidden band for RS at room temperature, determined from the intrinsic absorption edge, is 5.23 eV. Above and below the Curie point there is a linear dependence of the magnitude of E_g on temperature with the corresponding coefficients $(\partial E_g/\partial T)_p^p = -(6 \pm 1) \cdot 10^{-4}$ eV/deg and $(\partial E_g/\partial T)_p^s = -(3 \pm 1.2) \cdot 10^{-4}$ eV/deg. The magnitude of the coefficient is greater in the paraelectric phase than in the ferroelectric. The phase transition at 24°C is accompanied by the significant discontinuity $\Delta(\partial E_g/\partial T)_p = +(3 \pm 2.2) \cdot 10^{-4}$ eV/deg. The course of the temperature dependence with heating and cooling coincide. Because of the low decomposition temperature of RS (55°C) and the great sensitivity of the crystal to the surrounding conditions of temperature and humidity, measurements could not be made above +40°C. In contrast to TGS crystals, the ferroelectric domains in RS crystals are optically distinguishable. Scattering of light by the domain walls in the X section of the RS crystal can have an effect on the optical transmission of the crystal measured by a spectrophotometer. In order to eliminate this effect, the investigation of the temperature dependence of the intrinsic absorption edge in RS was carried out on samples of Y and Z sections. Figure 4.5 presents the results of investigation of the width of the forbidden band of RS in the region of both phase transitions for samples of X, Y, and Z sections cut from a single crystal. As is seen in the figure, all three dependences display the same change with temperature: breaks occur at -18 and $+24$°C. In the paraelectric phase below -18°C, the value of the coefficient $(\partial E_g/\partial T)_p$ is $-(3.1 \pm 0.9) \cdot 10^{-4}$ eV/deg. In the ferroelectric phase near the lower phase transition, E_g changes more strongly with temperature than in this phase near the upper Curie point. The value of the coefficient $(\partial E_g/\partial T)_s$ in the ferroelectric region near -18°C is $-(6 \pm 1.2) \cdot 10^{-4}$ eV/deg, and the discontinuity of the temperature coefficient $\Delta(\partial E_g/\partial T)_p$, which accompanies the lower phase transition, is $-(3 \pm 2.2) \cdot 10^{-4}$ eV/deg. The temperature depen-

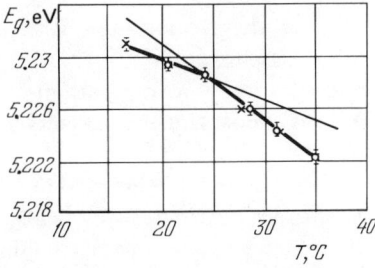

Figure 4.4. Temperature dependence of the width of the forbidden band E_g of Rochelle salt for an X-section sample in the region of the upper Curie point.

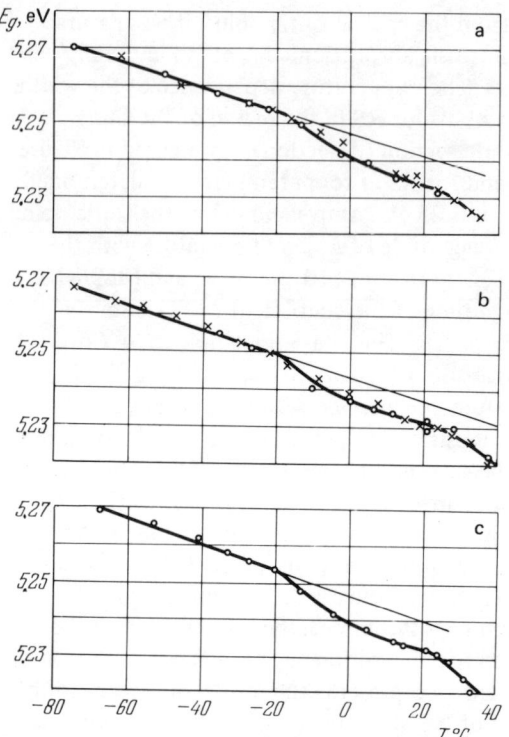

Figure 4.5. Temperature dependence of the width of the forbidden band of RS in the region of the lower and upper Curie points for X-, Y-, and Z-section samples (a, b, and c, respectively).

dence of the dielectric constant ϵ_x and spontaneous polarization P_0 measured on the same samples displayed the typical anomalies at the temperatures -18 and $+24°$C. These data indicate that the breaks in the curves $E_g = E_g(T)$ coincide with the Curie points determined from independent dielectric measurements. The investigations carried out for the three sections of the RS crystal showed an analogous behavior of the width of the forbidden band in the region of the phase transitions for all sections. These data also indicate that the domain structure for RS crystals does not have a significant effect on the measurement of the intrinsic absorption edge.

Thus, the nature of the observed anomalies in E_g in the region of the second-order phase transitions in RS also corresponds to the conclusions of the thermodynamic theory. We draw attention to the fact that the discontinuity $\Delta(\partial E_g/\partial T)_p$ in RS is almost three times greater than in TGS, whereas the discontinuity in the heat capacity in RS is so small that there are no reliable data on it (1-7%, according to the data of [103]). This confirms the assumption made above that the measurement of the anomalies in E_g at first- and second-order phase transitions can be a more sensitive method of recording and identifying the phase transitions in a number of cases than measurement of the anomaly in the heat capacity.

Measurement of the intrinsic optical absorption was also carried out for such a little studied crystal as potassium iodate, which is a ferroelectric below 212°C [104]. According to the data of quadrupole resonance investigation in KIO_3 crystals, there occur several phase transitions: a first-order transition at 75°C, and two phase transitions at temperatures below room temperature. According to x-ray data, KIO_3 has a pseudocubic perovskite-like cell with $a = 8.92$ Å. According to other data, KIO_3 has trigonal symmetry with $a = 4.41$ Å and $\alpha = 89.41°$. The KIO_3 crystals investigated in [95] were octuplets in the form of a cube with significant cloudiness at the growth junctions of the individual octants [105]. The structure, dielectric, and piezoelectric properties of these crystals was studied in [106]. According to these data, the KIO_3 crystals have a pseudocubic lattice with $a = 8.94$ Å, while the anomalies in ϵ and $\tan \delta$ are observed at temperatures less than 212°C. There are no data at all in the literature on the heat capacity of KIO_3.

Investigation of the intrinsic optical absorption edge of KIO_3 was carried out in the range of temperatures from room temperature to 250°C (Figure 4.6). The width of the forbidden band at room temperature is 4.02 eV. Four samples of KIO_3 were cut from the most transparent parts of the octants of single crystals perpendicular to the different directions of the pseudocube edges. Identical results were obtained for all the samples. Three sections of the linear change in E_g with temperature can be isolated in the temperature dependence $E_g = E_g(T)$ for KIO_3, presented in Figure 4.6. The temperature dependence of the width of the forbidden band displays two anomalies at 65 and 180°C. At 65°C, E_g de-

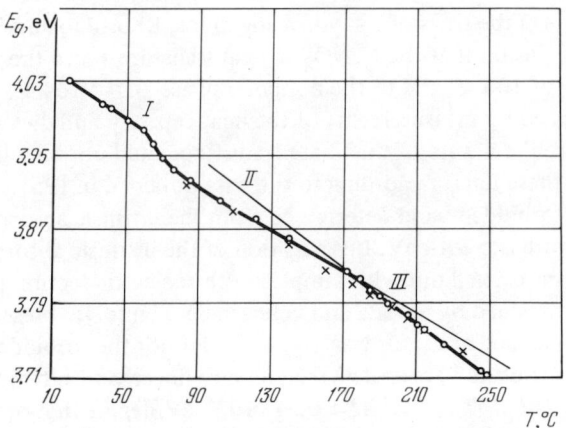

Figure 4.6. Temperature dependence of the width of the forbidden band for KIO_3 in the region of the two phase transitions (close to 65 to 180°C).

creases continuously by an amount ~0.03 eV. At 180°C, a break is observed in the course of the temperature dependence $E_g = E_g(T)$. Above and below 180°C, E_g changes linearly with temperature with coefficients $(\partial E_g/\partial T)_p^s = -(11.6 \pm 0.25) \cdot 10^{-4}$ eV/deg and $(\partial E_g/\partial T)_p^p = -(14.4 \pm 0.35) \cdot 10^{-4}$ eV/deg, respectively. A second-order phase transition evidently occurs at 180°C, accompanied by a discontinuity in the temperature coefficient $\Delta(\partial E_g/\partial T)_p = +(2.8 \pm 0.6) \cdot 10^{-4}$ eV/deg, while the discontinuity in E_g at 65°C is caused by a first-order phase transition. Investigation of the temperature dependences of the dielectric constant and the electric conductivity with the same crystals of KIO_3 revealed anomalies near 65 and 180°C, which confirms the presence of the phase transitions. Measurements of ϵ in KIO_3 crystals at high temperatures are difficult because of high dielectric losses. Thus, data from investigations of the intrinsic absorption edge not only permit one to determine the temperatures of the phase transitions in KIO_3 (65 and 180°C), but also uniquely indicate the nature of these transitions.

Since the results of investigations of intrinsic optical absorption of $BaTiO_3$ and compounds of the $A^V B^{VI} C^{VII}$ group, which are typical ferroelectric semiconductors, with first-order phase transitions are discussed in the following separate sections, we will discuss here several other classes of ferroelectric crystals and individual ferroelectrics. It is typical of all the cases considered below that the study of the dependence $E_g = E_g(T)$ is a reliable method for identifying the phase transition, including singular ferroelectrics.

Until recently, the nature of the phase transition in the ferroelectric $NaNO_2$ had not been established with sufficient reliability.

According to the data of [107], $NaNO_2$ has a first-order phase transition at $T_1 = 163°C$. Above the Curie point, $NaNO_2$ has the rhombic structure (*mmm* group); the crystal belongs to the $2m$ group in the polar phase. The b axis is the ferroelectric axis. On the basis of x-ray investigations, Khoshino and Shibuyya arrived at the conclusion that the $NaNO_2$ crystal transforms into the antiferromagnetic phase at 163°C, and to the nonpolar phase at $T = 164°C$. These results are confirmed by measurements of the heat capacity and dielectric constant, although the nature of the phase transition could not be reliably established from these data. According to the data obtained in [95], the value of the width of the forbidden band determined from the intrinsic absorption edge at room temperature is ~3.14 eV. Investigation of the intrinsic absorption of $NaNO_2$ crystals was carried out with samples of ferroelectric section (Y section in the derivation assumed by Savada and colleagues). Figure 4.7 presents the temperature dependence of E_g for $NaNO_2$. The width of the forbidden band E_g in the ferroelectric and paraelectric regions vary linearly with temperature with a coefficient $(\partial E_g/\partial T)_p = -(7.3 \pm 0.2) \cdot 10^{-4}$ eV/deg. A first-order phase transition occurs at 160°C, accompanied by $\Delta E_g \simeq +0.02$ eV. Thus, investigation of the optical absorption of $NaNO_2$ confirms the data of Savada on the presence of the first-order phase transition at +160°C.

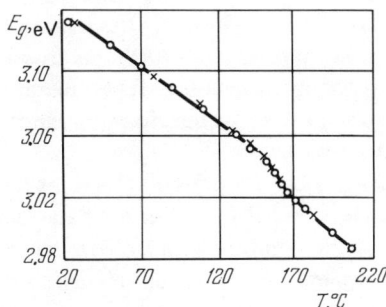

Figure 4.7. Temperature dependence of E_g for $NaNO_2$ in the region of the phase transition.

We also present the result of investigations obtained in [97, 101] for $E_g = E_g(T)$ for a new class of ferroelectrics: the trihydroselenites and their deuterated analogs.

We will discuss first of all the hydroselenites of lithium, sodium, and potassium: $LiH_3(SeO_3)_2$ (HSeL), $NaH_3(SeO_3)_2$ (HSeN), and $KH_3(SeO_3)_2$ (HSeK). As is well known, ferroelectricity in the first two crystals of this group was discovered in 1959 by Pepinsky and Vedam [108]. Intensive investigation of several crystals of this isomorphic series was carried out by Makita, Gavrilov-Podol'skaya, Blints, and others. A broad investigation of the properties of HSeN was carried out by Shuvalov and Ivanov [109]. According to these data, the HSeN crystals display the following ferroelectric phase transitions: a second-order, close to a first-order, phase transition from the paraelectric phase α to the ferroelectric phase β with a change in symmetry $2/m \to 1$ occurs at $-78.6°C$, and a first-order transition from the β phase to a ferroelectric phase γ with a change in symmetry $1 \to m$ occurs at $-172.5°C$. Blints published the results of dielectric measurements of HSeN in the region of the Curie point, but these results cannot give a clear answer about the nature of the phase transitions in view of the narrow temperature interval and the failure of Blints to take into account the domain contribution. Detailed investigations of the dielectric properties and domain structure in HSeN crystals near both phase transitions were carried out in [109]. Measurements of ϵ showed clearly the presence of the two phase transitions in HSeN. The magnitude of the dielectric constant decreased discontinuously at $T = -172.5°C$; temperature hysteresis occurs. Consequently, the transition at $-172.5°C$ is a first-order transition. It was shown in [109] that $KH_3(SeO_3)_2$ experiences a second-order phase transition at $-61.6°C$, accompanied by a dielectric anomaly and the generation of a twinned (probably domain) structure. It was established that in the "paraelectric phase" HSeK belongs to the rhombic class mmm, while in the low-temperature phase the crystal symmetry is 1. The data in the literature on the nature of the low-temperature phase are still insufficient to judge finally the ferroelectric or antiferroelectric nature of the phase transition of this crystal. According to the data obtained in [97], the width of the forbidden band for the investigated isomorphic crystals of the hydroselenites is about the same, and is about 4.9 eV at room temperature. Its temperature behavior, however, appears different for all three crystals, in accordance with the peculiarities of their phase transitions.

Figure 4.8 presents the temperature dependence of the width of the forbidden band for HSeN. In the paraelectric phase, E_g increases linearly with decreasing temperature with a coefficient $(\partial E_g/\partial T)_p = -(7.1 \pm 0.3) \cdot 10^{-4}$ eV/deg. The temperature coefficient changes to the value $-(4.2 \pm 0.3) \cdot 10^{-4}$ eV/deg at the Curie point at $-79°C$. Moreover, the change in the value of E_g by about 0.01 eV occurs in the region of the Curie point. A clearly expressed discontinuity in E_g by about 0.06 eV is observed at $-172.5°C$, where the temperature hysteresis of the lower phase transition is about 13°. The data presented confirm once again the conclusions of Makita about the presence of the two phase transitions in HSeN, and the conclusions reached in [109] about the nature of these phase transitions (the second-order, close to first-order, transition at $-78.6°C$, and a first-order transition at $-172.5°C$). Two systems of antiparallel domains with noncollinear vectors P_0 in the different systems of domains are formed with the transition of the HSeN and HSeK crystals into the ferroelectric triclinic class 1. Scattering of light by the domain walls in HSeN and HSeK can have a significant effect on the optical transmission measured by a spectrophotometer; thus sections for which this effect is reduced to a minimum were taken in [97]. These are the Y section for HSeN and the Z section for HSeK (in the derivation assumed in [109]). In investigation of the intrinsic absorption of HSeN crystals carried out on samples of the Z section and arbitrarily oriented sheet, the same dependences for $E_g = E_g(T)$ were obtained as for the samples of the Y section. Thus, reflection and scattering of light by the domain boundaries is not a significant effect in HSeN. The effect of scattering at the domains for HSeK, on the contrary, must be taken into account [97].

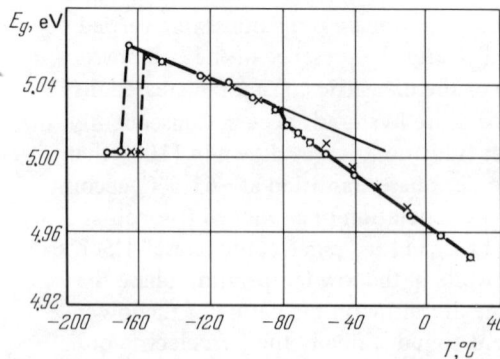

Figure 4.8. Temperature dependence of the width of the forbidden band for a Y section of sodium hydroselenite.

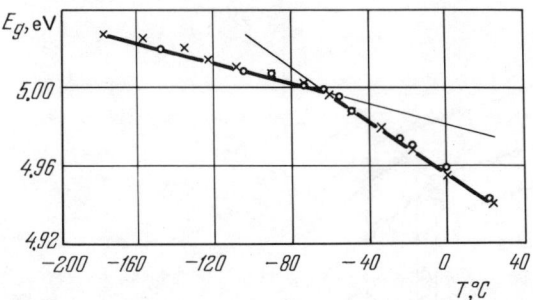

Figure 4.9. Temperature dependence of the width of the forbidden band of a Z section of potassium hydroselenite.

Investigation of the temperature dependence of the intrinsic absorption edge of HSeK crystals was carried out on samples of Z section and Y section. Figure 4.9 presents the dependence $E_g = E_g(T)$ for the Z-section samples of HSeK. The HSeK crystals, as has already been indicated, experience a second-order phase transition at $-61.6°C$. The coefficient $(\partial E_g/\partial T)_p$ is $-(6.8 \pm 0.3) \cdot 10^{-4}$ eV/deg in the "paraelectric" phase of HSeK. The temperature coefficient changes to $-(2.4 \pm 0.3) \times 10^{-4}$ eV/deg at $-61.6°C$. Thus, the magnitude of the discontinuity $\Delta(\partial E/\partial T)_p$ for HSeK, equal to $+4.4 \cdot 10^{-4}$ eV/deg, exceeds the magnitude of the discontinuity for HSeN at $T = -79°C$, equal to $+2.9 \cdot 10^{-4}$ eV/deg, which indicates the large distortion of the HSeK crystal lattice as compared to the HSeN at the phase transition. Shuvalov et al. [109] arrived at the same conclusion in comparing the results of measurements of disorientation of the optical indicators of the domains in these crystals. The monoclinic HSeL crystals do not display a ferroelectric phase transition right up to the melting temperature 110°C. The change of the width of the forbidden band of the HSeL crystal with temperature is presented in Figure 4.10. In the range of temperatures from room temperature to 40°C a linear change in $E_g = E_g(T)$ with a coefficient $(\partial E_g/\partial T)_p = -(20 \pm 0.6) \cdot 10^{-4}$ eV/deg occurs, which is larger in magnitude than that for the HSeN and HSeK crystals.

Figure 4.10. Temperature dependence of the width of the forbidden band of an HSeL crystal.

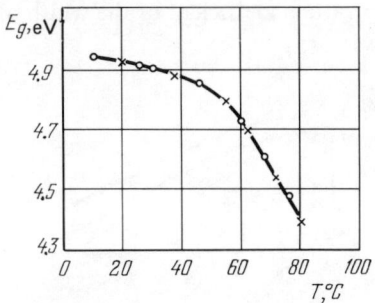

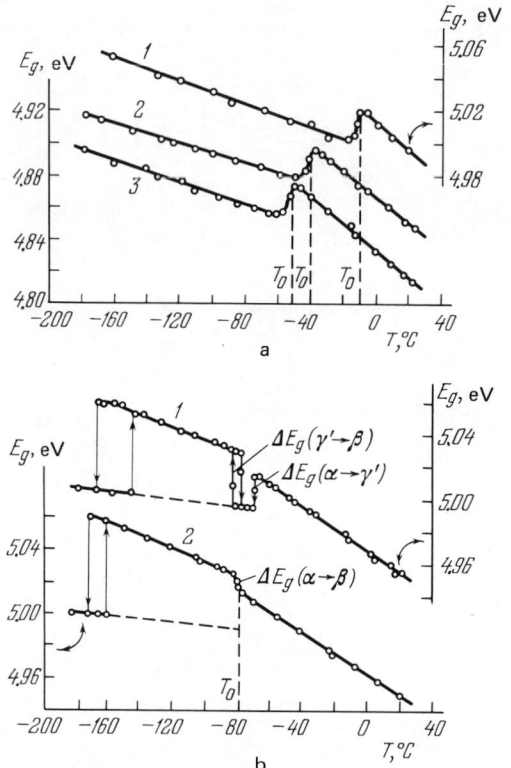

Figure 4.11. Temperature dependence of the width of the forbidden band of crystals of the system $Na(D_xH_{1-x})_3(SeO_3)_2$. a: $x > 0.5$; (1) $x = 0.94$; (2) $x = 0.7$; (3) $x = 0.6$. b: $x < 0.5$; (1) $x = 0.27$; (2) $x = 0$.

The temperature dependences of the intrinsic absorption edge of the crystals $Na(D_xH_{1-x})_3(SeO_3)_2$, $RbH_3(SeO_3)_2$, $RbD_3(SeO_3)_2$, $KD_3(SeO_3)_2$, $NH_4H_3(SeO_3)_2$, and $CsH_3(SeO_3)_2$ were obtained in [101].

The results of the experiments are presented in Figures 4.11–4.13, and in Table 2. The temperatures of the phase transition found from the anomalies in E_g agree well with the results of dielectric and other measurements.

All of the phase transitions are accompanied by a discontinuity in the temperature coefficient of the width of the forbidden band $\Delta(\partial E_g/\partial T)_p$. The only

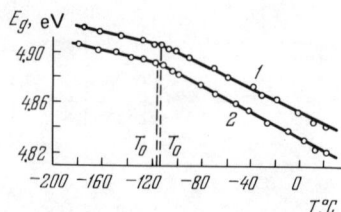

Figure 4.12. Temperature dependence of the width of the forbidden band of crystals of $RbH_3(SeO_3)_2$ (curve 1) and $RbD_3(SeO_3)_2$ (curve 2).

OPTICAL ABSORPTION AND FERROELECTRICS

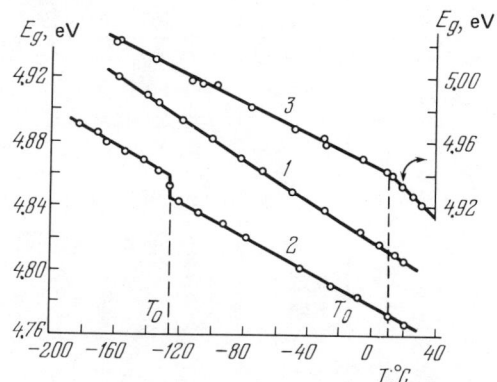

Figure 4.13. Temperature dependence of the width of the forbidden band of crystals of $NH_4H_3(SeO_3)_2$ (curve 1) $CsH_3(SeO_3)_2$ (curve 2) and $KD_3(SeO_3)_2$ (curve 3).

exception is evidently the $CsH_3(SeO_3)_2$ crystal, for which $\Delta(\partial E_g/\partial T) \simeq 0$ within the accuracy of the measurements ($\pm 0.3 \cdot 10^{-4}$ eV/deg) at the antiferroelectric transition. Moreover, first-order transitions are accompanied by a discontinuity in the width of the forbidden band, $\Delta E_g = E_g|_{T>T_1} - E_g|_{T<T_1}$. The $NH_4H_3(SeO_3)_2$ crystal is evidently a linear dielectric over the entire range of its existence and does not have an anomaly in E_g: $(\partial E_g/\partial T)_p \simeq -6.4 \cdot 10^{-4}$ eV/deg = const.

The behavior of E_g for crystals of the solid solutions $Na(D_xH_{1-x})_3(SeO_3)_2$, $NaH_3(SeO_3)_2$, and $NaD_3(SeO_3)_2$ merits special consideration. As was established, the symmetry of the ferroelectric phases in these crystals depends on the deuterium content: the α phase has the symmetry $2/m$, the γ' and γ phases have the symmetry m, and the β phase has the symmetry 1. Study of the dependence $E_g = E_g(T)$ (Figure 4.11a, b; Table 2) shows that the transitions $\alpha \to \gamma'$ and $\gamma' \to \beta$ (occurring for a deuterium concentration $x < 0.5$) are accompanied by discontinuities of E_g of opposite signs. The absolute value of ΔE_g is greater for the transition $\gamma' \to \beta$, and it determines the sign of ΔE_g for the transition $\alpha \to \beta$ in $NaH_3(SeO_3)_2$, if this transition can be considered as a superposition of the first two. Actually, one can write in this case, for $T \simeq T_0$,

$$\Delta E_g(\alpha \to \beta) = \Delta E_g(\alpha \to \gamma') + \Delta E_g(\gamma' \to \beta)$$

$$\Delta \frac{\partial E_g}{\partial T}(\alpha \to \beta) = \Delta \frac{\partial E_g}{\partial T}(\alpha \to \gamma') + \Delta \frac{\partial E_g}{\partial T}(\gamma' \to \beta).$$

It is not difficult to confirm by using the data of Table 2 and Figure 4.11b that these equalities are well satisfied. Since the transition $\alpha \to \gamma'$ is related to the generation of the spontaneous polarization component P_{0x} (in the m plane), the transition $\gamma' \to \beta$ is related to the generation of the component P_{0y} (along the second-order axis), while the transition $\alpha \to \beta$ is related to the generation of both components simultaneously, then the behavior of E_g, considered within the framework of the additive scheme, can serve as an illustration of the assump-

TABLE 2. Temperature Dependence of E_g for Several Hydroselenites

Crystal	T^0, °C	E_g at 20°C, eV	ΔE_g, eV	$\left.\dfrac{\partial E_g}{\partial T}\right|_{T>T^0} \cdot 10^4$, eV/deg	$\left.\dfrac{\partial E_g}{\partial T}\right|_{T<T^0} \cdot 10^4$, eV/deg	$\Delta \dfrac{\partial E_g}{\partial T} \cdot 10^4$, eV/deg
Na$(D_xH_{1-x})_3(SeO_3)_2$						
$x = 0.94$	$-11(1 \to \gamma)$	4.99	+0.023	$-8.4(\alpha)$	$-3.5(\gamma)$	$-4.9(\alpha \to \gamma)$
$x = 0.7$	$-40(\alpha \to \gamma)$	4.85	+0.023	$-8.3(\alpha)$	$-3.2(\gamma)$	$-5.1(\alpha \to \gamma)$
$x = 0.6$	$-53(\alpha \to \gamma)$	4.81	+0.023	$-8.3(\alpha)$	$-3.2(\gamma)$	$-5.1(\alpha \to \gamma)$
$x = 0.27$	$-68(\alpha \to \gamma')$	4.95	+0.023	$-7.0(\alpha)$	$-1.2(\gamma, \gamma')$	$-5.8(\alpha \to \gamma')$
	$-83(\gamma' \to \beta)$	—	-0.035	—	$-3.8(\beta)$	$+2.6(\gamma' \to \beta)$
	$-165(\beta \to \gamma)$	—	$+0.058$	—	—	$-2.6(\beta \to \gamma)$
$x = 0$	$-79(\alpha \to \beta)$	4.95	-0.011	$-7.1(\alpha)$	$-1.2(\gamma)$	$-3.3(\alpha \to \beta)$
	$-173(\beta \to \gamma)$	—	$+0.06$	—	$-3.8(\beta)$	$-2.6(\beta \to \gamma)$
RbH$_3$(SeO$_3$)$_2$	~-118	4.84	—	-4.7	-2.1	-2.6
RbD$_3$(SeO$_3$)$_2$	~-118	4.82	—	-5.0	-2.0	-3.0
KH$_3$(SeO$_3$)$_2$[a]	-62	4.88	—	-7.0	-2.4	-4.6
KD$_3$(SeO$_3$)$_2$	$+12$	4.93	—	-9.3	-4.8	-4.5
CsH$_3$(SeO$_3$)$_2$	-128	4.77	-0.013	-5.4	-5.4	—
NH$_4$H$_3$(SeO$_3$)$_2$	—	4.81	—	-6.4	—	—
LiH$_3$(SeO$_3$)$_2$	—	4.9	—	—	$-20(20°C)$	—
LiD$_3$(SeO$_3$)$_2$	—	4.92	—	—	$-20(20°C)$	—

[a] The deuterium content in the investigated samples was about 80%. The transition temperature is +24°C for almost complete deuteration.

tion of the independence of the P_{0x} and P_{0y} components of the polarization in $NaH_3(SeO_3)_2$.

Finally, we consider the peculiarities of the behavior of $E_g = E_g(T)$ in $RbH_3(SeO_3)_2$ and $RbD_3(SeO_3)_2$ crystals. Of greatest interest here evidently is the fact that a shift in the transition temperature is practically nonexistent with deuteration, although some decrease in the transition temperature possibly occurs.

Thus, the peculiarities of the behavior of the width of the forbidden band in the family of trihydroselenites can serve as an additional illustration of the anomalies in the physical properties at phase transitions, and as the method of identifying phase transitions.

In concluding this section, we discuss the results of investigations of the temperature dependence of $E_g = E_g(T)$ for singular ferroelectrics: gadolinium molybdate $Gd_2(MoO_4)_3$ and certain crystals in the family of boracites. As is well known, singular ferroelectrics display a number of interesting peculiarities in their physical properties near phase transitions as compared to ordinary ferroelectrics. Thus, it was of interest to investigate the temperature dependences of the width of the forbidden band of singular ferroelectrics and to compare the experimental results with the conclusions of the thermodynamic theory (cf. Chapter 1).

Experimental investigations of the temperature dependence $E_g = E_g(T)$ was carried out in [102] for the following boracites: $Fe_3B_7O_{13}Br(Fe-Br)$, $Fe_3B_7O_{13}Cl \cdot (Fe-Cl)$, $Mg_3B_7O_{13}(Mg-Cl)$, $Co_3B_7O_{13}Cl(Co-Cl)$, $Zn_3B_7O_{13}Cl \cdot (Zn-Cl)$, and $Cr_3B_7O_{13}Cl(Cr-Cl)$. As is well known, a majority of boracites experience a number of successive phase transitions, whose characteristics from the results of investigations of the dielectric and thermal properties are summarized in Table 3. For example, it was observed that the boracite Fe-Br experiences an anomaly of $\epsilon = \epsilon(T)$ at the point of transition $mm2-3m$, while the other trigonal boracites Co-Cl, Zn-Cl, and Fe-Cl do not have anomalies of $\epsilon = \epsilon(T)$ at the transitions $mm2-m$ and $m-3m$. The m phase in the boracites Co-Cl and Zn-Cl also does not appear in thermograms (it was discovered with the help of the optical polarization method). The transition in the boracite Cr-Cl is little studied; it is known only that the crystal becomes birefringent at the phase transition.

The temperature behavior of E_g differs for all the investigated boracite crystals in accordance with the peculiarities of their phase transitions. Figure 4.14 and 4.15 show the dependences $E_g = E_g(T)$ for the boracites Mg-Cl, Cr-Cl, and Fe-Br. For all these crystals, E_g in the ferroelectric and paraelectric regions decreases linearly with increasing temperature. In the boracites Mg-Cl and Cr-Cl, the phase transitions $\overline{4}3m-mm2$ are accompanied by discontinuities of E_g in the opposite directions. The discontinuities of the temperature coefficient $(dE_g/dT)_p$ at the transition from the ferroelectric region to the paraelectric for these

TABLE 3. Phase Transitions in Boracites

Boracite	Phase transitions	T_0, °K	Order of transition	Anomaly	Temperature effect	ΔE_g, eV	$\Delta\left(\dfrac{dE_g}{dT}\right)\cdot 10^4$, eV/deg
Mg-Cl	$\bar{4}3m-mm2$	538	1, from data of measurement of $E_g = E_g(T)$	Yes	Not measured	$+(0.05 \pm 0.003)$	$+(0.9 \pm 0.6)$
Cr-Cl	$\bar{4}3m-mm2$	264	The same	Not measured	Not measured	$-(0.023 \pm 0.003)$	$-(4 \pm 1.8)$
Fe-Br	$\bar{4}3m-mm2$	495	1	Yes	Yes	$+0.017$	0
	$\bar{4}3m-3m$	403	1	Yes	Yes	$+0.027$	0
	$\bar{4}3m-mm2$	593	1	Yes	Yes	—	—
Fe-Cl	$\bar{4}3m-m$	498	1	Yes	Yes	$+0.01$	—
	$\bar{4}3m-3m$	476	1	No	Yes	$+0.02$	—
	$\bar{4}3m-mm2$	623	1	Yes	Yes	0	—
	$\bar{4}3m-m$	538	2, from data of measurement of $E_g = E_g(T)$	No	No	0	$-(7.5 \pm 1)$
Co-Cl	$\bar{4}3m-3m$	468	The same	No	No	0	$-(1.9 \pm 1.1)$
	$\bar{4}3m-mm2$	703	1	Yes	Yes	—	—
	$\bar{4}3m-m$	564	1, from data of measurement of $E_g = E_g(T)$	No	No	—	—
Zn-Cl	$\bar{4}3m-3m$	472	The same	No	No	—	—

OPTICAL ABSORPTION AND FERROELECTRICS

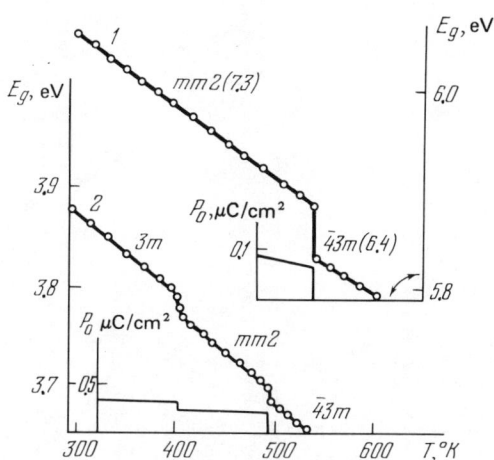

Figure 4.14. Temperature dependences $E_g(T)$ and $P_0(T)$ for the boracites Mg-Cl (1) and Fe-Br (2). The temperature coefficients $(dE_g/dT)_p \cdot 10^4$, eV/deg, are indicated in parentheses here and in the following figures.

boracites also have differing sign (cf. Table 3). The nature of the transitions in the boracites Mg-Cl and Cr-Cl at 538 and 264°K has not been previously determined. The anomalies of E_g, the discontinuity in ΔE_g, and the hysteresis for the boracite Cr-Cl observed by us indicate that these transitions are first order. For the boracite Fe-Br (Figure 4.14), both first-order phase transitions $\bar{4}3m$-$mm2$-$3m$ are accompanied by discontinuities in E_g. It should be noted that the magnitudes of the discontinuities and the dependences $E_g = E_g(T)$ obtained in polarized and unpolarized light coincide. Within the limits of the measurement errors, the coefficients $(dE_g/dT)_p$ for the boracites Fe-Br have the same values in all phases.

We note the following peculiarities of the temperature behavior of the spontaneous polarization $P_0 = P_0(T)$ for the various boracites (Figure 4.14). The temperature dependence $P_0 = P_0(T)$ was investigated by measuring the pyroelectric coefficient. The spontaneous polarization P_0 for the boracite Mg-Cl arises discontinuously at 538°K, and then increases linearly with decreasing

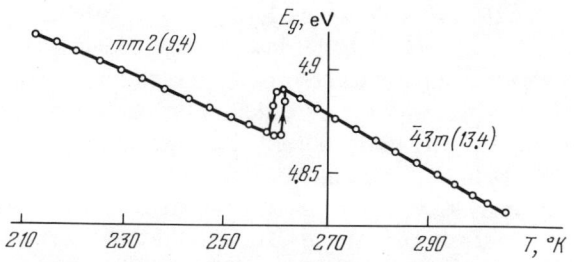

Figure 4.15. Temperature dependence $E(T)$ for the boracite Cr-Cl.

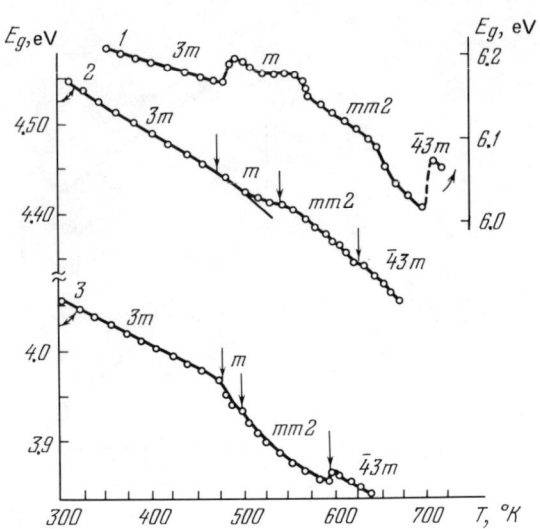

Figure 4.16. Temperature dependence $E_g(T)$ for the boracites Zn-Cl (1), Co-Cl (2), Fe-Cl (3).

temperature. The polarization P_0 for the boracite Fe-Br changes only at the temperatures of the phase transitions. The value of P_0 in the boracite Cr-Cl is very small ($\sim 3 \cdot 10^{-10}$ C/cm^2), and measurements of the temperature dependence $P_0 = P_0(T)$ were not made.

Investigation of the dependences $E_g = E_g(T)$ for the trigonal boracites Zn-Cl, Co-Cl, and Fe-Cl experiencing a number of successive transitions $\bar{4}3m$–$mm2$–m–$3m$ were of interest in connection with the fact that the phase m in the boracites Zn-Cl and Co-Cl is not observed from dielectric and thermal measurements. As is seen in Figure 4.16, the phase m is well displayed in the temperature dependence of E_g for all the boracites. The transitions $3m$–m and m–$mm2$ for Fe-Cl are accompanied by a discontinuity ΔE_g. However, the anomalies are diffuse. This behavior is found in accordance with the presence of a latent heat of transition. For the high-temperature transitions $\bar{4}3m$–$mm2$ in the boracites Zn-Cl and Fe-Cl, anomalies in E_g are observed, but the magnitude of the true discontinuity ΔE_g could not be determined, since the transitions are accompanied by a change in the form of the intrinsic absorption edge. The domain structure possibly leads to additional absorption or scattering of light and affects the form of the edge with the transition to the paraelectric phase. The transitions $3m$–m and m–$mm2$ in the boracite Zn-Cl are accompanied by a discontinuous change in E_g, which indicates that these are first-order transitions. However, the magnitude of the discontinuity ΔE_g at 473°K varies for the samples, which evidently can also be related to the domain structure. As for the discontinuities ΔE_g in Mg-Cl, Fe-Br, and Cr-Cl, their magnitude was the same

OPTICAL ABSORPTION AND FERROELECTRICS

for boracites of differing thickness, but the disappearance of the domain structure in the paraelectric phase could not lead to a change in the form of the edge. An anomalous behavior of E_g was observed for all the transitions $\overline{4}3m{-}mm2{-}m{-}3m$ in the boracite Co-Cl, but discontinuities ΔE_g were not observed (if they in fact exist, then their magnitude cannot exceed $3 \cdot 10^{-3}$ eV). This result contradicts the thermogram of the boracite Co-Cl, which displays a latent heat at the transition $\overline{4}3m{-}mm2$.

Figure 4.17 presents the temperature dependence $E_g = E_g(T)$ for gadolinium molybdate. Identical results were obtained for samples of the different X, Y, and Z sections. As is well known, investigations of the dielectric properties of $Gd_2(MoO_4)_3$ did not give a unique answer to the problem of the nature of the phase transition at 432°K. In the ferroelectric phase, E_g decreases linearly with increasing temperature with a temperature coefficient $(dE_g/dT)_p = -(6.6 \pm 0.2) \cdot 10^{-4}$ eV/deg. In the paraelectric region, $(dE_g/dT)_p = -(8.6 \pm 0.6) \cdot 10^{-4}$ eV/deg. Thus, a discontinuity in the temperature coefficient of $\Delta(dE_g/dT)_p = (2 \pm 0.8) \cdot 10^{-4}$ eV/deg is observed in the absence of a discontinuity of E_g, which corresponds to a second-order phase transition or a strongly diffused first-order phase transition. We note that the temperatures of the anomalies of $E_g = E_g(T)$ coincide well in all the investigated crystals with the temperatures of the phase transitions known from the literature.

The data of Figure 4.14 indicate the agreement between the experimental dependence $E_g = E_g(P)$ for the boracites Fe-Br and Mg-Cl and relation (1.57) presented in Chapter 1 for singular ferroelectrics.

These data indicate that the dependence $E_g = E_g(P)$ for the boracite Mg-Cl is linear, where the coefficient $a_1 = 0.7 \cdot 10^6$ eV $\cdot$ cm^2/C. The experimental data for the boracite Fe-Br also agrees with (1.57) for $a_1 = 0.1 \cdot 10^6$ eV-cm^2/C. This comparison could not be carried out for the other boracites because of the absence of reliable results on the temperature dependence of the spontaneous polarization. The data of Figure 4.15 for the boracite Cr-Cl indicates that $a_1 < 0$, since $\Delta E_g < 0$, which also agrees with the sign of $\Delta(dE_g/dT)_p$. The electrical absorption was investigated only for the boracite Cr-Cl. In accordance

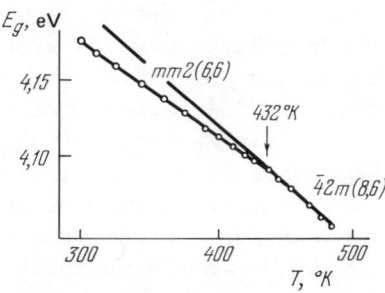

Figure 4.17. Temperature dependence $E_g(T)$ for gadolinium molybdate.

with (1.58), a significant Kern-Harbeke effect is not observed either in the region of the phase transition or far from it.

Thus, investigation of the fundamental optical absorption near phase transitions in singular ferroelectrics is also useful from the point of view of identifying the phase transitions. Moreover, it permits one to verify some conclusions predicted by the thermodynamic theory with respect to the dependence of the width of the forbidden band of the crystal on the spontaneous polarization.

4.2. Intrinsic Optical Absorption of Single Crystals of $BaTiO_3$

Barium titanate, from the time of the discovery of its ferroelectric properties [1, 2], and also other ferroelectrics with the perovskite structure, have been investigated in great detail [5, 7, 11].

The nonpolar phase of $BaTiO_3$ above 120°C belongs to the cubic centrosymmetric point group $\bar{6}/4$. The polar phase, stable at temperatures below 120°C, has tetragonal symmetry (point group $4mm$). A typical first-order phase transition of the displacement type occurs at $T_1 \simeq +120°C$ in $BaTiO_3$. Two other phase transitions at +5 and -90°C are observed with a decrease in temperature in barium titanate, and the newly arising phases have the rhombic and rhombohedral symmetry, respectively. A complex domain structure is formed in $BaTiO_3$ crystals: the polar axis (spontaneous polarization) in the tetragonal phase can be parallel to any of the six directions [100] of the initial cubic cell, to any of the 12 directions [110] of the cubic cell in the rhombic phase, or to any of the eight directions [111] of the initial cubic cell in the low-temperature rhombohedral phase.

An investigation was carried out in [93] on the intrinsic absorption edge of $BaTiO_3$ in the region of the phase transition from the tetragonal phase to the cubic. The results of these measurements for unpolarized light are presented in Figure 4.18. The temperature dependence of the intrinsic absorption edge of $BaTiO_3$ was investigated previously [110], but anomalies in E_g were not observed at the phase transition. As is seen in Figure 4.18 [93], a discontinuity of the width of the forbidden band of barium titanate was first observed for the transition from the tetragonal phase to the cubic. A linear dependence of the width of the forbidden band on temperature occurs in the ferro- and paraelectric regions with close values of the coefficients $(dE_g/dT)_p = -(7 \pm 0.5) \cdot 10^{-4}$ eV/deg. This value is close to the temperature gradient $(\partial E_g/\partial T)_p$ obtained previously in [110]. Near the phase-transition temperature (for the crystal investigated in [93], $T_1 \simeq +105°C$), a discontinuity $\Delta E_g \simeq 0.02$ eV was observed. The dependence of E_g on temperature in the region of the phase transition displays a temperature hysteresis of $\sim 2°$. As is seen in Figure 4.18, the discontinuity of the width of the forbidden band is somewhat broadened in temperature, which is evidently caused by broadening of the transition region because of domain and

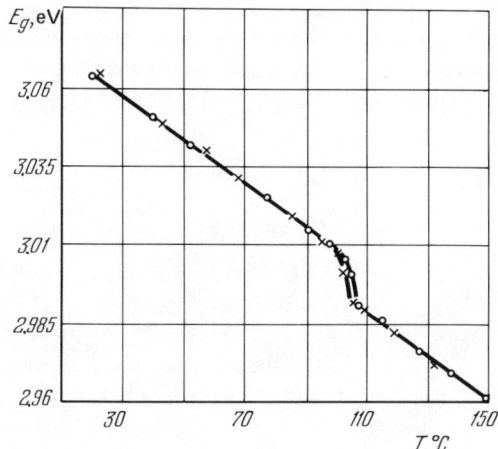

Figure 4.18. Temperature dependence of the width of the forbidden band of barium titanate in the region of the upper phase transition.

surface effects, internal stresses, and inhomogeneous temperature distribution. Some scatter $\Delta E_g = (2-3) \cdot 10^{-2}$ eV was obtained for all the crystals investigated in [93]. Measurement of $E_g = E_g(T)$ for barium titanate in polarized light was performed by Gähwiller [111], independently of [93]. Figure 4.19 presents the dependence $E_g = E_g(T)$ obtained by Gähwiller for a single-domain crystal in the region of the phase transition from the cubic phase to the tetragonal. In the paraelectric region, $(dE_g/dT)_p = -4 \cdot 10^{-4}$ eV/deg, which is close to the value obtained in [93]. A discontinuity $\Delta E_g \simeq 0.01$ eV occurs at the point of the phase transition ($T_1 \simeq +115°C$). Independent measurements of the spontaneous polarization made by Gähwiller showed that for $BaTiO_3$ near the phase transi-

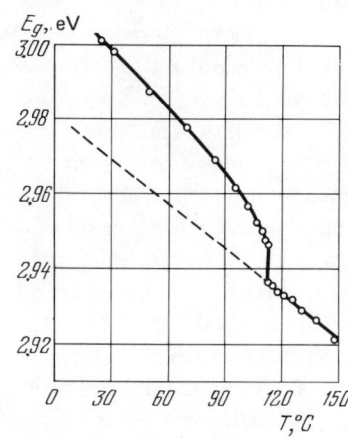

Figure 4.19. Temperature dependence of the width of the forbidden band in barium titanate in polarized light (at $\mathcal{E} \perp c$) [111].

tion, relation (1.46) is fulfilled, where the expansion coefficients $a = 0.46 \pm 0.02$ eV/(C/m^2)^2, $b = 13.2 \pm 0.8$ eV/(C/m^2)^2, and $c \simeq 0$. The maximum value of the spontaneous polarization in BaTiO$_3$ is $P_0 \simeq 20 \cdot 10^{-6}$ C/cm^2.

All the dependences $E_g = E_g(T)$ presented above were obtained for single-domain crystals of BaTiO$_3$. We now turn to the problem of the role of the domain structure of BaTiO$_3$ in measuring the optical absorption edge near the phase transitions.

A number of experiments were carried out in [93, 98] to investigate the effect of the domain structure on the position of the intrinsic absorption edge of BaTiO$_3$ and its behavior with temperature. One of these consisted of first making the barium titanate samples single domains. The observed dependences $E_g = E_g(T)$ for the single-domain samples were analogous to those obtained for multidomain crystals. In another experiment to completely eliminate the effect of the domain structure on the intrinsic absorption edge of BaTiO$_3$, the samples were placed under the effect of a constant electric field intensity of 0.6 kV/cm, applied in the [001] direction during the entire measurement process. The field provided a transition of the crystal from the paraelectric phase to the single-domain state with a decrease in temperature, while the nature of the observed dependence $E_g = E_g(T)$ was analogous to that previously obtained. The experiments performed, which eliminated the effect of the domain structure, led to the conclusion that the true value of ΔE_g, independent of the domain structure, was measured for BaTiO$_3$ in the region of the upper phase transition. This conclusion also follows from the measurements of $E_g = E_g(T)$, which give identical results for the multidomain samples of BaTiO$_3$ of varying thickness and with varying domain structure.

Thus, the domain structure in the tetragonal phase of BaTiO$_3$ does not give a significant additional absorption or scattering of light and thus does not affect the measurement results. The data of [112] also indicates this, showing that the transparency of the crystal does not change at the high-temperature phase transition in BaTiO$_3$. We have another pattern in the behavior of the width of the forbidden band and the effect of the domain structure on it for the low-temperature phase transitions in BaTiO$_3$.

Absorption measurements were carried out in [110] for BaTiO$_3$ over a wide interval of temperatures covering all three phase-transition points. The so-called apparent absorption edge was measured in contrast to the true intrinsic absorption edge. A large discontinuity of the edge of ~ 0.23 eV and temperature hysteresis were observed in the region of the phase transition at $-70°$C, where the width of the hysteresis and the magnitude of the discontinuity varied depending on the thickness of the various samples. The fact that the relatively crude measurements performed in [110] revealed the discontinuity of the edge only at $-70°$C and did not reveal it at $+5$ and $+120°$C can be explained by the fact that when the crystal changes from the orthorhombic phase to the rhombo-

OPTICAL ABSORPTION AND FERROELECTRICS

hedral, there arises a significantly more complex and thinner domain structure than for the other phase transitions in $BaTiO_3$. More detailed investigations of the absorption for the low-temperature phase transitions in $BaTiO_3$ were performed in [93, 98]. These investigations indicated that the anomaly of the absorption edge and the hysteresis phenomena in the region of low-temperature phase transitions in $BaTiO_3$ depend to a strong degree on the nature of the domain structure. For the multidomain sample, the discontinuity of the width of the forbidden band in the region of the phase transitions always consists of a true and additional effect. The latter is due to the domain structure. In crystals of TGS and RS, as has been indicated, the domain structure does not produce additional absorption and scattering. In the case of the upper phase transition in $BaTiO_3$, the possible effect of the domain structure was eliminated and the true effect of the change in the width of the forbidden band was investigated. Moreover, it was shown in [93, 98] that the effect of the domains in $BaTiO_3$ crystals at the transition from the cubic to the tetragonal phase is small and does not introduce additional effects into the spectral measurements in contrast to the low-temperature phase transitions.

Crystals of $KNbO_3$ were investigated in [113] as another ferroelectric with the perovskite structure. These crystals undergo a ferroelectric first-order phase transition from the tetragonal phase to the cubic at +430°C. Figure 4.20 presents the dependence $E_g = E_g(T)$ for $KNbO_3$ in a wide temperature interval also encompassing the phase transition at +200°C (to the ferroelectric

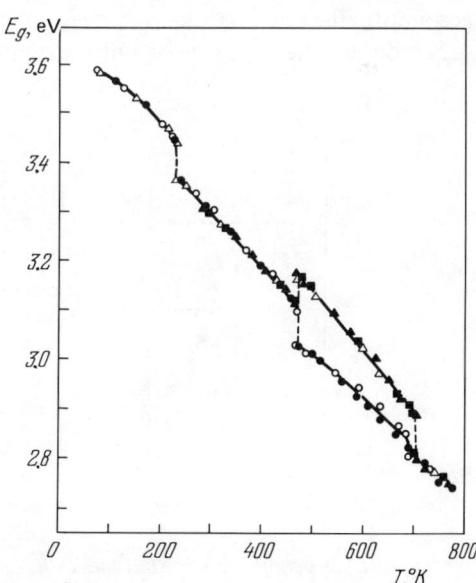

Figure 4.20. Temperature dependence of the width of the forbidden band in $KNbO_3$ with heating and cooling [113].

phase). The ferroelectric phase transition at +430°C is accompanied by a large discontinuity $\Delta E_g \simeq 0.1$ eV.

We will consider the band structure of perovskite ferroelectrics in more detail in Section 4.6. We will note only here that sections of direct interband transitions in perovskite ferroelectrics are related to a transition between the $2p$ level of oxygen and the conduction band corresponding to the d level. This transition is denoted $\Gamma_{15} \to \Gamma_{25'}$ in the cubic phase. The transition from the cubic to the tetragonal phase, by changing the band structure, increases the energy of the direct transition [114], which also leads to the discontinuity ΔE_g.

It was indicated above that determination of E_g for BaTiO$_3$ and other ferroelectrics was carried out from the steepest section of the absorption edge. We turn now to the complex and still unexplained problem of determining the form of the intrinsic optical absorption edge in BaTiO$_3$. Two contradictory groups of works exist in the literature. According to the first group of works, for example [98, 115], the absorption edge in BaTiO$_3$ is due to direct and indirect interband transitions. An exponential form of the edge was discovered in other works [116, 117], and it was shown that it satisfies the Urbach rule [118]. These latter results also satisfactorily agree with the calculation of the band structure of perovskite ferroelectrics, which will be discussed in Section 4.6.

A thorough investigation of the intrinsic optical absorption edge of single crystals of BaTiO$_3$ at room temperature was carried out by Cox [115]. The reflection of light from the crystal surface of varying thickness was taken into account by the known technique. The results of this work, presented in Figure 4.21, confirm the existence of a direct transition with a threshold energy $E_d =$

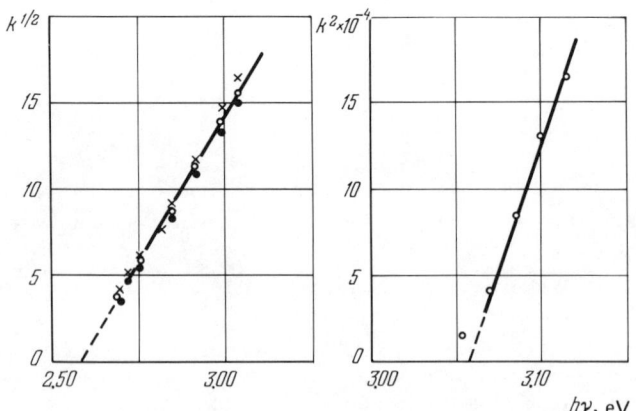

Figure 4.21. Dependences $k^{1/2}$ ($h\nu$) and k^2($h\nu$) for BaTiO$_3$ at room temperature [115].

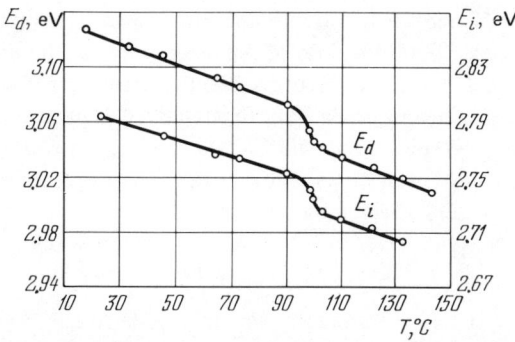

Figure 4.22. Temperature dependences of the energies E_d and E_i corresponding to the "direct" and "indirect" optical transitions in $BaTiO_3$ in the region of the transition from the tetragonal phase to the cubic phase.

3.05 eV, and indicate the presence of an indirect interband transition with a threshold energy $E_i = 2.66$ eV. The fact that this result was reproduced for all the samples led the authors to the conclusion that the form of the absorption edge under investigation characterizes interband transitions in $BaTiO_3$ and does not depend on the impurity content in the crystal. The form of the intrinsic absorption edge was also investigated in [120] for single crystals of $BaTiO_3$ at various temperatures in the region of the upper phase transition, where the dependences $k^2 = k^2(h\nu)$ and $k^{1/2} = k^{1/2}(h\nu)$ were constructed for temperatures below and above the Curie point. As in [115], reflection of light from the surface of the crystal of varying thickness was taken into account to obtain correct values of the absorption coefficient, and the calculations were carried out according to the known technique [121]. The values of the energies E_d and E_i, corresponding to the direct and indirect optical transitions, were determined from the dependences $k^2(h\nu)$ and $k^{1/2}(h\nu)$ by extrapolating them to the value $k = 0$. According to the data obtained in [120], $E_d = 3.13$ eV and $E_i = 2.77$ eV at room temperature. It would appear that the results of these measurements confirmed the existence of the direct and indirect optical transitions in $BaTiO_3$, although the average values of the energies E_d and E_i were greater than the corresponding values obtained in [115] by ~0.1 eV. According to [120], the slope of the sections of the direct and indirect transitions does not depend on temperature, while the energies E_d and E_i decrease linearly with temperature both in the paraelectric and in the ferroelectric phase. However, as is seen in Figure 4.22, an anomalous shift of E_d and E_i occurs in the region of the upper phase transition in $BaTiO_3$. It is significant that, according to [120], the form of the intrinsic absorption edge and the nature of the interband transitions in $BaTiO_3$ do not change over a wide temperature interval, including the transition from the tetragonal phase to the cubic. Investigations of the form of the edge were not performed for the low-temperature phase transitions in BaTiO$_3$, since they are made difficult because of the large effect of the domain structure.

We now turn to [116, 117]. Figure 4.23 presents the spectral distribution of the absorption coefficient of BaTiO$_3$ for polarized light near the edge. The curves obtained by Wemple [116] and Gähwiller [111] at room temperature are presented for comparison on the same figure. The experimental curves of Wemple follow well the Urbach exponent:

$$k = k_0 \exp\left[(h\nu - E_0)/k_B(T+T_0)\right], \qquad (4.1)$$

where $T_0 = (140 \pm 10)°K$, $h\nu$ is the photon energy, k_B is the Boltzmann constant, and k_0 and E_0 are constants. In the opinion of Wemple, the presence of the long-wavelength tail on the absorption curves obtained by Gähwiller [111] and other authors is caused by impurities introduced into the crystal during the process of growth from the melt and are not related to the interband optical transitions. According to [116], this also determines the divergence between the two groups of works indicated above with respect to the interpretation of the nature of the absorption edge. Since the exponential form of the edge, strictly speaking, does not permit one to determine the width of the forbidden band E_g, the latter is determined in [116] in an indirect manner for $k \simeq 10^3$ cm^{-1}. According to [116], $E_g^{\|} = 3.38$ eV and $E_g^{\perp} = 3.27$ eV, where $E_g^{\|}$ and $E_g^{\perp}$ are the energies for the plane of polarization parallel and perpendicular to the direction of spontaneous polarization, respectively.

The exponential form of the intrinsic absorption edge (4.1) was observed in other perovskite ferroelectrics, for example, in SrTiO$_3$ [121], KTaO$_3$ [117], and also in ferroelectrics with other structures (cf. Section 4.3), and one could assume, following [116], that this mechanism of optical absorption is typical

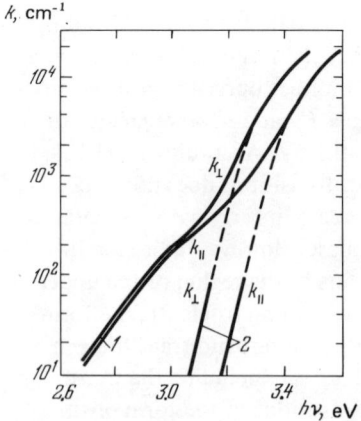

Figure 4.23. Intrinsic absorption edge in BaTiO$_3$ according to Gähwiller (1) and Wemple (2) [116].

in general for ferroelectrics, if not for a number of difficulties. Thus, if it is assumed, in accordance with the generally assumed mechanism, that the Urbach rule is due to the interband electron-phonon interaction, then it is natural to assume, in accordance with the pseudo-Jahn-Teller mechanism of phase transitions discussed in Chapter 2, that this interaction is due to the transverse optical phonon at the center of the Brillouin zone (the soft model). This same conclusion follows from the fact that, according to [122], the interaction of electrons with the soft mode is the basic mechanism of carrier scattering in $BaTiO_3$. At the same time, the value of the characteristic temperature T_0 in the Urbach law (4.1) does not agree with the energy of the soft phonon in the perovskite ferroelectrics [116]. Thus, $T_0 = 140°K$ [117] in $KTaO_3$, and $T_0 = 50°K$ [121] in $SrTiO_3$, whereas the frequencies of the soft mode in them approximately coincide; on the contrary, $T_0 = 140°K$ in $BaTiO_3$, although the frequency of the soft mode in $BaTiO_3$ is significantly higher than in $SrTiO_3$. One should add to this that the mechanism itself responsible for satisfaction of the Urbach rule is not precisely defined and that the experimental investigation of the form of the optical absorption edge should be continued and refined.

4.3. Intrinsic Optical Absorption of Ferroelectrics of the Groups $A^V B^{VI} C^{VII}$ and $A_2^V B_3^{VI}$

The last two sections reported on the optical absorption of ferroelectric crystals, which are dielectrics with a wide forbidden band $E_g \gtrsim 3$ eV. Crystals of the type $A^V B^{VI} C^{VII}$ and $A_2^V B_3^{VI}$, the subject of this section, are high-resistance semiconductors with a relatively narrow forbidden band (~ 2 eV). The discovery in 1962 of the ferroelectric properties in single crystals of SbSI and other semiconductors of the type $A^V B^{VI} C^{VII}$ served as an impetus for further investigation of these and other ferroelectric semiconductors [3, 4]. In particular, a number of works, which we will discuss in the following chapters, was devoted to an investigation of single crystals of SbSI, and other ferroelectrics of this group as new semiconductors and photoconductors, and to the study of the behavior of semiconductor parameters in the region of the ferroelectric phase transition.

Crystals of SbSI display a first-order ferroelectric phase transition of the displacement type at a temperature $T_1 \simeq 22°C$ [3, 4]. Structural investigations have established that the SbSI crystals and their isomorphic analogs are rhombic and belong to the class *mmm* (space group D_{2h}^{16}) above T_1, and to the class *mm* (C_{2v}^9) below T_1 [123, 124]. Additional phase transitions are observed in ferroelectrics of the type $A^V B^{VI} C^{VII}$ besides the main phase transition to the ferroelectric phase. Thus, a phase transition at a temperature of $-40°C$ is observed in SbSI with a change in symmetry $mm \to 2$ ($C_{2v}^9 \to C_2^2$) [100, 125-127].

Intrinsic optical absorption of single crystals of SbSI was first investigated by Kern [31] and Harbeke [32]. According to [32] the width of the forbidden

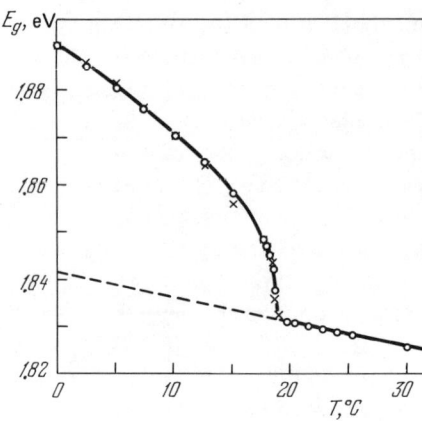

Figure 4.24. Temperature dependence of the width of the forbidden band in SbSI [128]; & [001]; circles are experimental values; crosses are calculated from (1.46).

band for SbSI is $E_g^\perp = 1.95$ eV and $E_g^\parallel = 1.88$ eV, where $E_g^\perp$ and $E_g^\parallel$ are the energies corresponding to the perpendicular and parallel direction of light polarization (with respect to the direction of spontaneous polarization). The temperature coefficients $(dE_g/dT)_p$ in the ferro- and paraelectric regions are, respectively, $-(2.2 \pm 0.2) \cdot 10^{-3}$ eV/deg and $-(0.9 \pm 0.2) \cdot 10^{-3}$ eV/deg. A discontinuity $\Delta E_g \simeq 0.02$ eV is observed at the Curie point. Later measurements [92, 120, 128] showed good agreement with these data. Figure 4.24 represents the temperature dependence $E_g = E_g(T)$ near the phase transition in the paraelectric region [128]. The width of the forbidden band was determined from the steep portion of the dependence of the absorption coefficient k on the energy ($10^2 < k < 10^3$ cm^{-1}). Independent measurements of the temperature dependence of the spontaneous polarization $P = P(T)$ in the same temperature region show good agreement with (1.46), as well as for BaTiO$_3$, where the expansion coefficient $a = 2.8$ eV/(C/m^2)^2 from the data of [128] and 1.7 eV/(C/m^2)^2 from the data of [120]. Reliable values for the coefficients b and c were not obtained for SbSI. The maximum value of the spontaneous polarization P_0 in SbSI is close to the corresponding value in BaTiO$_3$ ($P_0 \simeq 20 \cdot 10^{-6}$ C/cm^2).

Since the group of ferroelectric semiconductors of the type $A^V B^{VI} C^{VII}$ are relatively new, we will present here, following [126], a summary of their physical properties, including the dependence $E_g = E_g(T)$ in the temperature interval where the phase transitions occur. Figure 4.25 presents the temperature dependences of the dielectric constants ϵ, the logarithm of the electrical conductivity $\ln \sigma$, the width of the forbidden band E_g, and the heat capacity C_p for SbSi, SbSBr, and SbSeI. Table 4 presents some basic parameters for six ferroelectrics of this class (notation: σ is the electrical conductivity at $T = 300°$K, T_1 is the temperature of the phase transition, T_0 is the Curie–Weiss temperature, ΔE_g and Δu are the discontinuities of the width of the forbidden band and the

OPTICAL ABSORPTION AND FERROELECTRICS

donor activation energy at the phase transition, respectively, ΔS is the discontinuity of the entropy, ΔQ is the latent heat at the first-order phase transition, ΔC_p is the discontinuity of the heat capacity for second-order phase transitions, C is the Curie–Weiss constant, P_0 is the spontaneous polarization, and β and γ are the Devonshire coefficients). It is seen from Figure 4.25 and Table 4 that crystals of SbSI, SbSBr, BiSBr, and BiSI display an anomaly of the dielectric constant $\epsilon = \epsilon(T)$ at the temperature corresponding to the ferroelectric phase transition. A discontinuity of ΔE_g and the latent heat of transition ΔQ typical of a first-order phase transition are observed at these same temperatures. Along

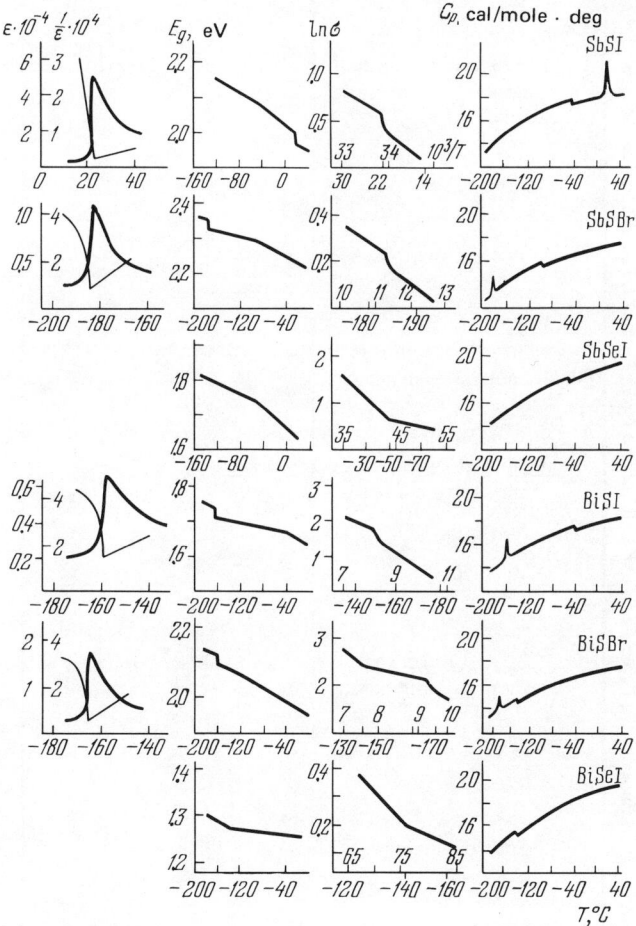

Figure 4.25. Temperature dependences $\epsilon(T)$, $E_g(T)$, $\sigma(T)$, and $C_p(T)$ for some ferroelectrics of the type $A^V B^{VI} C^{VII}$.

TABLE 4. Some Experimental Data Typical of First- and Second-Order Phase Transitions in Compounds of the Type $A^V B^{VI} C^{VII}$

Compound	Electrical conductivity at room temperature σ, ohm$^{-1}\cdot$cm^{-1}	Forbidden band width E_g, eV	Phase-transition order	Temperature of phase transition T_1, °K	Curie–Weiss temperature T_0, °K	Forbidden band width discontinuity ΔE_g, eV	Change in activation energy of donor levels Δu, eV	Change in entropy ΔS, cal/mole·deg	Latent heat of phase transition ΔQ, cal/mole	Heat capacity discontinuity at second-order phase transition ΔC_p, cal/mole·deg	Curie–Weiss constant $C \cdot 10^{-5}$ °K	Spontaneous polarization $P_0 \cdot 10^6$ C/cm^2	Coefficient $\beta \cdot 10^{13}$, cgs esu	Coefficient $\gamma \cdot 10^{23}$, cgs esu
SbSI	10^{-8}	1.92	1	295	281	0.03	−0.34	0.06	15	—	3.8	20	−3.2	7.4
SbSBr	10^{-9}	2.22	2	233	—	0	—	0	0	0.9	—	—	—	—
SbSeI	10^{-7}	1.63	1	91	80	0.02	−0.016	0.04	3.6	—	1.2	7.5	−54	480
			2	178	—	0	—	0	0	0.03	—	—	—	—
BiSI	$2 \cdot 10^{-2}$	1.64	2	223	—	0	0.20	0	0	0.01	—	—	—	—
			1	115	88	0.02	−0.05	0.02	2.1	—	1.9	7.0	−110	1300
			2	233	—	0	—	0	0	0.1	—	—	—	—
BiSBr	$3 \cdot 10^{-2}$	1.96	1	108	99	0.02	−0.005	0.03	3.2	—	1.5	8.5	−4	48
			2	133	—	0	0.015	0	0	0.2	—	—	—	—
BiSeI	$7 \cdot 10^{-2}$	1.28	2	133	—	0	0.006	0	0	0.08	—	—	—	—

with the first-order phase transition, the representatives of this class of ferroelectrics display second-order phase transitions. The value $\Delta S \simeq 0.06$ cal/mole · deg was obtained from the temperature dependence $C_p = C_p(T)$ for SbSI. Independent determination of $P_0 \simeq 20 \cdot 10^{-6}$ C/cm^2 and $C \simeq 3.8 \cdot 10^5$ °K from dielectric measurements showed that the values of $\Delta S, P_0$, and C for SbSI will satisfy the thermodynamic relation (1.34) $\Delta S = (2\pi/C) P_0^2$. It was not possible to measure P_0 directly for SbSBr. However, by measuring ΔS and C for these crystals and using relation (1.34), one could calculate the value of the spontaneous polarization (cf. Table 4). The value of P_0 for SbSBr and BiSI is almost three times less than for SbSI.

Crystals of SbSI and BiSeI display only second-order phase transitions at the temperatures 233 and 133°K, respectively. As is seen from Figure 4.25 and Table 4, these phase transitions are accompanied by a discontinuity of the heat capacity, a discontinuity of the temperature coefficient $(dE_g/dT)_p$, and anomalies of the electrical conductivity. Anomalies of ϵ were not observed in either case. Second-order phase transitions are also observed for the crystals of SbSI, SbSBr, BiSI, and BiSBr at the temperatures indicated in Table 4. These phase transitions were recorded in [126] from data of optical absorption by measuring $\Delta(dE_g/dT)$. Unfortunately, the mechanism of second-order phase transitions in the ferroelectrics of the type $A^V B^{VI} C^{VII}$ is still unclear and the appropriate structural data are absent in the literature.

We turn now to the problem of the form of the intrinsic absorption edge and the nature of the interband optical transitions in ferroelectrics of the type $A^V B^{VI} C^{VII}$. This problem was investigated in most detail for SbSI. The form of the absorption edge of SbSI in polarized light was investigated in [129] from the dependence of $k^{1/2}$ on the photon energy $h\nu$. The author was led to the conclusion of the existence of two linear sections indicating the presence of indirect optical transitions related to emission and absorption of phonons with energies 0.09 and 0.14 eV, respectively, at the temperatures 24 and -164°C. It was also shown in [126] that the form of the absorption edge and the nature of the transitions do not change over a wide temperature interval including the phase transitions in SbSI. Further investigation of the nature of the optical transitions in the ferroelectric SbSI was aimed at comparing the energies of the phonons obtained from the section of indirect transitions with data from a study of the vibration structure of SbSI [130]. An analogous approach was taken in [131] to study the edge of the intrinsic absorption of SbSBr. Another reason stimulating continuation of these investigations was the absence of analytical data on the indirect transitions in SbSI obtained by the most consistent method, for example, that described in [132].

Determination of the width of the forbidden band at various temperatures by extrapolation from the steepest section to $k \to 0$ (k is the absorption coefficient) or description of the temperature dependence $E_g = E_g(T)$ from temperature changes of the point with a specific absorption level, for example $k = 500$ cm^{-1} [133], cannot provide sufficient accuracy of the results obtained and can generally lead to inaccurate results in the case of the indirect transitions. This follows from these arguments. We assume that there are two semiconductors with the same forbidden bands, but with different phonon spectra. The absorption edge in this case for indirect transitions can have different slopes, and the two methods described above will lead to different values of E_g in these semiconductors, which is incorrect. On the other hand, analysis of the indirect transitions in SbSI is complicated by the presence of phase transitions. Thus, measurements in the absorption spectrum in the intermediate infrared portion of the spectrum with changes in the absorption edge at the phase transitions were compared in [130].

Figure 4.26 presents the transmission curves of SbSI crystals for the polarization $\mathcal{E} \parallel c$ and $\mathcal{E} \perp c$ at temperatures 293 and 95°K. A strongly polarized absorption band III is observed in the region 0.08 eV with $\mathcal{E} \perp c$.

Measurement of the transmission in the paraelectric phase of SbSI at 323°K also confirmed the presence of this band. The short-wavelength portion of a wide absorption band with high absorption for $\mathcal{E} \parallel c$ is observed in the region 0.06 eV at temperatures of 323 and 293°K. Thus, there are no essential differences between the transmission spectrum of SbSI in the phases D_{2h}^{16} and C_{2v}^{9} in the region 0.055–0.085 eV. Absorption in the region of long wavelengths beginning with 0.07 eV is evidently related to lattice vibrations, while the lower

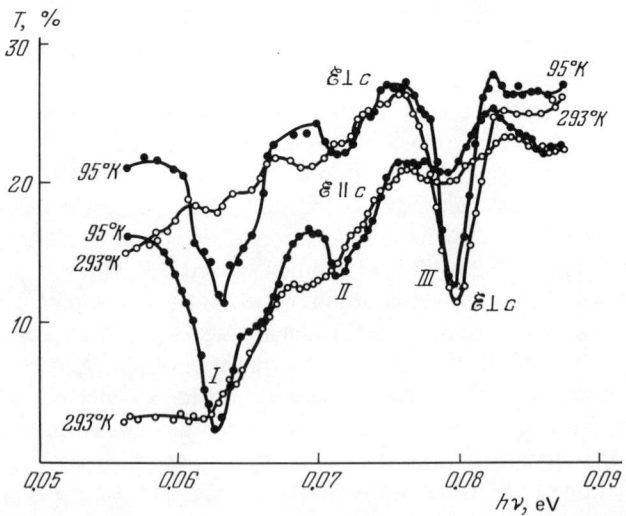

Figure 4.26. Transmission of SbSI crystals for various polarizations and temperatures in the IR region of the spectrum.

transmission at $\mathcal{E} \parallel c$ agrees with the data of [134], where reflection was observed in the region of the vibrational bands of SbSI and the reflection for $\mathcal{E} \parallel c$ is greater than the reflection for $\mathcal{E} \perp c$ in the same region. The fact that there are no changes in frequency of the highest-frequency vibrational bands with the transitions from the paraelectric phase to the ferroelectric is explained by the fact that absorption in the soft ferroelectric mode can be observed only in the far IR region of the spectrum. At 95°K, i.e., in the ferroelectric phase C_2^2, the absorption selectivity in the region 0.6–0.7 eV is enhanced at 0.063 eV (bands I) and weak bands II (0.0715 eV) appear for both polarizations. If the bands I, II, and III were caused by transitions of electrons from the valence band to local impurity levels or from local impurity levels to the conduction band, then the integral intensity of the bands must increase with decreasing temperature, which is not observed. Thus, the observed bands may be caused by absorption at the highest-frequency optical lattice vibration (longitudinal optical modes at $k = 0$) or selective absorption at free carriers (intraband absorption).

By comparing the transmission curves for SbSI crystals with the analogous curves for SbSBr we conclude that the bands I and II are vibrational in nature, while the band III is electronic. The position of band I agrees well with the values of the reduced masses m^*_{SbSI} and m^*_{SbSBr}, which would be expected for the highest-frequency absorption by the lattice at k = 0. We assume that the appearance of the band II at low temperatures is related to a change in symmetry

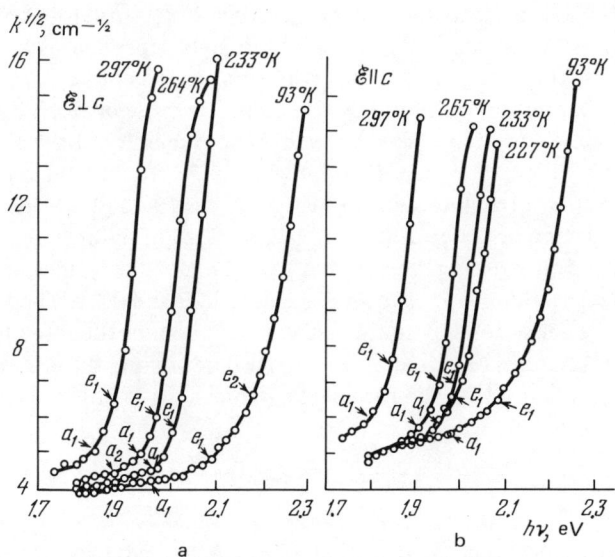

Figure 4.27. Indirect transitions in SbSI for polarizations $\mathcal{E} \perp c$ (a) and $\mathcal{E} \parallel c$ (b).

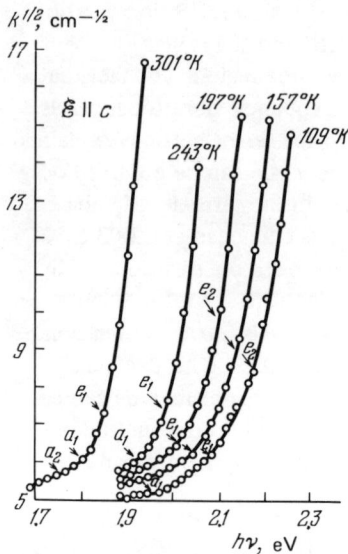

Figure 4.28. Indirect transitions in SbSI for $\mathcal{E} \parallel c$.

of the crystals of SbSI ($C_{2v}^9 \rightarrow C_2^2$) and with the removal of degeneracy of the vibrations similar to that which occurs for the highest-frequency longitudinal optical modes of BaTiO$_3$ with a decrease in temperature and change in symmetry of the crystals at the phase transitions.

Figure 4.27a, b presents the dependences $k^{1/2} = k^{1/2}(h\nu)$ for SbSI crystals at various temperatures and polarization. Analogous dependences taken from [129] are presented in Figure 4.28 for comparison. The choice of the main section of the indirect transitions ($\epsilon_{a_1} \leq h\nu \leq \epsilon_{e_1}$) is made difficult by the presence of phase transitions. The degree of reproducibility of the abscissas of the observed breaks can be judged from Figures 4.28 and 4.27b from the curves for $k^{1/2} = k^{1/2}(h\nu)$ at almost identical temperatures 297 and 301°K. By analyzing the temperature behavior of the separate sections of indirect transitions and the temperature changes in absorption in the IR region, Gerzanich et al. [130] chose the main sections as are shown in Figures 4.27 and 4.28. The arrows indicate the breaks corresponding to the emission of phonons (denoted by e_n) and absorption of phonons (denoted by a_n). After determining the abscissas of the breaks ϵ_{a_n} and ϵ_{e_n} with the help of the expressions [132]:

$$E_g = \frac{\varepsilon_{e_n} + \varepsilon_{a_n}}{2}, \quad k\theta_n = \frac{\varepsilon_{e_n} - \varepsilon_{a_n}}{2}, \quad (4.2)$$

the authors found E_g, $k\theta_1$, and $k\theta_2$ at various temperatures for the polarizations $\mathcal{E} \parallel c$ and $\mathcal{E} \perp c$ (see Table 5). The behavior of the $k\theta_1 = 0.06$ eV phonons in indirect transitions of the phase C_2^2 is typical, which agrees with enhanced

TABLE 5. Results of Analyzing Indirect Transitions in SbSI Crystals (Values in eV)

		Space group								
		D_{2h}^{16}		C_{2v}^9					C_2^2	
		301°K	297°K	265°K	243°K	233°K	227°K	197°K	157°K	109°K	93°K
$\mathcal{E} \parallel c$	E_g	1.82	1.82	1.93	1.96	1.97	1.96	1.95	1.99	2.02	2.02
	$k\theta_1$	0.02(5)	0.02(5)	0.02	0.02	0.02	0.06	0.06	0.06	0.06	0.06
	$k\theta_2$	0.06(5)	—	—	—	—	—	—	—	0.15	0.15
$\mathcal{E} \perp c$	E_g	—	1.85	1.95	—	1.99	—	—	—	—	2.02
	$k\theta_1$	—	0.02	0.02	—	0.02	—	—	—	—	0.06
	$k\theta_2$	—	—	—	—	—	—	—	—	—	0.16

selectivity of absorption at $h\nu_\text{I} = 0.063$ eV and the appearance of absorption at $h\nu_\text{II} = 0.0715$ eV (Figure 4.26). The fact that $h\nu_\text{II} \simeq 0.0715$ eV $> k\theta_1 \simeq 0.06$ does not contradict the fact that the IR absorption is due to direct transitions at the center of the Brillouin zone ($\mathbf{k} = 0$), while phonons of the same branch participate in the indirect transitions (if this is allowed by the selection rules), but with $\mathbf{q} = \Delta\mathbf{k} = \mathbf{k}_\text{min} - \mathbf{k}_\text{max}$, where $\mathbf{k}_\text{min}$ is the point of the minimum of the conduction band and $\mathbf{k}_\text{max}$ is the point of the maximum of the valence band. Since it follows from the dispersion of the optical branch that $d(\hbar\omega)/d|\mathbf{k}| < 0$, then $\hbar\omega(\Delta\mathbf{k}) < \hbar\omega(0)$. The comparatively higher energy $k\theta_2$ can, in principle, correspond to the total phonon processes or can be related to the fact that transitions from another main branch are superimposed onto the indirect transitions from the one main branch because of the complexity of the structure.

The points plotted in Figure 4.29 determine E_g at various temperatures and polarizations. The values of E_g thus determined are certainly more accurate than the values found from extrapolation. It is interesting that in the region of the second-order phase transition, the obtained values of E_g indicate the passage of the dependence $E_g = E_g(T)$ through a minimum. Such a minimum had not been observed in this region according to the data published in other works [100, 126] obtained without analyzing the indirect processes. Thus, strictly speaking, the behavior of such a minimum completely agrees with the thermodynamic relation (1.46). Actually, it follows from (1.46) that the passage of the dependence $P_0(T)$ through a minimum, which was observed in the region of the second-order phase transition in [125, 127], must lead to the appearance of a minimum in $E_g = E_g(T)$. Since the investigations in [130] were carried out on crystals obtained by the same technology as in [127], it is possible to make a quantitative evaluation of the magnitude of the drop in $(\delta E_g)_2$ in the region of the second-order phase transition and the discontinuity $(\Delta E_g)_1$ in the region of the first-order phase transition, and to make a comparison with the drop in δP_0^2 according to [127]. Actually, keeping the quadratic term in (1.46), we have

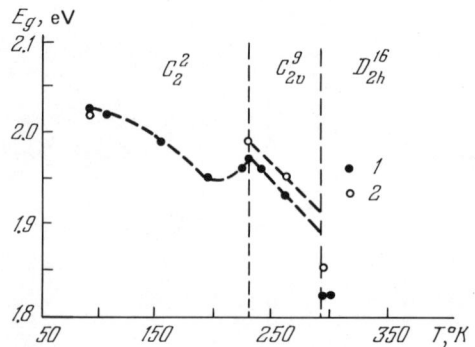

Figure 4.29. Values of E_g in different phases of SbSI found from analyzing indirect transitions: (1) $\mathscr{E} \parallel c$; (2) $\mathscr{E} \perp c$.

$$(\Delta E_g)_1 = aP_0^2(T), \tag{4.3}$$

$$(\delta E_g)_2 = a\delta P_0^2. \tag{4.4}$$

Then we have, from (4.3) and (4.4),

$$(\delta E_g)_2 = (\Delta E_g)_1 \frac{\delta P_0^2}{P_0^2}. \tag{4.5}$$

According to [127], the spontaneous polarization of SbSI decreases from $29.4 \cdot 10^{-6}$ C/cm^2 to $26.0 \cdot 10^{-6}$ C/cm^2 in the region of the second-order phase transition; thus, $\delta P_0^2/P_0^2 \simeq 0.22$.

We find from Figure 4.29 that $(\Delta E_g)_1 = 0.075$ eV (we assume that E_g depends linearly on T in the phase C_{2v}^9 [126, 100]). Thus, according to (4.5), $(\delta E_g)_2 \simeq 0.016$ eV, while it is seen from Figure 4.29 that $(\delta E_g)_2 \simeq 0.02$ eV. Such a qualitative and quantitative agreement of the obtained values of $(\delta E_g)_2$ in terms of δP_0^2 and from the analysis of the indirect transitions indicates that the interpretation presented in [130] of the minimum in the dependence $E_g = E_g(T)$ in the region of the second-order phase transition of SbSI raises no doubts. On the other hand, this once again speaks in favor of the pseudo-Jahn–Teller model, according to which the general causes in the change of E_g and P_0 at phase transitions lead to correlations of these quantities described by relation (1.46).

Following [130], we briefly discuss the band model of SbSI, taking into account the presence in all three phases of the band $h\nu_{III} \simeq 0.08$ eV for $\mathcal{E} \perp c$. Gerzanich et al. [130] assume that the polarized band III is caused by intraband transitions, similar to the polarized band ($\mathcal{E} \parallel c$) at 0.11 eV in tellurium [119]. According to [135], the minimum of the valence band of SbSI is located at point Γ of the Brillouin zone, while the minimum of the conduction band is on the k_z axis (point Λ, cf. Figure 4.50). Thus, depending on whether SbSI is an n- or p-type semiconductor, the separate band in the energy model must be located at a distance of 0.08 eV above the bottom of the conduction band (point Λ) or below the top of the valence band (point Γ). According to the data of many works, SbSI is an n-type semiconductor [136, 137], although data in [138, 139] from an investigation of the Zeebeck effect for SbSI leads to the conclusion of p-type conductivity. The selection rules for direct transitions established in [135] made it possible to determine the allowed polarizations in all three phases of SbSI at different points of the Brillouin zone. Since the polarization $\mathcal{E} \parallel c$ is forbidden and the polarization $\mathcal{E} \perp c$ is allowed by the selection rules only at point Γ for all three phases (D_{2h}^{16}, C_{2v}^9, and C_2^2), the subband mentioned above must evidently be located at 0.08 eV below the top of the valence band. Since data for only two polarizations ($\mathcal{E} \parallel c$ and $\mathcal{E} \perp c$) were available to Gerzanich et al. [130], they made use of the one-dimensional model [135]. Then, in this model 0.08 eV = $E_{\Gamma_2^+} - E_{\Gamma_2^-}$. On the other hand, the distance of 0.08 eV between subbands found in [135] correlates not too badly with the reflection peaks of SbSI in the visible region of the spectrum [135]. Actually, if $E_2 = E_{\lambda_1} - E_{\lambda_2}$ ($\mathcal{E} \perp c$) and $E_1 = E_{\lambda'_2} - E_{\lambda_2}$ ($\mathcal{E} \parallel c$), then the distance between the peaks $E_2 - E_1 = E_{\lambda_1} - E_{\lambda'_2}$ is 0.08 eV in the phase D_{2h}^{16}, 0.09 eV in the phase C_{2v}^9, and 0.13 eV in the phase C_2^2, while the transition $E_{\lambda'_2} - E_{\lambda_1}$ is allowed only for $\mathcal{E} \perp c$. We will discuss this problem once again as applied to SbSI in Section 4.6, which is devoted to the band structure of several ferroelectrics. Here, we note in accordance with [130] that only more detailed investigations can refine the parameters and reliability of one band model or another. In particular, detailed investigations of the indirect transitions in all phases for all three polarizations, with the help of instruments which would permit isolating an energy interval of $\sim 10^{-3}$ eV, are necessary for SbSI. Moreover, it is necessary to establish the selection rules for the indirect transitions in all phases of SbSI. Establishment of the more precise position of bands located above the bottom of the conduction band and below the top of the valence band is evidently possible with a detailed combined analysis of the results of investigations of reflection [135], edge absorption, and electrical reflection [140] of SbSI crystals.

The nature of the interband optical transitions in SbSI, SbSBr, and other ferroelectrics of the type $A^V B^{VI} C^{VII}$ cannot be considered as finally established. Thus, for example, there are a group of works [141, 142] in which the authors are led to the conclusion that the absorption edge in $A^V B^{VI} C^{VII}$ is caused by direct transitions subject to the Urbach rule, although these same authors were subsequently led to the conclusion of the presence in SbSI of indirect transitions on the basis of analysis of the band structure (cf. Section 4.6). Thus, the spectral and temperature dependences of the absorption coefficient k in the interval of energies $h\nu = 1.6$–2.1 eV ($k = 10$–10^3 cm^{-1}) and temperatures 83-373°K were investigated for SbSI in [142]. As a result, Zeinally et al. [142] were led to the conclusion that in the range $k = 40$–10^3 cm^{-1} there is an exponential dependence of k on $h\nu$ subject to the Urbach rule:

$$k = k_0 \exp\left[\frac{\sigma}{kT^*}(h\nu - E_0)\right], \tag{4.6}$$

where σ, k_0, and E_0 are constants; T^* is the effective temperature:

$$T^* = \frac{\hbar\omega}{2k} \coth\left(\frac{\hbar\omega}{2k_B T}\right), \tag{4.7}$$

$\hbar\omega$ is the energy of the phonon interacting with the electron; k_B in (4.7) is the Boltzmann constant. As in [130], the absorption is anisotropic. The edge of the intrinsic absorption for $\mathcal{E} \perp c$ corresponds to higher energies than for $\mathcal{E} \parallel c$. Figure 4.30 presents the dependences of k on $h\nu$ on a semilogarithmic scale, illustrating agreement with (4.6). The widths of the forbidden band $E_g^\perp = 1.96$ eV

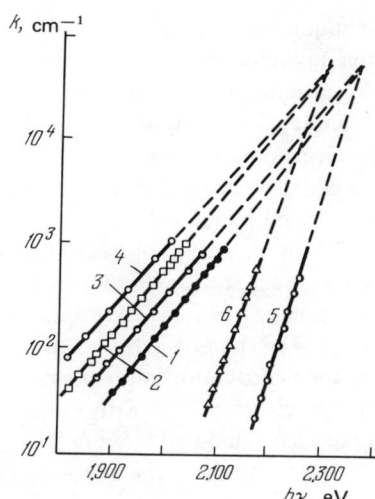

Figure 4.30. Spectral dependence of the absorption coefficient of SbSI in polarized light at different temperatures [142]. (1, 3, 5) & ⊥c; (2, 4, 6) & ∥ c. Temperatures (°K): (1, 2) 300; (3, 4) 373; (5, 6) 83.

and $E_g^{\parallel}$ = 1.84 eV were determined in [142] for SbSI at $k \simeq 10^2$ cm^{-1}. The phonon frequency $\omega \simeq 210$ cm^{-1}, which agrees well with the frequency $\omega \simeq 215$ cm^{-1} determined from analogous measurements in [141], was calculated by comparing the experimental curves with (4.6) and (4.7). The phonon frequency ω, which was calculated in [141, 142] from data on optical absorption (assuming direct Urbach transitions), agrees well with $\omega \simeq 212$ cm^{-1} [143] and ω = 215 cm^{-1} [144] independently determined from combination scattering spectra. It was shown in [143, 144] that this frequency corresponds to the frequency of vibrations of atoms along the z direction and, consequently, it is this vibration that interacts with the electrons and causes the Urbach nature of the intrinsic absorption edge. In the opinion of Chisler [143], the frequency of the vibrations $\omega \simeq 212$ cm^{-1} characterizes valence vibrations of the Sb$^+$-S bond and, not being a vibrational frequency of the soft (ferroactive) mode, is observed in both phases of SbSI, not changing in the region of the phase transition. Thus, in SbSI, as in perovskite ferroelectrics, direct Urbach transitions, if they actually occur, are evidently caused by interaction of the electrons with longitudinal optical phonons and do not have a significant role in the transverse optical mode of vibration (the soft mode), which, it would appear, contradicts the vibron model of ferroelectric phase transitions. All this once again indicates the necessity of further experimental study of the nature of the interband optical transitions in ferroelectrics.

In concluding this section, we discuss the problem of phase transitions in single crystals of Sb_2S_3. It was emphasized above that the change in the temperature dependence of the intrinsic absorption edge of ferroelectrics is a convenient technique in a number of cases for identifying phase transitions. This technique was applied in [99] to the study of possible phase transitions in semiconductor crystals of antimony trisulfide Sb_2S_3. As is well known, the orthorhombic single crystals of Sb_2S_3 are structurally the same as the paraelectric phase of the ferroelectric SbSI and belong to the space group D_{2h}^{16}. A number of dielectric properties of Sb_2S_3 single crystals have been investigated. Anomalies of the temperature dependence of the dielectric constant ϵ, tan δ, dark current, and photocurrent have been obtained as a result [145, 146]. The anomalous dispersion of ϵ has been observed and investigated along with the resonance absorption of microwaves in Sb_2S_3, which is evidently related to the excitation of low-frequency optical vibrations in the region of the phase transition [147]. The possibility of the existence of such vibrations in ferroelectrics near the phase-transition temperature has already been discussed in Chapter 2. The pyroelectric effect with a maximum value of the pyrocurrent at $T = +52°C$ was observed and measured in Sb_2S_3 single crystals [148].

Investigations on the dielectric properties of Sb_2S_3 and comparison of the crystal-chemical structure of Sb_2S_3 crystals and the ferroelectric SbSI have permitted one to assume the presence of ferroelectric phase transitions in Sb_2S_3. However, the data obtained have been insufficient to determine the Curie temperature and the nature of the phase transitions. In this connection, the temperature dependence of the width of the forbidden band $E_g = E_g(T)$ in single crystals of Sb_2S_3 was investigated in [99]. The spectral measurements were carried out with the illumination directed perpendicular to the spontaneous polarization axis in unpolarized light in the temperature interval from -178 to $+150°C$. The values of E_g from the literature for Sb_2S_3, determined by various methods, differ greatly. According to the data obtained in [99], the width of the forbidden band for Sb_2S_3 determined from the steep section of the intrinsic absorption edge is 1.63 eV at room temperature. For all the Sb_2S_3 crystals investigated in [99], the temperature dependence $E_g = E_g(T)$ displayed an anomaly near $T_0 = +20°C$ (Figure 4.31). The value of T_0 varied in the range $+10$ to $+20°C$ for the investigated crystals. A linear change of E_g with temperature occurred above and below T_0. For $T_0 = +10$ to $+20°C$, the temperature co-

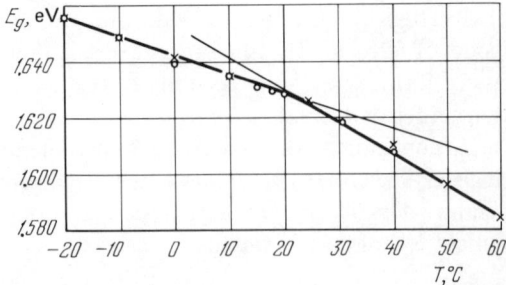

Figure 4.31. Temperature dependence of the width of the forbidden band for Sb_2S_3.

efficient $(\partial E_g/\partial T)_p$ has a discontinuity $\Delta(\partial E_g/\partial T)_p \simeq -(3.2-4.8) \cdot 10^{-4}$ eV/deg. Judging from the nature of the anomaly, a second-order phase transition occurs in the temperature interval +10 to +20°C in the Sb_2S_3 crystals investigated in [99].

We should note that the temperature coefficient $(\partial E_g/\partial T)_p$ for Sb_2S_3 is greater in absolute value in the temperature region $T > T_0$ and correspondingly smaller in the region $T < T_0$, whereas the opposite occurs for SbSI. Moreover, the first-order phase transition in SbSI at 22°C is accompanied by a discontinuity $\Delta E_g \simeq +0.02$ eV. As is seen in Figure 4.31, a discontinuity ΔE_g is not observed for Sb_2S_3 crystals. (If it actually exists, then its value does not exceed $3 \cdot 10^{-3}$ eV.) Two breaks in the course of the temperature dependence $E_g = E_g(T)$ at +10 and +60°C, respectively, were observed in one of the Sb_2S_3 crystals in [99]. The discontinuity of the temperature coefficient at +60°C is smaller in magnitude than at +10°C and is $(1.4 \pm 1) \cdot 10^{-4}$ eV/deg.

Recent investigations of the structure, dielectric properties, and infrared spectra of natural and grown crystals of Sb_2S_3 have permitted one to refine the nature of the phase transitions in Sb_2S_3 [149]. It has been shown as a result of these investigations that two phase transitions occur in Sb_2S_3, in agreement with the optical measurements [99]: a low-temperature transition in the region 290-310°K, and a high-temperature transition whose temperature depends strongly on the stoichiometry and varies in the range 420-490°K. The phase transition in the region 290-310°K is related only to the change in the magnitude of the spontaneous polarization and evidently occurs without a change in the symmetry of the crystal or with a change $C_2^2 \rightarrow C_{2v}^9$, the same as in SbSI in the region of the low-temperature second-order phase transition. The change in symmetry in $C_{2v}^9 \rightarrow D_{2h}^{16}$ in Sb_2S_3 evidently occurs at 420-490°K. Detailed structural investigations over a wide temperature range are necessary to confirm these assumptions. The results of investigations of the phonon spectra of artificial and natural crystals of Sb_2S_3 confirm the polarity of the crystals up to the high-temperature phase transition. For the polarization $\mathcal{E} \parallel c$, the number of active peaks of infrared absorption does not change at the transition through the temperature of the low-temperature phase transition and corresponds in the temperature interval from 100 to 400°K to the number of single-phonon absorption peaks for the C_{2v}^9 symmetry. For the polarization $\mathcal{E} \perp c$, several peaks disappear above the temperature of the low-temperature phase transition, which may be related to small structural changes at this temperature. The results of investigations of the temperature dependence of the NQR frequency confirm the existence of structural changes in the first coordination sphere of Sb atoms in the temperature range 290-310°K. The ferroelectric properties of the crystals and the phase transitions are caused by weak coupling in the first coordination sphere of the Sb atoms [146]. This agrees with the results of [150], according to which the low-frequency mode of vibrations of Sb_2S_3 responsible for the high static dielectric constant is related to vibrations of the Sb atoms.

4.4. Polarization Fluctuations and the Intrinsic Absorption Edge of Ferroelectrics

In this section, we will discuss in somewhat more detail one of the possible mechanisms causing the Urbach form of the intrinsic absorption edge of ferroelectrics.

In Section 4.3, we discussed a mechanism based on electron–phonon interaction which, as a detailed analysis has shown [151], leads to an exponential form of the edge under the assumption of a high electron–phonon coupling coefficient in the polar crystal. Emphasizing once again that the problem of the form of the intrinsic absorption edge of ferroelectrics requires additional experimental study, we discuss in this section the theory proposed by Bonch-Bruevich and Burtsev [118], which permits one to explain the exponential form of the edge and its behavior near the phase transition on the basis of a fluctuation mechanism specific for ferroelectrics.

As was shown by Bonch-Bruevich [152], modulation of the bands by a Gaussian random field leads to the generation of a long-wavelength absorption tail having an exponential form. Random fields in semiconductors can be due to various causes. These can be fluctuations caused by strong alloying (amorphous semiconductors) or random distribution of impurities of various kinds. However, even "ideal" crystals at temperatures differing from absolute zero can display a tail in the density of states due to fluctuations of the volume density of carriers and correspondingly a tail on the curve of the intrinsic optical absorption. This kind of fluctuation can be considered as a static electric profile if the electron relaxation time is small as compared to the period of change in this random electric field.

From this point of view, the ferroelectric semiconductor near the phase transition is an analog of a disordered (and even amorphous) semiconductor. Fluctuations of the internal field in this case are then caused by fluctuations of the order parameter, i.e., spontaneous polarization. According to [153], these fluctuations make a contribution to the scattering of free carriers in ferroelectrics by analogy to scattering at a deformation potential in linear semiconductors (carrier scattering on acoustical vibrations).

As follows from thermodynamics, the mean-square fluctuations of the field $\mathcal{E}_i$ or polarization P_i in the volume of the crystal V are, respectively,

$$\overline{\Delta \mathcal{E}_i^2} = \frac{4\pi kT}{V\varepsilon},$$

$$\overline{\Delta P_i^2} = \frac{\varepsilon kT}{4\pi V}. \tag{4.8}$$

Using (4.8) and the method proposed in [152], Burtsev obtained an expression for the interband absorption coefficient α in an anisotropic ferroelectric:

$$\alpha = \sum_i \alpha_{0i} \exp\left[(h\nu - E_g)/\varphi_i^{1/3}\right], \tag{4.9}$$

$$\varphi_i = \frac{\hbar^2}{12m_i^*} \left\langle \left(\frac{\partial U}{\partial r_i}\right)^2 \right\rangle, \quad i = x, y, z. \tag{4.10}$$

Here, m_i^* are the diagonal components of the effective mass tensor, and $U(\mathbf{r})$ is the potential of the random internal field at point $\mathbf{r}$. The brackets in (4.10) denote averaging over all possible configurations of the random potential. The effective number of exponents in (4.9) varies from one to three and depends on the anisotropy of the crystal. If the medium is close to isotropic ($\varphi_x \simeq \varphi_y \simeq \varphi_z$), then the Urbach absorption tail is one exponent. Denoting the field averaged over the volume of the unit cell by $\overline{\mathscr{E}(\mathbf{r}, t)}$ and the corresponding fluctuation by $\Delta\mathscr{E}(\mathbf{r}, t)$, we have

$$\mathscr{E}(\mathbf{r}, t) = \overline{\mathscr{E}(\mathbf{r}, t)} + \Delta\mathscr{E}(\mathbf{r}, t). \tag{4.11}$$

The following correlation function for the fluctuations $\Phi_{ij}(\mathbf{r})$ is introduced later in [118]:

$$\Phi_{ij}(\mathbf{r}) = \Phi_{ij}(\mathbf{r}' - \mathbf{r}'') = \overline{\Delta\mathscr{E}_i(\mathbf{r}', t) \Delta\mathscr{E}_j(\mathbf{r}'', t)},$$
$$i, j = x, y, z. \tag{4.12}$$

It follows from (4.10) and (4.12) that

$$\Phi_{ij}(0) = \overline{\Delta\mathscr{E}_i^2} = \frac{12m_i^*}{q^2\hbar^2}\varphi_i. \tag{4.13}$$

We isolate a volume V in the crystal, much larger than the volume of the unit cell, and calculate the fluctuation of the intensity of the macroscopic field in the volume V:

$$\Delta\mathscr{E}(t) = \frac{1}{V}\int_V \Delta\mathscr{E}(\mathbf{r}, t)\, d\mathbf{r}. \tag{4.14}$$

We have, for the mean square of the macroscopic fluctuation, according to (4.14),

$$\overline{(\Delta\mathscr{E}_i)^2} = \frac{1}{V^2}\int_V d\mathbf{r}' \int_V d\mathbf{r}'' \overline{\Delta\mathscr{E}_i(\mathbf{r}', t)\Delta\mathscr{E}_i(\mathbf{r}'', t)} = \frac{1}{V}\int_V \Phi_{ii}(\mathbf{r})\, d\mathbf{r}. \tag{4.15}$$

The definition of the correlation function (4.12) is used in (4.15). Comparison of (4.15) and (4.8) leads to

$$\int_V \Phi_{ii}(\mathbf{r})\, d\mathbf{r} = kT(\varepsilon\varepsilon_0)^{-1}. \tag{4.16}$$

We denote the characteristic linear correlation dimension of the macroscopic inhomogeneity of the field by r_0. The following form of the correlation function is postulated in [118]:

$$\Phi_{ii}(\mathbf{r}) = \Phi_{ii}(0) \exp(-r/r_0). \tag{4.17}$$

Substituting (4.17) into (4.16) and integrating, we find

$$\varphi_i = \frac{q^2 \hbar^2}{12 m_i^*} \frac{kT}{8\pi r_0^3} (\varepsilon \varepsilon_c)^{-1}. \tag{4.18}$$

As is emphasized in [118], the choice of the analytic form of the correlation function (4.17) does not affect the result (4.18), with an accuracy to a numerical factor, only if the function (4.17) tends to zero sufficiently rapidly for $r > r_0$. Substitution of (4.18) into (4.9) gives the final expression for the energy dependence of the intrinsic absorption edge.

We analyze the result obtained in [118] and compare it with the data presented in Sections 4.2 and 4.3 from [116, 117, 142]. First of all, as follows from the consideration presented above, fluctuation of the polarization (or field) leads to an exponential form of the intrinsic optical absorption tail. As a phase transition, the form of the tail can change, in principle, both because of the change in the number of exponents in (4.9) and because of the dependence of φ_i on the correlation volume r_0^3 and the dielectric constant ϵ. In such triaxial ferroelectrics as BaTiO$_3$ and other perovskites, the absorption tail must be exponential both above and below the Curie point. If one assumes, following [154], that the correlation volume r_0^3 in the region of the phase transition depends weakly on the temperature, then it would appear that the slope of the exponent of (4.9) at the transition through the Curie point must change significantly because of the temperature dependence of ϵ. However, this change was not observed in [116]. It is still difficult to judge the causes of this divergence. It is possible that the fluctuation volume r_0^3 actually has a temperature maximum at the Curie point, compensating the maximum of ϵ in the same region.

It is no less interesting to compare the results of (4.18) with the experimental data, according to which the slope of the Urbach exponent can be identical for various ferroelectrics with the perovskite structure (for example, for BaTiO$_3$ and KTaO$_3$; cf. Section 4.2) and having a different phonon spectrum. From the point of view of (4.18), this fact becomes clear if one considers the close values of r_0^3, ϵ, and the effective mass m_i^* for them. According to (4.1), the experimental value for BaTiO$_3$ is $\varphi^{1/3} \simeq k(140° + T_1) \simeq 45 \cdot 10^{-3}$ eV. Substituting the parameters for $kT \simeq 36 \cdot 10^{-3}$ eV, $\epsilon \simeq 8000$ [116], $m_i \simeq 0.1 m_0$, and $r_0 \simeq 0.2 \cdot 10^{-6}$ cm [155] into (4.18), we find $\varphi_i^{1/3} \simeq 0.6 \cdot 10^{-3}$ eV. As is seen, this value differs by more than an order of magnitude from the experimental value obtained in [116]. This divergence can first of all be related to inhomogeneity in evaluating the correlation length r_0 and, moreover, to the analytical form of the approximation (4.17) assumed in [118]. In uniaxial ferroelectrics, for example SbSI, $\varphi_x, \varphi_y \gg \varphi_z$ occurs and, consequently, the

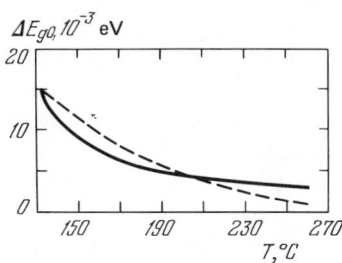

Figure 4.32. Dependence $\Delta E_{g_0}(T)$ in the cubic phase of $BaTiO_3$.

Urbach absorption tail must be represented by two exponents according to [118]. Thus, the meaning of the approach taken in [118] involves not the quantitative comparison with experiment, but the fundamental possibility of explaining the exponential form of the intrinsic absorption edge of ferroelectrics on the basis of polarization fluctuations.

The role of polarization fluctuations in forming the intrinsic absorption edge of a ferroelectric can be verified directly from (4.8). Keeping the quadratic term in the polarization in (1.46), we have, for the paraelectric region;

$$\overline{E}_{g0} \simeq E_{g0} + \frac{a}{2}\overline{\Delta P^2}, \qquad (4.19)$$

where $\overline{E}_{g0}$ is the width of the forbidden band in the paraelectric region with consideration of polarization fluctuations, $\overline{\Delta P^2}$ satisfies (4.8), and a is the expansion coefficient known from (1.46). Verification of (4.19) was performed in [116] for the cubic phase of $BaTiO_3$, where a deviation from linearity in the dependence $E_{g0} = E_{g0}(T)$ was observed. Figure 4.32 presents the experimental dependence $\Delta E_{g0} = \Delta E_{g0}(T)$ obtained in [116] for the cubic phase of $BaTiO_3$ (dashed curve) and the corresponding theoretical curve constructed in accordance with (4.19) and (4.8) (solid curve). For the temperature dependence $\epsilon = \epsilon(T)$, borrowed from [116], and $V \simeq 3 \cdot 10^{-20}$ cm^3 [156], agreement with these experimental curves is obtained with $a \simeq 1$ eV/(C/m^2)^2, which agrees well in order of magnitude with [111, 120] where the value of a was obtained from a comparison of (1.46) with the experimental dependence $E_g = E_g(T)$ for the tetragonal phase of $BaTiO_3$.

Of course, this correspondence does not definitely speak in favor of the mechanism of interband absorption proposed in [118], but indicates conclusively the effect of polarization fluctuations on the position of the intrinsic absorption edge of a ferroelectric.

4.5. Effect of an Electric Field on the Intrinsic Absorption Edge of Ferroelectrics

A number of works have been devoted to a theoretical and experimental investigation of the effect of an electric field on the width of the forbidden band of semiconductors. A rigorous theoretical consideration of the problem of absorption of light by a crystal placed in a uniform electric field was carried out by Keldysh and Franz [157, 158]. They predicted a shift of the intrinsic absorption edge of semiconductors toward the long-wavelength region in a strong electric field. In fields of the order of 10^5-10^6 V/cm, the magnitude of the shift can reach $\sim 10^{-2}$ eV. Experimental verification of the shift of the intrinsic absorption edge was obtained, for example, with single crystals of $A^{II}B^{VI}$ and $A^{III}B^{V}$. On the basis of numerous investigations performed for films and single crystals of various materials (CdTe, CdS, PbS, Se, ZnTe), it was shown that the shift in the intrinsic absorption edge with an electric field applied to the semiconductor is caused by various mechanisms, which cannot be reduced to just a single Franz-Keldysh effect. For example, an estimate was made for the change of the width of the forbidden band under the effect of electrostriction, which for linear semiconductors and dielectrics gives in a field of 10^4 V/cm a value of 10^{-7} eV; this is four or five orders of magnitude smaller than the Franz-Keldysh effect. In the case of ferroelectrics, the change in the width of the forbidden band due to electrostriction can be 0.003 eV.

It was shown in Chapter 1 on the basis of the phenomenological theory of ferroelectric semiconductors that in the region of a ferroelectric phase transition, a shift in the intrinsic absorption edge occurs under the effect of an external electric field, whose sign is determined by the sign of the electron–phonon coupling constant. This fact follows directly from the expansion of E_g in even powers of the polarization and is the result microscopically of the application of the pseudo-Jahn-Teller effect to the mechanism of ferroelectric phase transitions. Since the expansion constant $a > 0$ for SbSI and for BaTiO$_3$ (this follows from an analysis of the intrinsic absorption edge, cf. Sections 4.2 and 4.3), then at

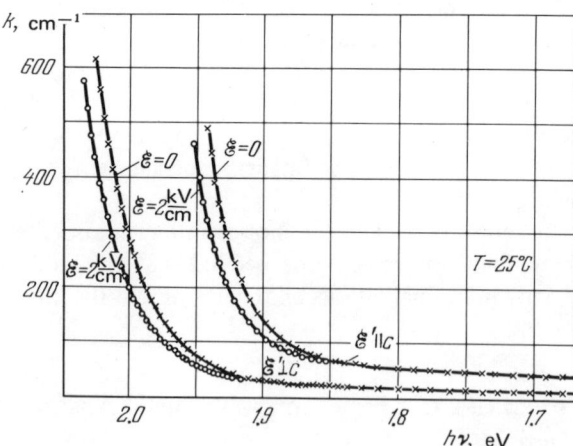

Figure 4.33. Shift of the intrinsic absorption edge in SbSI under the effect of a constant electric field [32].

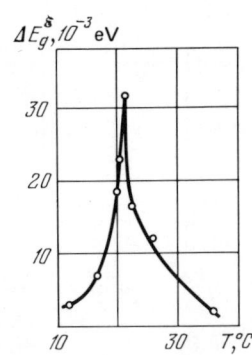

Figure 4.34. Temperature dependence of the shift of the absorption edge $\Delta E_g^{\mathscr{E}}$ under the effect of an electric field $\mathscr{E} = 2$ kV/cm in the region of the phase transition of SbSI [32].

least for these ferroelectrics the field must shift the intrinsic absorption edge toward the shorter wavelengths [cf. (1.52)]. Thus, the sign of the effect for BaTiO$_3$ and SbSI must be opposite the sign of the Franz-Keldysh effect.

Actually, in investigating SbSI crystals, Kern [31] discovered that a constant electric field applied along the spontaneous polarization axis of the crystal shifts the intrinsic absorption edge toward shorter wavelengths. As compared to the Franz-Keldysh effect, the shift of the edge was observed in smaller fields (of the order of 10^3-10^4 V/cm), and the sign of the shift was opposite the sign of the Franz-Keldysh effect. The absorption edge in SbSI was shifted toward shorter wavelengths independent of the direction of the field. Detailed investigations of the shift in the intrinsic absorption edge of SbSI under the effect of a field was carried out by Harbeke [32]. The results obtained by him are presented in Figures 4.33-4.36. As is seen from Figure 4.33, the electric field $\mathscr{E} = 2$ kV/cm shifts the absorption edge toward higher energies. The measurements are carried out in polarized light. For the two different polarizations $\mathscr{E} \perp c$ and $\mathscr{E} \parallel c$ ($\mathscr{E}$ is the polarization vector), the values of the absorption edge differed by 0.07 eV, while the magnitude of the shift is the same within the measurement error. The shift

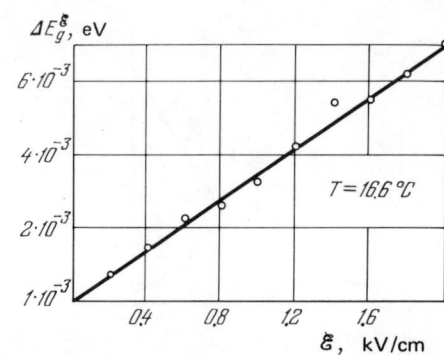

Figure 4.35. Dependence of the shift of the absorption edge $\Delta E_g^{\mathscr{E}}$ on the electric field intensity $\mathscr{E}$ in the ferroelectric phase for SbSI [32].

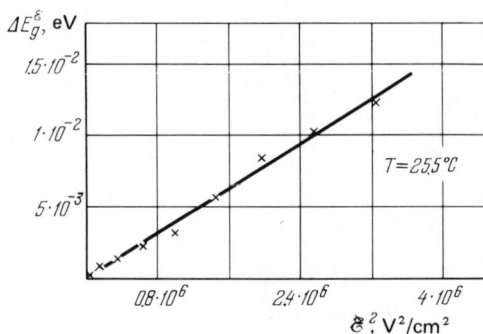

Figure 4.36. Dependence of the shift of the absorption edge $\Delta E_g^{\mathcal{E}}$ on the square of the field intensity in the paraelectric phase for SbSI [32].

of the absorption edge ΔE_g in SbSI was measured as a function of temperature and field intensity. It is seen from Figure 4.34 that the shift ΔE_g reaches a maximum value at the phase transition. Figures 4.35 and 4.36 present the field dependence $\Delta E_g = \Delta E_g(\mathcal{E})$. In accordance with the predictions of thermodynamic theory, this dependence is linear in the ferroelectric region and quadratic in the paraelectric region.

In accordance with (1.52), the temperature dependence of the field shift $\Delta E_g^{\mathcal{E}} = \Delta E_g^{\mathcal{E}}(T)$ is determined by the temperature dependence of the dielectric constant $\epsilon = \epsilon(T)$ and the spontaneous polarization $P_0 = P_0(T)$ near the Curie point. The temperature dependence $\Delta E_g^{\mathcal{E}} = \Delta E_g^{\mathcal{E}}(T)$ for SbSI was investigated, in particular, by Toyoda [128]. Figure 4.37 presents the dependence $\Delta E_g^{\mathcal{E}} = \Delta E_g^{\mathcal{E}}(T)$ in polarized light $\mathcal{E} \perp c$ obtained for SbSI in [128]. Since the Curie-Weiss law is satisfied in the paraelectric region for SbSI, substitution of $\epsilon = C/(T - T_0)$ and $P_0 = 0$ into (1.51) leads to the temperature dependence $(\Delta E_g^{\mathcal{E}})^{-1/2} \sim$

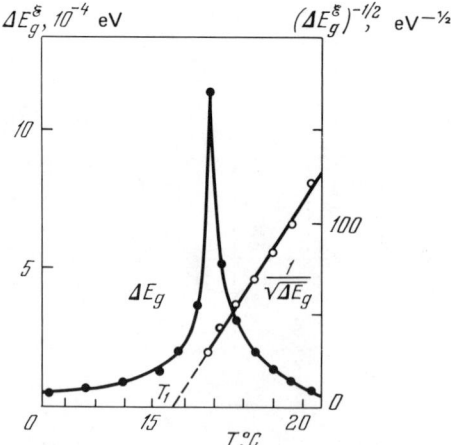

Figure 4.37. Temperature dependence of the shift $\Delta E_g^{\mathcal{E}}$ in SbSI in a variable field $\sim 23 \cdot 10^3$ V/cm [128].

OPTICAL ABSORPTION AND FERROELECTRICS

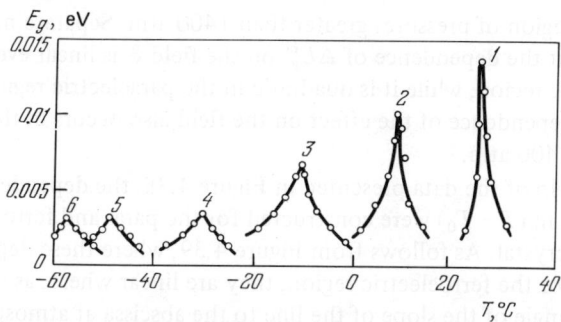

Figure 4.38. Shift of the intrinsic absorption edge of single crystals of SbSI in a field of 2 kV/cm, depending on the temperature for various values of the hydrostatic pressure p.

$(T - T_0)$ shown in Figure 4.37 and at the same time the quadratic dependence of $\Delta E_g^{\mathcal{E}}$ on the external field $\mathcal{E}$, presented in Figure 4.36. By substituting into (1.51) the value of the Curie-Weiss constant $C \simeq 1.67 \cdot 10^5$ °K for SbSI from the linear dependence of $(\Delta E_g^{\mathcal{E}})^{-1/2}$ on $(T - T_0)$, the expansion coefficient was obtained in [128]: $a = 2.68$ eV/(C/m²)², close to the values of a determined from the spontaneous discontinuity in the width of the forbidden band ΔE_g for the phase transition in SbSI.

Another approach to verifying the temperature dependence of the Kern-Harbeke effect was undertaken in [159]. As has already been indicated in Chapter 1, $(\Delta E_g^{\mathcal{E}}) \sim (T - T_0)^{-1}$ near first-order phase transitions in the ferroelectric region, while $\Delta E_g^{\mathcal{E}} \sim (T - T_0)^{-1/2}$ for second-order phase transitions in the ferroelectric region. Investigation of the state diagram for the ferroelectric SbSI showed the existence of a triple point with coordinates $T = -40°C, p = 1500$ atm, near which the lines of the first-order phase transitions arrive at the line of the second-order phase transitions [160-162]. Thus, it was of interest to investigate the Kern-Harbeke effect in SbSI over a sufficiently wide interval of hydrostatic pressures in order to compare the temperature dependences of the effect corresponding to phase transitions far from the point where the nature of the phase transition changes. Figure 4.38, borrowed from [159], presents six curves of the temperature dependence of the intrinsic absorption edge of SbSI in a field of 2 kV/cm corresponding to six different pressures in the region from 1 to 2000 atm and to the Curie temperatures from 25 to -57°C. As is seen from Figure 4.38, a maximum in the shift of the intrinsic absorption edge is observed at atmospheric pressure at the Curie point ($T_1 = 25°C$). The magnitude of this shift $\Delta E_g^{\mathcal{E}} \simeq 0.013 \pm 0.001$ eV agrees well with the data presented above from the literature. A decrease of this effect is observed with increasing pressure, and at $p = 1400$ atm, its value at the maximum is 0.003 eV. Broadening of the temperature maximum of the effect occurs simultaneously with increasing pressure. The magnitude and nature of the temperature dependence of the effect do not change in the region of pressures greater than 1400 atm. Separate measurements have shown that the dependence of $\Delta E_g^{\mathscr{E}}$ on the field $\mathscr{E}$ is linear everywhere in the ferroelectric region, while it is quadratic in the paraelectric region. The same nature of the dependence of the effect on the field also occurs in the region of pressures $p > 1400$ atm.

With the help of the data presented in Figure 4.38, the dependences of $\ln(1/\Delta E)_g^{\mathscr{E}}$ on $\ln(T - T_0)$ were constructed for the para- and ferroelectric regions of the crystal. As follows from Figure 4.39, where these dependences are presented for the ferroelectric region, they are linear where, as should be expected, the angle of the slope of the line to the abscissa at atmospheric pressure is close to 45°. This angle decreases with approach to the triple point, and is ~20° near it. It does not change significantly with a further increase in pressure.

We add that at temperatures sufficiently far from the Curie point, inversion of the sign of the effect was observed, caused by the fact that the magnitude of the Kern-Harbeke effect is comparable to the magnitude of the Franz-Keldysh effect.

Harbeke proposed in [32] that electrostriction is at least partially the basis of the field shift of the absorption edge of SbSI. This assumption is based on the closeness of the time dependences of the deformation and displacement of the absorption edge under the effect of the field. This assumption found indirect verification in [91, 92] from measurements of the dependence of E_g for SbSI on the hydrostatic pressure and temperature in the region of the phase transition.

A detailed investigation of the effect of a constant electric field on the intrinsic absorption edge in single crystals of $BaTiO_3$ near the upper Curie point was carried out in independently performed works [98, 163]. It was shown as a result that equally good agreement with the thermodynamic relations (1.52) occurs for crystals of SbSI and $BaTiO_3$. The investigations in [98] were car-

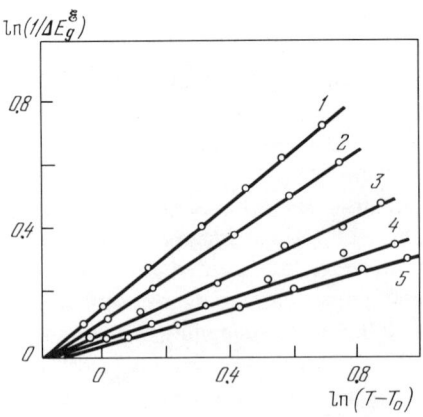

Figure 4.39. Dependence of $\ln(1/\Delta E_g^{\mathscr{E}})$ on $\ln(T - T_0)$ for single crystals of SbSI in the ferroelectric region at various values of hydrostatic pressure.

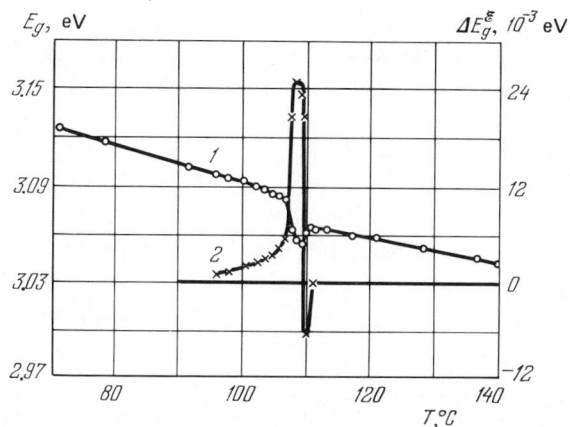

Figure 4.40. Temperature dependence of $E_g(T)$ for $BaTiO_3$ in the region of the phase transition from the tetragonal phase to the cubic phase (curve 1). Temperature dependence of the shift of the absorption edge $\Delta E_g^{\mathcal{E}}$ under the effect of an electric field of 1.1 kV/cm in the region of the phase transition (curve 2).

ried out with samples of $BaTiO_3$ having the form of a parallelepiped with sides parallel to the pseudocubic directions [100]. The measurements were performed on mechanically free crystals, the field was applied in the direction of the tetragonal axis, and illumination was perpendicular to the direction of the applied electric field. The $BaTiO_3$ crystals investigated in [98] contained a large amount of uncontrolled impurities, since technically pure $BaCO_3$ that contained a certain amount of Sr was used to prepare them. It was discovered in investigating the absorption edge of these crystals that with the transition from the ferroelectric phase to the paraelectric, a direct discontinuity in the width of the forbidden band toward lower energies $\Delta E_g \simeq +3 \cdot 10^{-2}$ eV was first observed, and then an inverse discontinuity $\Delta E_g \simeq -1 \cdot 10^{-2}$ eV (Figure 4.40, curve 1). As for SbSI, a constant electric field applied to the crystal shifts the absorption edge toward shorter wavelengths, i.e., leads to a decrease in the width of the forbidden band by an amount $\Delta E_g^{\mathcal{E}}$. The sign and magnitude of $\Delta E_g^{\mathcal{E}}$ do not change with a change in the direction of the field. The dependence of the shift $\Delta E_g^{\mathcal{E}}$, which arises under the effect of a field of intensity 1.1 kV/cm, on temperature (Figure 4.40, curve 2) was plotted in [98]. It was seen from this dependence that $\Delta E_g^{\mathcal{E}}$ has a maximum value at the temperature of the phase transition $T_1 = 108°C$. The magnitude of $\Delta E_{g\,max}^{\mathcal{E}}$ is $+2.5 \cdot 10^{-2}$ eV and is close to the magnitude of the spontaneous discontinuity in the width of the forbidden band $\Delta E_g \simeq +3 \cdot 10^{-2}$ eV at the phase transition. The phase transition was simultaneously fixed by observation of the hysteresis loop and other dielectric anomalies. In the paraelectric region at $T = 110°C$, the shift of $\Delta E_g^{\mathcal{E}}$ changes sign, and $\Delta E_{g\,max}^{\mathcal{E}} \simeq -6 \cdot 10^{-3}$ eV.

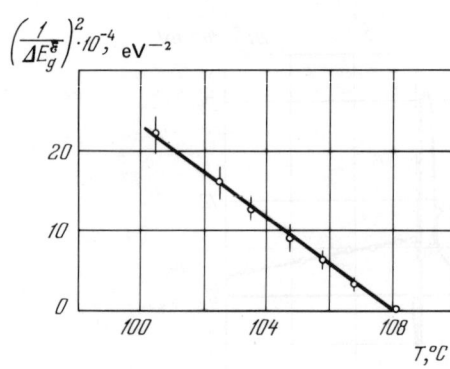

Figure 4.41. Dependence of $(1/\Delta E_g^{\mathcal{E}})^2$ on temperature for $BaTiO_3$; $\mathcal{E} = 1.1$ kV/cm.

A shift in the absorption edge is not observed at higher temperatures with an electric field intensity of 1 kV/cm (Figure 4.40, curve 2). The temperature dependence of $\Delta E_g^{\mathcal{E}}$ near the phase transition in the ferroelectric region satisfies the relation $(1/\Delta E_g^{\mathcal{E}})^2 \sim (T_1 - T)$ (Figure 4.41). The temperature of the phase transition is determined from Figure 4.41. The temperature dependence of $\Delta E_g^{\mathcal{E}}$ presented in Figure 4.41 does not contradict (1.52) because of the broadening of the phase transition $P_0 = P_0(T)$. The dependence of $\Delta E_g^{\mathcal{E}}$ on the intensity of the electric field applied to the crystal was measured in [98]. As for SbSI, the shift of the intrinsic absorption edge depends linearly on the field intensity in the ferroelectric region and quadratically in the paraelectric region (Figures 4.42 and 4.43). The coefficients of the dependence of $\Delta E_g^{\mathcal{E}}$ on the field intensity $\mathcal{E}$ in the ferroelectric region and on the square of the field intensity $\mathcal{E}^2$ in the paraelectric region are compared for $BaTiO_3$ and SbSI in Table 6. The data for SbSI are borrowed from [32].

A quadratic dependence of $\Delta E_g^{\mathcal{E}}(\mathcal{E}^2)$, similar to that presented in Figure 4.43, is observed in the paraelectric region at a sufficiently great distance from the Curie point. Superimposed on the dependence of $\Delta E_g^{\mathcal{E}}$ on the field in the same paraelectric region, but near the Curie point, is also the shift of the Curie point with the field, which must have a significant effect on the form of the ex-

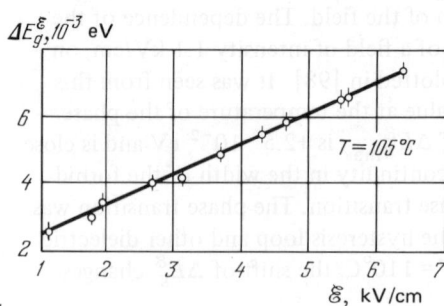

Figure 4.42. Dependence of the shift of the absorption edge $\Delta E_g^{\mathcal{E}}$ for $BaTiO_3$ on the electric field intensity $\mathcal{E}$ in the ferroelectric region at $T = 105°C$.

OPTICAL ABSORPTION AND FERROELECTRICS

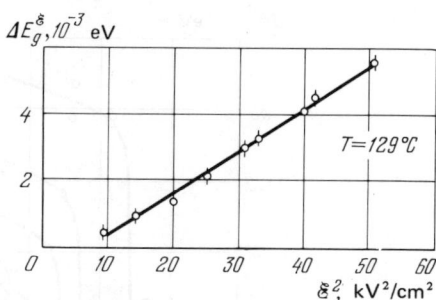

Figure 4.43. Dependence of $\Delta E_g^{\mathcal{E}}$ for $BaTiO_3$ on the square of the field intensity in the paraelectric region at $T = 129°C$ for $BaTiO_3$.

perimental curve. The corresponding results are presented in Figure 4.44, which represents the dependences of the shift $\Delta E_g^{\mathcal{E}}$ on the electric field intensity $\mathcal{E}$, taken near the Curie point $T_1 \cong 110°C$ at various temperatures $T > T_1$. All the obtained curves display a steep section at specific field intensities $\mathcal{E}_c$. A section of negative values of $\Delta E_g^{\mathcal{E}}$ is observed in the fields $\mathcal{E} < \mathcal{E}_c$. This form of the curves is explained by the fact that the electric field shifts the Curie point and thereby induces a transition of the crystal to the ferroelectric phase. In accordance with curve 1 of Figure 4.40, this transition of the crystal from the paraelectric to the ferroelectric phase is accompanied by an anomalous change in the width of the forbidden band E_g, where the latter first decreases and then increases sharply. The amplitudes of the shifts $+\Delta E_g^{\mathcal{E}}$ and $-\Delta E_g^{\mathcal{E}}$ in Figure 4.44 are close to the values of direct and inverse discontinuities observed in the temperature dependence of the width of the forbidden band represented by curve 1 in Figure 4.40. As is seen from Figure 4.44, the nature of the curves does not change with increasing temperature, but the amplitudes of the shifts $\Delta E_g^{\mathcal{E}}$ decrease and at temperatures where the electric field does not yet induce a transition to the ferroelectric state, a relatively weak change in $\Delta E_g^{\mathcal{E}}$ with field intensity according to a quadratic law is observed (Figure 4.43). From the dependence of the field intensity $\mathcal{E}_c$, corresponding to the steep section of the curves of Figure 4.44, on the temperature of the crystal, one can evaluate the coefficient of the shift in the Curie temperature with the field $dT_1/d\mathcal{E}$. As is seen from Figure 4.45, this dependence is linear, while the value of the coefficient $dT_1/d\mathcal{E} \simeq$

TABLE 6. Electrical Absorption in $BaTiO_3$ and SbSI

Crystal	T_1, °C	$\dfrac{d(\Delta E_g^{\mathcal{E}})}{d(\mathcal{E}^2)}$, eV·cm/V	$\dfrac{d(\Delta E_g^{\mathcal{E}})}{d(\mathcal{E}^2)}$, eV·cm²/V²
$BaTiO_3$	108	$(1 \pm 0.1) \cdot 10^{-6}$ at $T_1 - T = 3°$	$(0.12 \pm 0.012) \cdot 10^{-9}$ at $T - T_1 = 20°$
SbSI	21	$2.8 \cdot 10^{-6}$ at $T_1 - T = 4°$	$3.9 \cdot 10^{-9}$ at $T - T_1 = 4.5°$

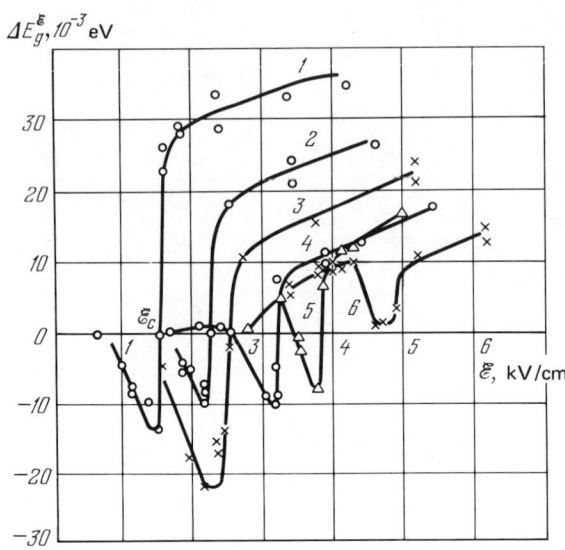

Figure 4.44. Dependence of the shift $\Delta E_g^{\mathscr{E}}$ on the electric field intensity in the paraelectric region near the phase transition in BaTiO$_3$. Curves 1-6 correspond to various temperature near the Curie point.

$1.2 \cdot 10^{-3}$ deg · cm/V determined from this dependence agrees with the values $(dT_1/d\mathscr{E})$ [$(1.2-1.4) \cdot 10^{-3}$ deg · cm/V] known for BaTiO$_3$ from the literature [15].

From the point of view of the microscopic mechanism, the experimental results obtained for BaTiO$_3$ do not contradict the piezodeformation and electrostriction of the crystal under the effect of the electric field. For BaTiO$_3$ in the tetragonal phase, the change in the width of the forbidden band E_g under the effect of the electric field can be represented in the following form:

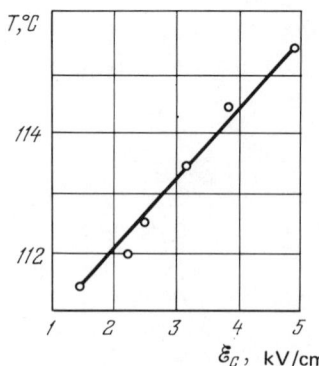

Figure 4.45. Dependence of the field intensity $\mathscr{E}_c$ on temperature.

$$\frac{dE_g}{d\mathscr{E}} = \sum_{i=1}^{3} \frac{\partial E_g}{\partial u_{ii}} \frac{du_{ii}}{d\mathscr{E}} + \left(\frac{\partial E_g}{\partial \mathscr{E}}\right)_u. \qquad (4.20)$$

Here, E_g is the width of the forbidden band, $\mathscr{E}$ is the electric field applied in the direction of the tetragonal axis (the axis of spontaneous polarization), and u_{ii} are the components of the deformation tensor. The first term on the right determines the change in width of the forbidden band under the effect of the field as a result of the inverse piezoeffect, while the second term characterizes the direct effect (for example, the Franz-Keldysh effect). It is seen from (4.20) that the dependence of the width of the forbidden band on the field intensity $\mathscr{E}$ is linear in the ferroelectric (tetragonal) phase. Since the sign of the components of the deformation tensor u_{ii} does not depend on the direction of the field $\mathscr{E}$, the sign of the change in E_g also does not depend on the field direction. The measurements were carried out in fields greater in magnitude than the coercive field. It should be noted that the mechanisms indicated above must also be realized in principle for linear piezoelectrics, for example, CdS. A change in the direction of the field to the opposite must then change the sign of $\Delta E_g^{\mathscr{E}}$. The magnitude of this effect in linear piezoelectrics must be small, and this effect for them can thus be overlapped by the Franz-Keldysh effect.

Barium titanate is not a piezoelectric in the paraelectric region. Only electrostriction deformation proportional to the square of the fields occurs in this case. A quadratic dependence of the change in width of the forbidden band with the field is observed accordingly.

We note in conclusion that for crystals of $NaNO_2$ and HSeN, an electric field with intensities up to 4 kV/cm did not cause a significant shift in the absorption edge, even near the Curie temperature [120].

4.6. Band Structure of Ferroelectrics

The nature of the energy spectrum of the electrons in the crystal determines not only the intrinsic optical absorption, but also the other properties of ferroelectrics, such as, for example, the electrical conductivity and photoconductivity. Investigation of the energy or band structure of the electrons in ferroelectric semiconductors is particularly important in both phases near the Curie point, since the semiconductor parameters of these crystals change sharply at the phase transition. Moreover, the changes resulting from the wave functions of the electrons at the phase transition are of fundamental interest for the theory of ferroelectricity.

The available works on the study of the band structure of ferroelectrics can be divided into two groups. The first consists of a group-theory analysis of the energy dispersion laws of the electrons in the crystal in the vicinity of singular points of the Brillouin zone. The second consists of the determination of the peculiarities of the band structure from experiments on optical reflection, absorption, and photoconductivity. The "weighted core presentation" method developed by Kovalev and Lyubarskii [164] is the basis of the first group. It was shown by these works in particular that for the groups D_{2h}^{16} and C_{2v}^{9}, which describe the symmetry of a number of ferroelectrics of the type $A^V B^{VI} C^{VII}$ in the paraelectric and ferroelectric phases, respectively (SbSI, SbSBr, BiSI, etc.), and also $A_2^V B_3^{VI}$ (Sb$_2$S$_3$, etc.), paired contact of the energy bands must occur at the boundary of the Brillouin zone [164]. For crystals with this symmetry, the general nature of the dispersion laws with and without consideration of the spin-orbital interaction was investigated by Gashimzade [165] and Karpus and Batarunas [166] (as applied to Sb$_2$S$_3$); Bercha and Tovstyuk [167, 168], Rashba [169], and Borets [170] showed the possibility of the existence in these crystals of "loops" of extrema and coupled minima on the isoenergy surfaces. Chepur and coauthors [171] studied in a group-theory investigation the anisotropy of these crystals, which is a result of their being chained or layered. This model leads to a twofold degeneracy of the energy levels at the center of the Brillouin zone.

We discussed above the experimental works on the study of intrinsic absorption of ferroelectrics. These works in combination show that the appearance of spontaneous polarization at the phase transition changes the width of the forbidden band of the crystal. On the basis of comparing the optical reflection spectra of $A_2^V B_3^{VI}$ compounds, the conclusion was reached in [172] about the predominance of the p nature of the states between which transitions occur, which are responsible for the peaks in the reflection spectrum of Sb$_2$S$_3$ in the range of energies to 9 eV. By analyzing the optical reflection spectra of SbSI at temperatures of 300, 273, and 90°K with D_{2h}^{16}, C_{2v}^{9}, and C_2^2 symmetries, respectively, Bercha and coauthors [135] isolated from the extremum points, with the help of the selection rules, those which could be responsible for the absorption edge with greatest probability. These are points of the type $\lambda(0, 0, k_z)$ for ferroelectrics of the $A^V B^{VI} C^{VII}$ type. It should be kept in mind only that, first, the absorption edge may correspond to a transition not between extremum points, and second, that extremum points themselves may not lie in the direction of high symmetry in the Brillouin zone.

More definite quantitative information about the electron energy spectrum can be obtained from direct quantum-mechanical calculations. As is well known, better results are given by the LCAO (linear combination of atomic orbitals) method for dielectrics and high-resistance semiconductors, which are ferroelectrics. Thus, a majority of calculations for ferroelectrics have been performed by this method [114, 173-179]. The empirical pseudopotential and combined plane wave method [180] have also been used along with the LCAO method.

A majority of the calculations have been carried out for crystals with the perovskite structure. Kahn and Leyendecker [181] calculated the band structure of SrTiO$_3$, which is similar to the structure of BaTiO$_3$, by using the Wolfsberg-Helmholtz approximation. This approximation is used in a majority of LCAO calculations. It is assumed that Sr ions do not participate in the chemical bond. The diagonal values of the Hamiltonian were constructed as a sum of the ionization potential of Ti or O, the Madelung potential, and the parameter of electrostatic split under the effect of an electrostatic field. The Madelung potentials were found as a combination of the known Madelung potentials of the structures of ReO$_3$, CsCl, and NaCl. For the ionic model of Sr^{2+}Ti^{4+}O$_3^{-2}$ generally used, the difference between the levels defining the width of the forbidden band—the 2p for oxygen and the 3d for titanium—was almost 15 eV instead of the required ~3 eV. By varying the charges of the O and Ti ions, a satisfactory value E_g = 3.25 can be achieved for a charge O (-1.67) and Ti ($+3$). It should be noted that in varying the charges of the ions, the authors did not take into account the shift of the energy eigenvalues in the isolated ions, but only the changes in the Madelung potentials. The values of the overlap integrals were taken from calculations for similar compounds. According to this calculation, the 2p orbitals of oxygen and the 3d orbitals of Ti predominate in the valence band. The width of the forbidden band, as has already been indicated in Section 4.2, is determined by the gap between the states at the center of the Brillouin zone $\Gamma_{15} - \Gamma_{25}$. The isoenergy surfaces corresponding to the lower conduction band are six ellipsoids lying along the [100] direction in wave vector space. The results of the calculation agree satisfactorily with the experimental data on the conductivity, Hall effect, and thermal emf. Michel-Calendini and Mesnard [176] calculated the band structure of BaTiO$_3$ in the paraelectric phase by using an analogous technique but rejecting a number of semiempirical estimates and calculating the overlap integrals numerically (Figure 4.46). The frequency dependence of the imaginary part of the dielectric constant $\epsilon_2 = \epsilon_2(\omega)$, which agrees well with experiment, was determined from the obtained band structure. An LCAO calculation of the band structure of SrTiO$_3$ was made in [179].

Gähwiller [111], on the basis of the calculation of the band structure of SrTiO$_3$ in the paraelectric phase [181], evaluated the change in E_g in BaTiO$_3$ at the ferroelectric phase transition. The ionic model Ba$^{1.65+}$Ti$^{3.3+}$O$_3^{1.65-}$ was used in [111]. As a result of the ionic shift, the levels determining the width of the forbidden band are also shifted and split because of the reduction in symmetry. Assuming the change in the quantum-mechanical integrals to be insignificant, i.e., taking into account only the change in the Madelung potentials, Gähwiller obtained a value of ΔE_g for BaTiO$_3$ exceeding the experimentally observed value by more than two times. Brews [114], also taking the calculation for SrTiO$_3$ in the paraelectric phase [181] as a basis, determined the band structure of BaTiO$_3$ in the ferroelectric phase. The changes in the matrix elements because of the ionic shift were assumed proportional to the changes in the corresponding overlap integrals. Along with the shift of the levels,

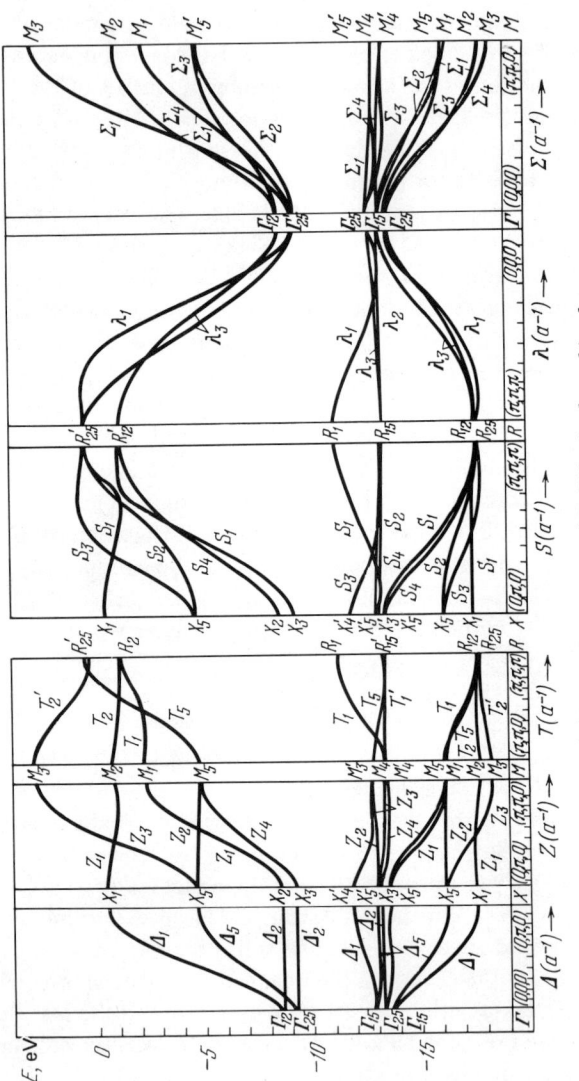

Figure 4.46. Band structure of BaTiO$_3$ in the cubic phase.

their splitting occurs, where the most important is the splitting of $\Gamma_{25'}$ into two levels. The obtained value of $\Delta E_g = 0.1$ eV is close to the experimental value for $KNbO_3$ (cf. Section 4.2). The band structure of $SrTiO_3$ in the paraelectric phase was calculated in [182], which agrees satisfactorily with experiments on optical reflection. Moreover, the Kern-Harbeke effect in perovskites is explained in this work on the basis of changes in the band structure with the ionic shift. Both the changes in the overlap integrals and the Madelung potentials were taken into account in [182]. The effect of the latter was predominant. It is shown that the Kern-Harbeke effect is determined by the magnitude of the polarization as a whole, and not by the value of the ionic shift or by the change in effective charges separately. This explains the closeness of the values of this effect in various perovskites.

An independent calculation of the band structure in the tetragonal phase for tetragonal $BaTiO_3$ by the LCAO method was made by Michel-Calendini and Mesnard [177]. In contrast to [114], the maximum of the valence band is shifted at the phase transition from the center of the Brillouin zone, which is explained by the insufficiently accurate calculation of the $pp\sigma$ and $pp\pi$ overlap integrals in [114]. The gap between the split levels $\Lambda_{2'}$ and Λ_5 is $30 \cdot 10^{-3}$ eV in contrast to $80 \cdot 10^{-3}$ eV in [114]. A positive dichroism $k_\perp/k_\parallel = 1.9$ ($k_\perp$ and $k_\parallel$ are the corresponding absorption coefficients for polarized light) was also revealed in [177]. This corresponds to splitting of the conduction band by $110 \cdot 10^{-3}$ in the a and c directions (the experimental value of this splitting is $120 \cdot 10^{-3}$ eV; cf. Figure 4.23). The conduction band is somewhat curved in the Δ [010] direction, which leads to a decrease in the transverse mass. The corresponding band structure is presented in Figure 4.47.

Significantly fewer calculations of band structure have been made for ferroelectric semiconductors of the $A^V B^{VI} C^{VII}$ and $A_2^V B_3^{VI}$ types. This is related, first, to the more complex chain or strip structure of these crystals (the presence of nontrivial translations in the symmetry space group, nonequivalent atoms of one type in the cell, for example, for crystals of the $A_2^V B_3^{VI}$ type, the presence of interstitial atoms) and, second, to the absence of a sufficient amount of experimental data. The overlap integrals and Madelung potentials for some of these crystals are presented in [183]. The band structure of Sb_2S_3 was considered only in the paraelectric phase [175] because of the absence of structural data in the ferroelectric phase. Since the coupling between strips is weak, the interaction between them was neglected, which led to a one-dimensional model ($k \parallel c$). The band structure of Sb_2S_3 obtained in [175] is presented in Figure 4.48. The value of the width of the forbidden band obtained in [175] was ~ 2.2 eV, which corresponds to a direct transition at $k_z = 0.8\ \pi/c$ (the transition is allowed for $E \parallel c$). Analysis of the wave functions obtained showed that p-symmetry electrons predominate both in the valence band and in the conduction band at distances up to ~ 5 eV from the edge of the bands [184].

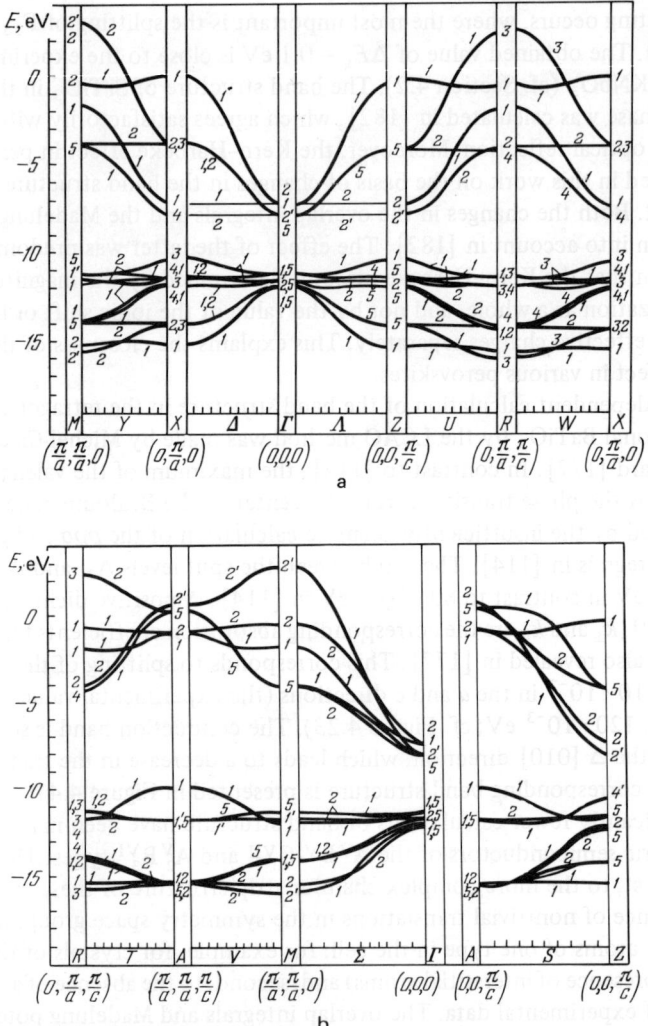

Figure 4.47. Band structure of $BaTiO_3$ in the tetragonal phase.

These are mainly the electrons of the sulfur in the valence band, and of antimony in the conduction band. The effective charges obtained from the calculation were +1.25 and +0.21, respectively, for Sb_1 and Sb_2, and -0.22, -0.64, and -0.6 for the three types of sulfur atoms S_1, S_2, and S_3, respectively (the designations for the nonequivalent atoms are the same as in [185]). As was shown in [184], the band structure obtained in [175] permits one to explain the origin of a majority of peaks in the reflection spectrum of Sb_2S_3 [172].

OPTICAL ABSORPTION AND FERROELECTRICS

The band structure of SbSI in the paraelectric phase was first calculated by Yamada and Chihara [173]. Only the p orbitals of sulfur and antimony were taken into account in the calculations; the overlap integrals and the interaction between chains were neglected. A qualitative band diagram without designation of an energy scale was presented in [173]. A more precise calculation within the framework of a one-dimensional model was made independently for the paraelectric and ferroelectric phases of SbSI by Khasabov and Nikiforov [174]. The width of the forbidden band E_g in both phases is determined by the direct transition between the points $\Gamma_{6,8}$, as in [173]. The charge of I was taken as −1, while it was varied from zero to −2 for S and Sb. The value of E_g in the paraelectric phase then varied insignificantly: from 2.36 to 2.08 eV,

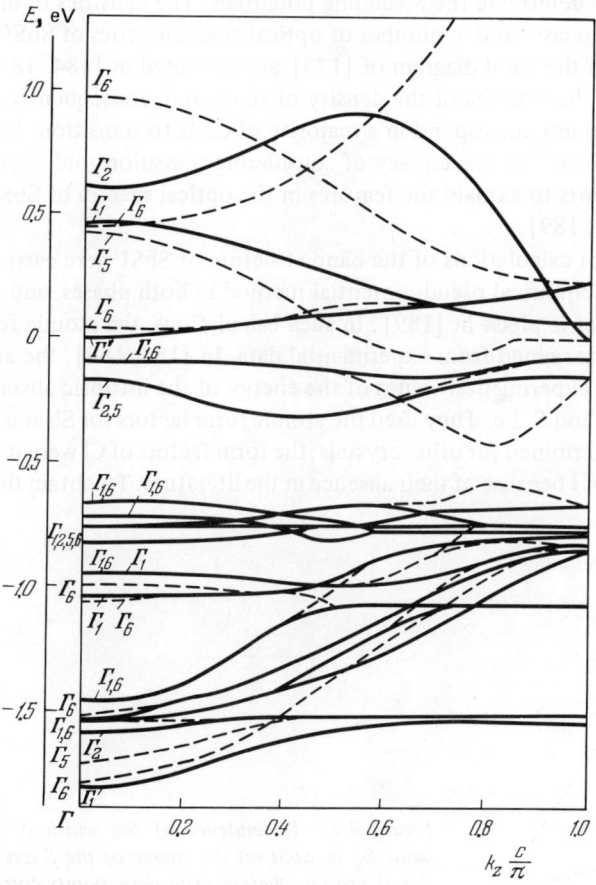

Figure 4.48. Band structure of Sb_2S_3. Solid and dashed lines correspond to the representations Δ_1 and Δ_2, respectively.

and significantly more strongly in the ferroelectric phase (from 3.57 to 2.07 eV) (cf. Figure 4.49). The best agreement with the experimental value of the discontinuity ΔE_g at the phase transition in SbSI is obtained for a charge of the sulfur of -1.8. The form of the bands at the phase transition varies insignificantly, the change amounting mainly to their shift. A larger (p_x-p_y) hybridization occurs in the ferroelectric phase than in the paraelectric phase. The polarity of the bonds was assumed fixed within the limits of the accuracy of the calculation. Since the values of the effective charges are determined from the calculation, one can obtain self-consistency in the charges. Such self-consistency was obtained for the paraelectric phase, and it was shown that coincidence of the calculated charges with the specified charges is achieved for the ionicities $Sb^{1.6+}S^{0.6-}I^{1-}$. This noncorrespondence can be explained by the limitation of the point ion model used to determine the Madelung potentials. The densities of electron states in both phases and a number of optical characteristics of SbSI obtained on the basis of the band diagram of [174] are presented in [184, 187]. It must be noted that the maxima of the density of states and, consequently, the peaks of $\epsilon_2 = \epsilon_2(\omega)$, do not correspond in a majority of cases to transitions between extremum points. The inadequacy of considering transitions only between extremum points to explain the features in the optical spectra of SbSI was also noted in [188, 189].

Subsequent calculations of the band structure of SbSI were carried out in [188] by the empirical pseudopotential method in both phases, and only in the paraelectric phase in [189]. In such calculations, the atomic form factors are usually determined from experimental data. In [188, 189], the authors used only the experimental values of the energy of the intrinsic absorption edge for $\mathscr{E} \parallel c$ and $\mathscr{E} \perp c$. They used the atomic form factors for Sb and S, previously determined for other crystals; the form factors of Cl were used for the form factors of I because of their absence in the literature. To obtain the values of

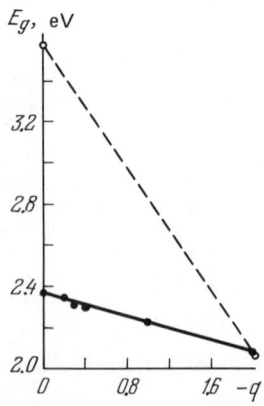

Figure 4.49. Dependence of the width of the forbidden band E_g in SbSI on the charge of the S ion (q in units of the electronic charge). The dark points correspond to the paraelectric phase; the dashed line corresponds to the ferroelectric phase.

OPTICAL ABSORPTION AND FERROELECTRICS 161

the energy of the direct transitions agreeing with the experimental position of the intrinsic absorption edge, Nakao et al. [188] were forced to select the form factors in such a way as to achieve coincidence in the paraelectric phase of the theoretical values of the energy gaps with the experimental values. Further calculations for both phases were carried out with the form factors thus chosen. According to [188], the direct transition in the paraelectric phase must occur for $\mathscr{E} \parallel c$ between $u_{5,6}$ and $u_{7,8}$ and for $\mathscr{E} \perp c$ between $u_{1,2}$ and $u_{7,8}$; and between $u_{3,4}$-$u_{3,4}$ and $u_{1,2}$-$u_{3,4}$, respectively, in the ferroelectric phase. [The notation of the irreducible representations taken in the original works is preserved here and subsequently; the directions of the coordinate axes coincide with those generally taken for SbSI; the point u has the coordinates $(\pi/a, \pi/b, 0)$.] The narrowest gap at the center of the Brillouin zone occurs between the points Γ_6 and Γ_8. The existence of an indirect transition $u_{5,6}$-Z_1 $(0, \pi/b, 0)$ with an

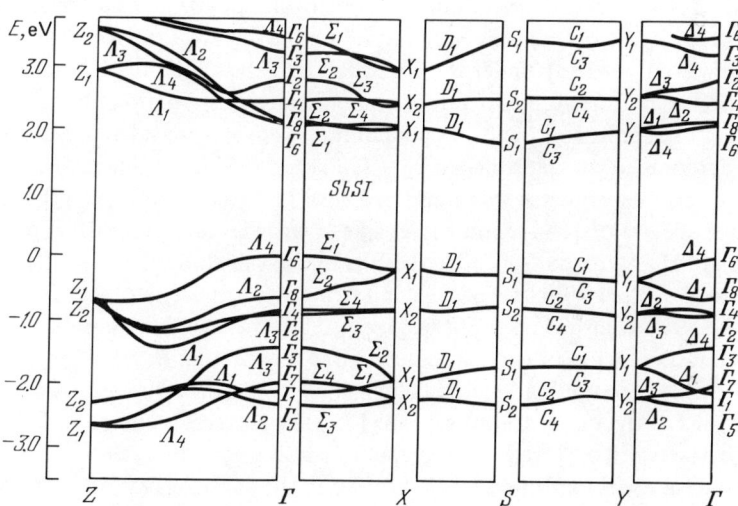

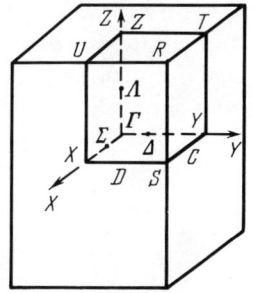

Figure 4.50. Band structure of SbSI in the paraelectric phase.

energy of 1.41 eV in the paraelectric phase, and $R_{3,4}$ (π/a, π/b, π/c)-Z_1 with an energy of 1.43 eV in the ferroelectric phase was an interesting result of the calculation in [188]. Although the absolute values of the energy of the direct transitions obtained in [188] do not coincide with those experimentally observed for SbSI (cf. Section 4.3), the fact itself of the presence of indirect transitions in the calculation of the band structure of SbSI by the pseudopotential method is of interest. Satisfactory values of the energy gaps in SbSI were successfully obtained in [189] without additional adjustment of the form factors. The band structure of SbSI in the paraelectric phase obtained in [189] is presented in Figure 4.50. The least energy of a direct transition, which occurs between the points Γ_6, is 2.08 eV. The width of the forbidden band is 1.82 eV, which corresponds to the indirect transition Γ_6-S_1 (π/a, π/b, 0), and agrees well with the experimental values for the paraelectric phase of SbSI (cf. Section 4.3). The adjustment of the form factors was evidently successfully eliminated in [189] because of the use of a very large number of plane waves (about 400) in the calculation, whereas there were only 50 for singular points and 40 for general points in [188]. It should also be noted that adjustment of the form factors was carried out in [188] under the condition that the experimentally observed absorption edge for two polarizations corresponded only to direct transitions at the points Γ or u, and the other possibilities were not considered. Nakao et al. [188] determined the dependence $\epsilon_2 = \epsilon_2(\omega)$ for SbSI from the band diagram obtained. This dependence was compared with that calculated in [187] from the reflection spectra of [135], and better agreement was obtained with experiment than in [174]. A comparison of the results of the calculation of the reflection spectrum $R = R(\omega)$ from data of the band structure with the natural experimental data for SbSI was also carried out in [189]. Agreement with experiment was obtained in the region of energies below 3 eV. The drop in $R(\omega)$ and $\epsilon_2(\omega)$ observed experimentally at energies above 4 eV contradicts the results of the calculations of the band structure [188, 189, 184]. It was assumed in [184] that the cause of the decrease in the experimental values of ϵ_2 and R with increasing energy is explained by absorption in the surface layer of the ferroelectric because of imperfections and the peculiar dielectric properties. Thus, although the band structure of SbSI and other ferroelectric semiconductors of complex composition has been insufficiently investigated up to now, the already available information about the nature of the interband transitions permits one to carry out in a number of cases a comparison with experiment and explain some of the general features of the electrical, photoelectric, and optical properties.

5

Photoferroelectric Phenomena and Photostimulated Phase Transitions

One of the peculiarities of ferroelectric semiconductors is the effect of electrons on the phase transition and the properties of the crystal near the Curie point. We considered in the second chapter the pseudo-Jahn–Teller effect as a possible micromechanism for this effect. The effect of nonequilibrium current carriers on the position of the Curie point and the ferroelectric properties near the phase transition has been called the photoferroelectric phenomena and the photostimulated phase transition in the literature [17-24]. As has already been noted in Chapter 1, photostimulated phase transitions are not a specific feature of ferroelectrics, but have been observed in a wide class of materials. One can assume in the general case that the mechanism of photostimulated phase transitions is related to the interaction of nonequilibrium electrons or electron excitations with one of the vibrational modes. In this respect, ferroelectric phase transitions are the most studied and experimentally accessible model. We will discuss this problem in more detail in Section 5.6 of this chapter.

Photoferroelectric phenomena are a large and varied class of phenomena. However, they can be divided into two basic groups. The first group includes phenomena related to the excitation of nonequilibrium carriers and optical recharged levels [46] in the neutral volume of the ferroelectric, in which there is no internal field. It is this group of phenomena which is described by the thermodynamic relations of Section 1.3 and the mechanism of the pseudo-Jahn–Teller effect. However, as was shown in Chapter 3, there always exists in a real ferroelectric an internal field, whose magnitude and distribution are determined by the conditions of the screening of the spontaneous polarization in the volume and at the surface of the crystal. Excitation of nonequilibrium carriers, on changing the screening conditions, simultaneously changes the magnitude and distribution of the internal field in the ferroelectric. Because of this, there can be a shift of the Curie point and a change in the whole set of properties of the ferroelectric, including the domain structure. This is the second group of phenomena caused by the effect of equilibrium or nonequilibrium carriers on screening of the spontaneous polarization, which are conventionally called screening phenomena. Thus, actually observed photoferroelectric phenomena can be related to two completely different mechanisms, whose experimental separation is a difficult problem in the majority of cases. The effect of "natural" illumination on the physical properties of a ferroelectric is most often caused by the appearance of both mechanisms. A typical example of this is the so-called "optical damage effect" of ferroelectrics, to which Section 5.5 is devoted.

In this chapter we will discuss primarily the first group of photoferroelectric phenomena, although many of them, as will be seen in the following, are a complex superposition of phenomena of both groups. The following two chapters are devoted to photoferroelectric phenomena of the second group—screening phenomena.

5.1. Thermodynamics of Photoferroelectric Phenomena

Near the first-order phase transition, the effect of the change in hysteresis ΔT_{hN} is also superimposed on the shift of the temperature of the phase transition ΔT_{1N} (the photohysteresis effect). Thus, for a first-order phase transition, $\Delta T_{1N} \neq \Delta T_N$, where ΔT_N is defined by expression (1.36). Moreover, the values of ΔT_{1N} and ΔT_{hN} can be close to the temperature intervals near the phase transition, where it is impossible to neglect fluctuations of polarization and, consequently, to use the expansion (1.14). All this requires additional refinement of the estimate of the magnitude of the effects described in Section 1.3.

In the case of first order-phase transitions, the coefficient α in the expansion of the thermodynamic potential does not go to zero at the transition point T_1. It is clear from the derivation of equation (1.36) that for first-order phase transitions it describes the photostimulated shift of the Curie–Weiss temperature T_0 defined from the condition $\alpha(T = T_0) = 0$ and appearing in the Curie–Weiss law (1.28). The Curie–Weiss temperature and its shift with illumination can, generally speaking, be determined experimentally from the temperature change of $1/\epsilon$ in the paraelectric region. However, such measurements require high accuracy in practice. Actually, it follows from the Curie–Weiss law (1.28) that the change in ϵ caused by the change in the Curie–Weiss temperature satisfies the relation

$$\frac{\Delta \epsilon}{\epsilon} = \frac{\Delta T_0}{T - T_0}.$$

If it is assumed that the difference $T_1 - T_0 \simeq 10°$ as, for example, in SbSI and BaTiO$_3$, while $\Delta T_0 \simeq 1°$, then it is clear that even in the immediate vicinity of the phase transitions, $\Delta \epsilon/\epsilon \simeq 0.1$ and decreases with increasing temperature.

For a first-order phase transition, the shift in the transition point with a change in concentration of nonequilibrium carriers can be obtained from (1.35):

$$\Delta T_1 = -\frac{C}{2\pi} aN + \left(\frac{2bN}{\beta} - \frac{cN}{\gamma}\right) \frac{1}{8\pi} \frac{\beta^2 C}{\gamma}. \tag{5.1}$$

Thus, if the shift ΔT_0 is determined only by the coefficient a, then the shift in temperature of the phase transition is also determined by the coefficients b and c. It was shown in Section 1.3 that for ferroelectric phase transitions, $a > 0$ and, consequently, $\Delta T_0 < 0$. At the same time, thermodynamics does not predict the sign of the coefficients b and c, and two situations are possible: $\Delta T_1 < 0$ and $\Delta T_1 > 0$. Thus, for first-order phase transitions, the photohysteresis effect should also be taken into account in evaluating the photostimulated shift of the phase-transition temperature.

Evaluation of the temperature hysteresis of the upper phase transition in BaTiO$_3$ from (1.35) with the use of experimental values of β, γ, and the Curie–Weiss constant C leads to a value eight times the observed value of SbSI. An analogous calculation for $\Delta T_h(14)$ with the use of the data of [190] gives a value $\Delta T_h = 6.5°$, whereas the experimentally observed value does not exceed $1°$. The inadequacy of the ordinary thermodynamic theory in considering the temperature hysteresis of the phase transition is explained by neglect of fluctuations. The problem of polarization fluctuations near the ferroelectric phase transition was first considered by Ginzburg [30]. According to [30], the thermodynamic theory of ferroelectrics based on the expansion (1.14) is valid only in some region near the point of the phase transition, excluding a temperature interval close to this point, that is,

$$\frac{\Delta T}{T_1} \gg \frac{k^2 T_1 \beta^2}{32\pi^2 \varkappa^3 \alpha'_T}. \tag{5.2}$$

Here ΔT is the magnitude of the interval where the expansion (1.14) becomes inapplicable; $\varkappa$ is the correlation parameter.

Estimates indicate that the expansion (1.14) can become incorrect at a distance of the order of a degree from the phase transition point.

Even more significant is consideration of the heterophase fluctuations near the first-order phase transition temperature. It is known that in BaTiO$_3$ nucleation centers of the new phase arise and disappear in a fluctuating manner with approach toward the upper phase transition point [14]. If the new phase is stable at the given temperature, then the nucleation centers of the new phase of sufficiently large size are stable and continue to grow in the course of time. The minimum size of the stable ("critical") nucleation centers is determined by the following condition. The negative change of the free energy at the phase transition caused by the formation of some volume of the new phase is compensated by the increase in free energy because of the appearance of the phase interface. Thus, there exists a minimum "critical" size which must be acquired by the nucleation center of the new phase in order that its growth become thermodynamically favored. The formation of such nucleation centers also causes the temperature hysteresis of the phase transition. It is clear that the temperature hysteresis ΔT_h^f with consideration of fluctuations is less than the value of ΔT_h

found from (1.35). If $\Delta T_h^f \ll \Delta T_h$, then the photostimulated charge ΔT_h, i.e., the decrease in the region of existence of metastable states, which is caused, for example, by illumination, may not even be observed in general and the conclusions of Section 1.3 are inapplicable in this case.

The temperature T_{max} corresponding to the maximum of the dielectric constant c obviously differs from the phase-transition temperature by the magnitude of the temperature hysteresis. Thus, the photostimulated shift of this temperature is

$$\Delta T_{max} = \Delta T_1 \pm \delta(\Delta T_h^f), \qquad (5.3)$$

where $\delta(\Delta T_h^f)$ is the photohysteresis effect. The different signs correspond to the different branches of the curve $\epsilon(T)$: the left (obtained with cooling of the sample) and the right (corresponding to heating). Figure 5.1 illustrates the two characteristic cases. In the first case, $|\Delta T_1| < |\delta(\Delta T_h^f)|$, and in spite of the decrease in the Curie–Weiss temperature and the phase-transition temperature, a shift of the left branch of $\epsilon = \epsilon(T)$ is observed toward higher temperatures (Figure 5.1a). On the other hand, the right branch of $\epsilon = \epsilon(T)$ can be shifted toward lower temperatures, even if $\Delta T_1 = 0$. In the second case when $|\Delta T_1| > |\delta(\Delta T_h^f)|$, both branches of $\epsilon = \epsilon(T)$ are shifted in one direction. One can then speak of the shift of the phase-transition temperature (Figure 5.1b). However, the effects of the shift in the Curie temperature and the change in the temperature hysteresis of the phase transition can be reliably separated only when they differ significantly in magnitude. In a number of cases, separation of these effects is related not only to technical, but also fundamental, difficulties, since the photohysteresis effect can be completely related to screening phenomena. We will discuss this in somewhat more detail.

The effect of nonequilibrium carriers on the temperature hysteresis of the phase transition ΔT_h^f can be caused by the following mechanism. The probabil-

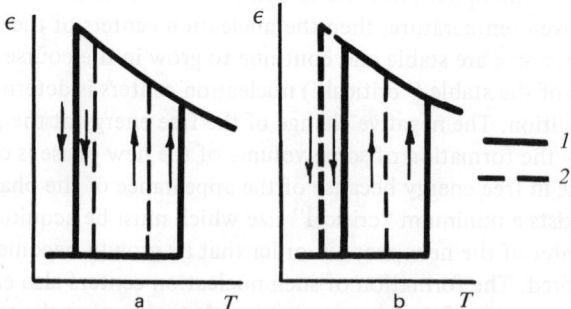

Figure 5.1. Diagram of the temperature variation of the dielectric constant in darkness (1) and with illumination (2). (a) $|\Delta T_1| < |\delta(\Delta T_h)|$; (b) $|\Delta T_1| > |\delta(\Delta T_h)|$.

ity of the formation of a spherical nucleation center of critical size for a fixed value of superheating (or supercooling) depends on the magnitude of the surface energy of the nucleation center W as $\exp(-W^3)$ [28]. In the case of the ferroelectric phase transition, when one of the phases is polar, W is the sum of the elastic energy $\sigma = W_1$ caused by distortion of the crystal lattice at the phase interface and the energy of the spontaneous polarization screening W_2:

$$W = W_1 + W_2. \tag{5.4}$$

The spontaneous polarization is then screened near the phase interface by free carriers (or those bound at trapping levels). The screening energy W_2 of the spontaneous polarization P by carriers with a density ρ is determined by integrating the free energy of the crystal F along the screening length l_D,

$$F = \frac{1}{2}\alpha P^2 + \frac{1}{4}\beta P^4 + \frac{1}{6}\gamma P^6, \quad W_2 = \int_0^{l_D} F\, dz,$$

with consideration of the Poisson equation

$$\frac{dP}{dz} = \rho$$

and the boundary conditions

$$P_{z=0} = 0, \quad P_{z=l_D} = P_0, \quad F(P_0) = 0, \quad \frac{dF}{dP}(P_0) = 0,$$

where α, β, and γ are the known coefficients in the expansion (1.14). As in Chapter 3, the correlation energy is neglected in F. Integration leads to the following expression for the screening energy W_2:

$$W_2 = 4 \cdot 10^{-2} \frac{\alpha}{\rho} P_0^3 \text{ ergs/cm}^2. \tag{5.5}$$

Here α is the inverse dielectric constant (1.26), $\rho = (N+n)q \simeq Nq$ is the density of screening charge, and n and N are, respectively, the concentration of free carriers localized at trapping levels ($N \gg n$ in ferroelectrics). It is clear from (5.5) that an increase in concentration of nonequilibrium carriers with illumination of the ferroelectric in the region of its natural photosensitivity decreases W_2 and, consequently, the temperature hysteresis. At high light intensities and corresponding strong recharging of levels, the screening energy W_2 can become less than the elastic energy W_1, and saturation must be observed in accordance with (5.4) in the dependence of the magnitude of the temperature hysteresis on the light intensity. This saturation, in principle, can also be caused by completely filled trapping levels. Conversely, illumination of the ferroelectric in the impurity spectral range, which empties the trapping levels thanks to transition of the captured carriers into the band, decreases W_2, and, accordingly, ΔT_h reaches the equilibrium value. It follows from the above discussion that the photohysteresis effect can be related partially or completely to the group of screening phenomena.

5.2. Photostimulated Shift of the Phase-Transition Temperature and the Photohysteresis Effect

The effect of electrons on the Curie–Weiss temperature and the temperature hysteresis of the phase transition, the topic of Section 1.3, is most easily observed experimentally with excitation of nonequilibrium carriers. The photostimulated shift of the phase transitions was first observed for crystals of SbSI [191, 192] and later in a series of works [193-195]. Figure 5.2 presents as an illustration the dependence of $\epsilon = \epsilon(T)$ for a BaTiO$_3$ crystal with cobalt impurity, which indicates that illumination of the crystal in the region of intrinsic absorption leads to a shift of the Curie point by $\sim 5°$ toward lower temperatures.[†] Unfortunately, Figure 5.2 does not present data on the temperature hysteresis and its change with illumination. Section 5.3 is devoted to the separation of photoferroelectric phenomena of various kinds in BaTiO$_3$. We will discuss here photoferroelectric phenomena in SbSI and in solid solutions based on $A^V B^{VI} C^{VII}$ and also in HgI$_2$.

In accordance with the conclusions of thermodynamic theory, illumination

---

[†]Unpublished work performed by L. Kai in the Laboratory of Dielectrics of Dijon University (France) and kindly given to the author by Professor Godfrua.

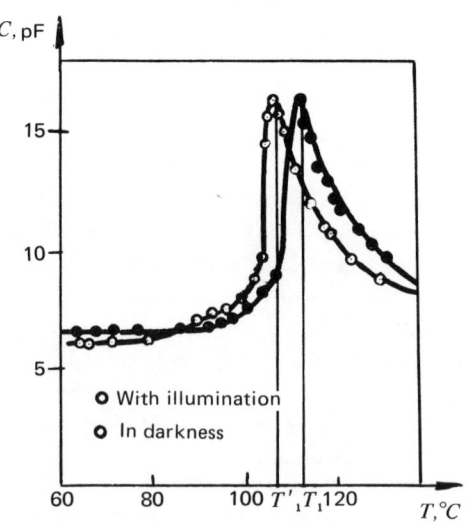

Figure 5.2. Effect of "natural" illumination on the temperature of the phase transition in BaTiO$_3$ (from the temperature dependence of the capacitance).

of SbSI crystals leads to a shift of the Curie temperature toward lower temperatures, where the maximum value of the shift is 1–1.5°. The magnitude of the shift of the Curie temperature with illumination ΔT_N depends on the sum $n + N$, where n and N are, respectively, the concentration of free carriers and carriers at trapping levels. Since for SbSI, as for other high-resistance photoconductors, the trapping coefficient $0 = n/N \ll 1$, the contribution of the free carriers to the photosensitive shift of the Curie temperature can be neglected and, consequently, ΔT_N depends only on N [cf. (1.36)], where N means the concentration of localized electrons. Measurements of ΔT_N performed in [191] for one of the SbSI crystals led to the value $N \simeq 3 \cdot 10^{17}$ cm^{-3}, which is plausible in principle. An attempt was made in [193] to establish a relation between the magnitude of the shift of the Curie temperature with illumination ΔT_N and filling of the trapping levels. Two groups of independent measurements were performed for this purpose. On the one hand, the dependence of the magnitude of the shift of the Curie temperature ΔT_N on the intensity of the illumination was measured. On the other hand, the activation energy of the trapping levels and their concentration were determined by the method of thermostimulated currents. We will show here that the results of these independent measurements satisfactorily agree with each other and are thus in good agreement with the conclusions of the theory of photosensitive phase transitions in ferroelectric semiconductors.

Single crystals of SbSI, grown from the gaseous phase and having the usual form of needles (the needle axis coincides with the direction of spontaneous polarization) were investigated in [193]. The average dimensions of the investigated crystals were 0.2 × 0.6 × 4 mm. Electrodes of silver paste were applied at the ends of the needles. The Curie temperature of the investigated crystals was near 23°C. The basic ferroelectric parameters of these crystals were in good agreement with the published data of [3, 126] and are presented in Table 4. The intrinsic absorption edge and the maximum of the photoconductivity of the SbSI crystals were in the region 620–660 nm. In accordance with the data of Table 4, the width of the forbidden band of SbSI experiences a discontinuity $\Delta E_g \simeq 10^{-2}$ eV at the phase transition.

In contrast to [191, 192], determination of the shift of the Curie temperature with illumination ΔT_N in [193] was carried out by measuring the pyroelectric current. It is well known that for SbSI, as for other ferroelectrics, a sharp peak of the pyroelectric current is observed at the temperature of the phase transition. Determination of ΔT_N was carried out in [193] by measuring the temperature of the maximum of the pyroelectric current with illumination of the crystal. The crystal under investigation was placed in a vacuum cryostat, which provided a vacuum of $\sim 10^{-5}$ mm Hg, with a window for illuminating the sample. The usual technique was used to measure the pyroelectric current. At a temperature above the Curie temperature, a constant electric field greater than the coercive field was applied to the sample, after which the sample was cooled in the external field to a temperature below the Curie point. After removal of the external field, the sample was heated and the peak of the pyroelectric current was recorded. These measurements were carried out successively in darkness and with illumination of the crystal, where the rate of heating of the crystal and all other experimental conditions were maintained strictly identical. Figure 5.3a presents two curves of the pyroelectric current taken, respectively, in darkness (curve 1) and with illumination of the crystal in the region of the maximum photosensitivity (curve 2), which illustrate the shift of the maximum of the pyrocurrent with illumination toward lower temperatures. Curve 2 was taken with continuous illumination of the crystal. The inertia of the shift of the Curie temperature with the light on and off was not investigated in this work. It should be noted only that preliminary illumination of the crystal does not affect the results of the subsequent measurement in darkness.

Measurement of the thermostimulated current in SbSI was carried out in [193] by the ordinary technique [196]. The crystal was cooled to −160°C in the same cryostat, after which illumination of the crystal with white light was carried out at this temperature. A voltage was applied to the crystal 10 min after cutoff of the illumination, and it was heated at a constant rate; the peak of the thermostimulated current was observed as a result. The temperature dependence of the dark and pyroelectric currents was kept in mind in measuring the thermostimulated current. The heating rate of the crystal in measuring the thermostimulated current was taken so that the temperature corresponding to the maximum of the thermostimulated current was significantly lower than the Curie temperature, i.e., so that the peaks of the thermostimulated and pyroelectric currents were well separated. Figure 5.3b presents the curve of the thermostimulated current, constructed with consideration of the contribution of the pyroelectric current, which displays one maximum at −20°C. The activation energy of the levels was determined from the maximum of the thermostimulated current with the help of the Bube formula, and their concentration was estimated by integrating the curves of the thermostimulated current and determining the photoelectric yield [196]. The results of these measurements did not depend on the intensity and time of illumination of the crystals, which indicates the complete filling of the levels with measurement of the thermostimulated current. Similar

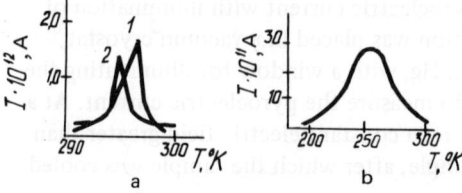

Figure 5.3. Temperature dependence of the pyroelectric and thermostimulated currents in SbSI. (a) Maxima of the pyroelectric current in darkness (1) and with illumination (2), $\Delta T_1 = 1°$; (b) curve of the thermostimulated current.

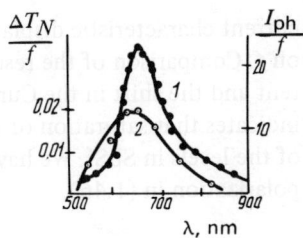

Figure 5.4. Spectral distribution of the photoconductivity (1) and the effect of the photosensitive shift of the Curie temperature (2) for SbSI (in relative units).

data for measurements of the thermostimulated conductivity were obtained in a number of independent works [197] for SbSI.

The spectral distribution of the effect of the photostimulated shift of the Curie temperature was investigated in [193]. Figure 5.4 presents the dependence of the shift of the Curie temperature ΔT_N per unit energy incident on the crystals on the wavelength of the incident light. The spectral distribution of the photoconductivity is presented for comparison in the same figure. In accordance with the theory, the maxima of both curves were located near 650 nm, i.e., the shift of the Curie temperature with illumination of SbSI occurs in the region of the intrinsic photosensitivity of the crystal.

Figure 5.5a presents the dependence of the shift of the Curie temperature ΔT_N (in degrees) on the intensity f of the monochromatic light incident on the crystal ($\lambda \simeq 630$ nm). The intensity-current characteristics of the photocurrent taken in the same interval of light intensity is presented in Figure 5.5b. It is seen by comparing the curves in Figure 5.5 that the crystal with a linear intensity-

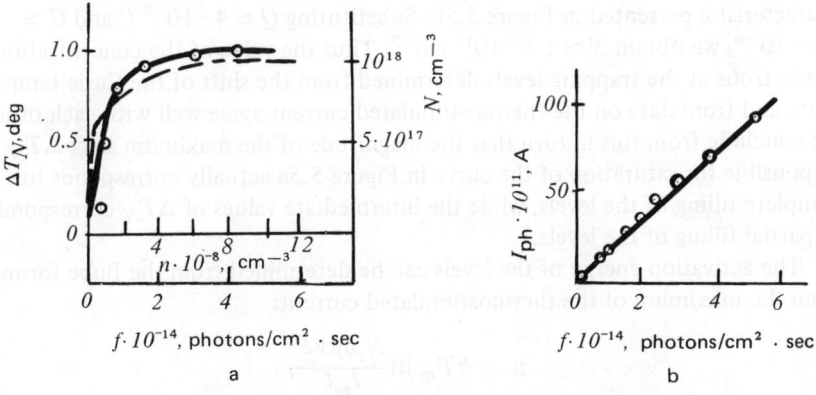

Figure 5.5. Dependence of the shift of the Curie temperature ΔT_N and photocurrent I_{ph} on the intensity of incident light on the crystal ($\lambda = 630$ nm). (a) Dependence of ΔT_N on f; solid curve is for experimental values of ΔN; dashed curve is for theoretical values calculated from (5.6); values of N calculated from equation (5.7) are plotted along the right vertical axis; (b) dependence of I_{ph} on f; crystal temperature 16°C.

current characteristic displays saturation of the curve of the dependence of ΔT_N on f. Comparison of the results of measurements of the thermostimulated current and the shift in the Curie temperature with illumination presented below indicates that saturation of the curve in Figure 5.5a is caused by complete filling of the levels in SbSI. We have from (1.36), keeping the quadratic term in the polarization in (1.46),

$$\Delta T_N = \frac{\Delta E_g C}{\pi P_0^2} N, \qquad (5.6)$$

where P_0 is the spontaneous polarization, C is the Curie-Weiss constant, ΔE_g is the discontinuity in the width of the forbidden band at the first-order phase transition, and N is the concentration of carriers at the trapping levels. Substituting into (5.6) the experimentally measured value of the shift $\Delta T_N \simeq 1°$ corresponding to the saturation of the curve in Figure 5.5a and also the independently measured values $P_0 \simeq 10\ \mu C/cm^2$, $C \simeq 1.1 \cdot 10^5$ deg, and $\Delta E_g \simeq 1.0 \cdot 10^{-2}$ eV, we obtain $N \simeq 1.2 \times 10^{18}\ cm^{-3}$, which agrees in order of magnitude with the results of measurements performed in [191].

On the other hand, the concentration N can be evaluated from data on the thermostimulated current. According to [198],

$$N = Q/qVG, \qquad G = I_{ph}/fVq.$$

Here Q is the total charge determined by integrating the curve of the thermostimulated current in Figure 5.3b, V is the volume of the crystal, q is the elementary charge, and G is the photoelectric yield depending on the ratio of the photocurrent to the light intensity I_{ph}/f, i.e., on the slope of the intensity-current characteristic presented in Figure 5.5b. Substituting $Q \simeq 4 \cdot 10^{-8}$ C and $G \simeq 4.6 \cdot 10^{-4}$, we obtain $N \simeq 1.1 \cdot 10^{18}\ cm^{-3}$. Thus the value of the concentration of electrons at the trapping levels determined from the shift of the Curie temperature and from data on the thermostimulated current agree well with each other. We conclude from this in turn that the magnitude of the maximum shift ΔT_N responsible for saturation of the curve in Figure 5.5a actually corresponds to complete filling of the levels, while the intermediate values of ΔT_N correspond to partial filling of the levels.

The activation energy of the levels can be determined from the Bube formula from the maximum of the thermostimulated current:

$$u = kT_m \ln \frac{N_c q \mu VS}{I_m l},$$

where T_m and I_m are, respectively, the temperature and thermostimulated current corresponding to the maximum of the curve in Figure 5.3b; S and l are, respectively, the cross-sectional area and the length of the crystal in the direction of the applied field; V is the voltage applied to the crystal; $\mu \simeq 50\ cm^2/V \cdot sec$ is the electron mobility in SbSI [199]; and $N_c \simeq 10^{19}$ cm^{-3} is the density of states. Substituting the values of the parameters corresponding to Figure 5.3b into the formula we find $u \simeq 0.6$ eV.

The dependence of the concentration of electrons at the trapping levels N on the light intensity f, as is usual, can be obtained from the steady-state solution of the kinetic equation (α-trapping [46]):

$$N = \frac{n}{N_{eM} + n} M, \qquad (5.7)$$

where M is the concentration of levels, n is the concentration of free electrons, and $N_{cM} = N_c \exp(-u/kT)$. The dependence of N on the light intensity appears in (5.7) in terms of n. For $n \gg N_{cM}$, saturation $N = M$ occurs, i.e., complete filling of the levels. Substituting the experimental values of n and N_{cM} into (5.7), one can obtain the dependence of N on f and compare it with the curve in Figure 5.5a. Since the values of N for the curve in Figure 5.5a are obtained from measurements of the shift ΔT_N and calculated from equation (5.6), this comparison serves as an additional verification of the theory. Substituting the values $N_{cM} \simeq 6 \cdot 10^7$ cm^{-3} and $M \simeq 10^{18}$ cm^{-3} determined from data on the thermostimulated current into (5.7) and also the values of the concentration of free carriers n determined from the photocurrent for $\mu \simeq 50$ cm^2/V · sec (Figure 5.5b), we find the dependence of N on f. This dependence is represented by the dashed curve in Figure 5.5a, and, as is seen from the figure, both curves are close to each other.

Thus the results obtained in [193] indicate good correspondence between the data of measurements of the photosensitive shift of the Curie temperature, photoconductivity, and thermostimulated conductivity of SbSI, which in turn serves as an experimental confirmation of the validity of equation (1.36) and the other conclusions of the theory. We remark that this correspondence could have been worse in principle. This is related not only to the necessity of strict determination of the lifetime of carriers in evaluating the concentration of levels from the data on the thermostimulated current [200]. Some divergence can be caused by the indeterminacy in evaluating the parameter P_0 and the discontinuity of the width of the forbidden band ΔE_g. The indeterminacy in evaluating the values of the spontaneous polarization P_0 becomes significant in the presence of poorly defined phase transitions. In the derivation of equation (1.36), the activation energies of the trapping levels were neglected as compared to the width of the forbidden band of the crystal. Consideration of the energies of the trapping levels both for electrons and for holes leads to a correction in evaluating ΔE_g, whose sign and magnitude depend on the shift of the trapping levels at the phase transition. An estimate using the results of [201] shows that this correction can be significant.

The photostimulated shift of the Curie point was evaluated in [193] without

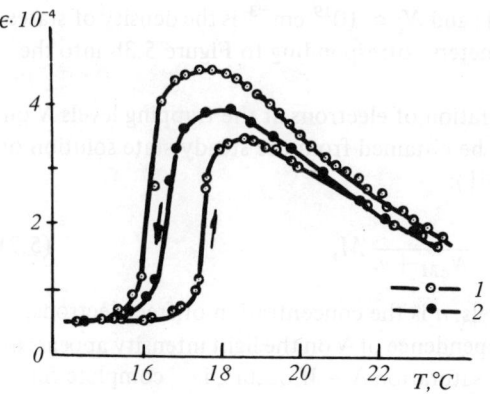

Figure 5.6. Temperature dependence of the dielectric constant of SbSI in darkness (1) and with illumination (2). Crystal No. 1.

consideration of the photohysteresis effect. A special investigation of the photohysteresis effect was performed in [202].

The temperature hysteresis of the phase transition in SbSI was recorded in [202] by measuring the temperature dependence of the dielectric constant ϵ. Figures 5.6–5.8 present typical situations observed with illumination of the crystal in the region of the natural photosensitivity. The temperature hysteresis of the phase transition with illumination usually decreased if the dark resistance of the crystal was sufficiently large ($\sigma \lesssim 10^{-9} \; \Omega^{-1} \cdot cm^{-1}$) and the crystal is without optical defects. In "low-resistance" crystals, the condition $W_2 \ll W_1$ is obviously satisfied in darkness and there is thus no photohysteresis effect. It is also possible that for high concentration of defects in the crystal, the process of nucleation formation is determined mainly by defects, and not by the screening

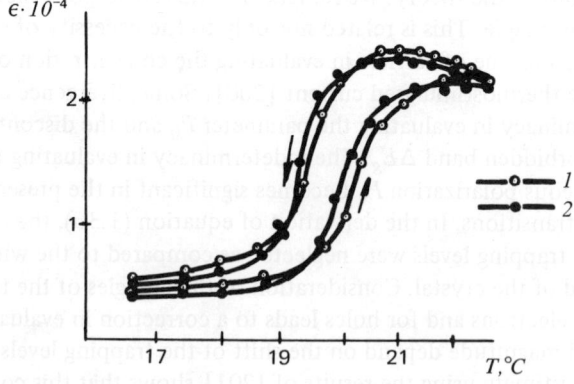

Figure 5.7. Temperature dependence of the dielectric constant of SbSI in darkness (1) and with illumination (2). Crystal No. 2.

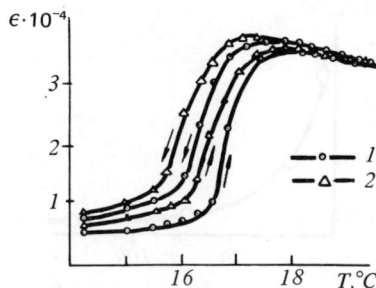

Figure 5.8. Temperature dependence of the dielectric constant of SbSI in darkness (1) and with illumination (2). Crystal No. 3.

conditions. The shift of the Curie point with illumination is definitely observed only in the case presented in Figure 5.8. This phenomenon is notable here because the photohysteresis effect is not observed for this crystal and the phenomena under consideration are not superimposed on each other. One cannot say anything so definite about the shift of the Curie point in the other two cases. However, for the sample whose characteristic is shown in Figure 5.6, one can indicate the shift of the Curie-Weiss temperature ($\Delta T_N \simeq 1°$) by considering the dependence $\epsilon = \epsilon(T)$ in the paraelectric phase.

Figure 5.9 illustrates the coincidence of the maximum of the spectral distribution of the photoconductivity and minimum of the magnitude of the temperature hysteresis. The measurements performed at equal energies confirm that the photohysteresis effect is related to the increase in photoconductivity of the crystal. Figure 5.10 presents the dependence of the temperature hysteresis on the photoconductivity, which was obtained with change in intensity of the illumination in the region of intrinsic absorption edge. This dependence displays saturation.

To compare the results of the experiment with theory, the spontaneous polarization P_0 was determined, and the concentration of electrons at the trapping levels N was determined from the thermostimulated conductivity. For the crystal for which the results are presented in Figures 5.6, 5.9, and 5.10, values of $P_0 \simeq 15\ \mu C/cm^2$, $M \simeq 10^{18}\ cm^{-3}$, and $u \simeq 0.55$ eV were obtained,

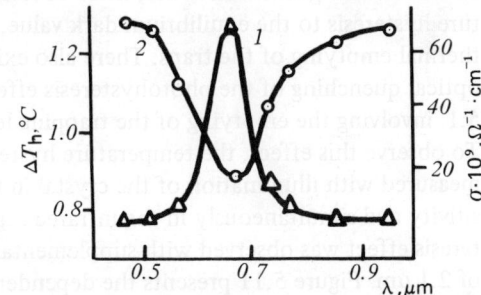

Figure 5.9. Spectral distribution of the photoconductivity (1) and photohysteresis effect (2) for SbSI.

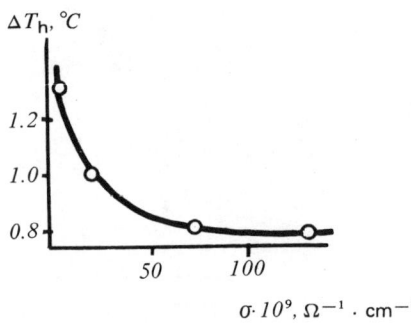

Figure 5.10. Dependence of the value of the temperature hysteresis on the photoconductivity for SbSI.

where M is the concentration of trapping levels and u is their activation energy. The concentration of electrons at trapping levels N was determined from equation (5.7), where the concentration of free electrons n determined from the steady-state conductivity appears. On the portion of saturation of the photohysteresis effect, $\sigma \simeq 10^{-7} \, \Omega^{-1} \cdot \text{cm}^{-1}$ (cf. Figure 5.10). As above, by taking the electron mobility in SbSI as $\mu \simeq 50 \, \text{cm}^2/\text{V} \cdot \text{sec}$, we have $n \simeq 10^{10} \, \text{cm}^{-3}$ and $N \simeq 10^{18} \, \text{cm}^{-3}$ from (5.7). Thus, on the section of saturation of the photohysteresis effect, $N \simeq M$ and, consequently, one of the possible causes of saturation can be complete filling of the traps. Evaluation of the screening energy from (5.5) under the same conditions on the saturation section gives $W_2 \simeq 10 \, \text{ergs/cm}^2$. As we see, this agrees well with other conditions of saturation of the photohysteresis effect $W_2 \ll W_1$, since the elastic energy of the nucleation boundary is $W_1 \simeq 100 \, \text{ergs/cm}^2$ for SbSI according to [203] (the value of W_1 was evaluated in [203] from other independent experiments, cf. Chapter 6). We note that the value of W_1 found in [203] coincides in order of magnitude with the value of W_1 obtained for perovskite ferroelectric semiconductors [204]. Thus the data obtained for SbSI in [202] do not permit separating the two possible mechanisms for saturation of the photohysteresis effect. Because of the absence in the literature of reliable values of the coefficient $b > 0$ [cf. (1.46)], comparison of the experimental values of ΔT_{hN} from (1.39) for SbSI is also impossible.

After the light is turned off, there occurs a gradual increase of the temperature hysteresis to the equilibrium dark value, which is naturally related to the thermal emptying of the traps. There also exists in SbSI the phenomenon of optical quenching of the photohysteresis effect, which was mentioned in Section 5.1, involving the emptying of the trapping levels by long-wavelength radiation. To observe this effect, the temperature hysteresis of the phase transition was measured with illumination of the crystal in the region of the natural photosensitivity and simultaneously in the infrared region. Quenching of the photohysteresis effect was observed with supplementary illumination with a wavelength of 2.1 μm. Figure 5.11 presents the dependence of the magnitude of ΔT_h on the wavelength of the infrared illumination. It is seen that the quenching actually occurs where the maximum of the quenching corresponds to the wavelength $\lambda \simeq$ 2.1 μm, where the magnitude of the temperature hysteresis almost reaches the "dark" equilibrium value. The activation energy of the trapping levels, determined from the spectral distribution of the quenching effect, is 0.6 eV, which agrees satisfactorily with data from the thermostimulated conductivity. Experiments on optical quenching of the photohysteresis effect indicate directly that the predominating role in photoferroelectric phenomena is played by carriers localized at the trapping levels.

The photostimulated shift of the Curie temperature and the photohysteresis effect are close in order of magnitude in SbSI crystals and cannot be strictly separated. The necessity of experimentally separating these phenomena led Grekov et al. [205] to study them in solid solutions of $SbSI_x Br_{1-x}$, which display low-temperature phase transitions. Illumination of these crystals at the maximum of the photoconductivity at low temperatures leads to a metastable recharging of the levels, which is preserved in darkness for a long time depending on the temperature and activation energy of the levels. According to the conclusion of Section 1.3, this must lead to a metastable shift of the Curie temperature, stable during the entire time of optical recharging of the levels ("frozen" shift). It is significant that the photohysteresis effect must not be observed under conditions of the "frozen" shift of the Curie temperature (if it is caused by a change in screening energy in accordance with the mechanism proposed above), since there is no thermal exchange between the deep trapping levels and the band for $u/kT \gg 1$.

Crystals of composition $SbSI_{0.35} Br_{0.65}$ displaying a low-temperature first-order phase transition at $T_1 \simeq 160°K$ were investigated in [205]. The experiment confirmed the assumptions. The phase transition in the crystal was recorded from the temperature variation of the dielectric constant, which was measured in darkness with heating and cooling. After this, the crystal was heated

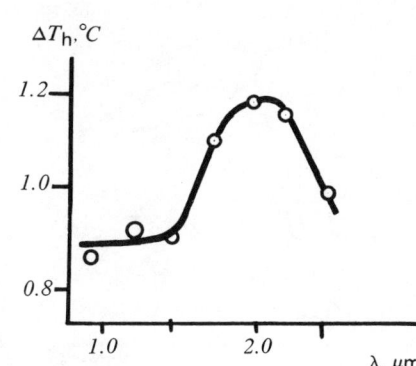

Figure 5.11. Spectral distribution of the quenching of the photohysteresis effect in SbSI.

to room temperature and again cooled with continuous "natural" illumination to a temperature exceeding the phase transition temperature by 20–25°. Since the illumination was produced in the paraelectric region, the photoelectric effect did not appear (cf. Section 5.3 and Chapter 7), and the distorted fields caused by the screening of the spontaneous polarization by nonequilibrium carriers did not arise. After cutoff of the illumination, whose intensity was varied, the sample was cooled in darkness below the temperature of the phase transition, and the temperature dependence of the dielectric constant was again measured with heating and cooling (Figure 5.12); the thermostimulated currents (Figure 5.13) for the excited crystal were then measured. A shift of both branches of the dielectric constant toward lower temperatures was observed after illumination, but the temperature hysteresis of the phase transition did not change. In accordance with the above discussion, this behavior of the $\epsilon = \epsilon(T)$ curves can be interpreted as a shift of the Curie temperature. The magnitude of the shift correlates with the degree of filling of the trapping levels, which is proportional to the area under the curves of the thermostimulated current in Figure 5.13. The maximum value of the shift was of the order of $\sim 0.5°$. The activation energy of the trapping levels, determined from the initial slope of the thermostimulated current curves [206], is ~ 0.6 eV. The measurements in [205] were performed for several crystals. The photohysteresis effect was observed in none of them. At the same time, the "frozen" shift of the Curie temperature could be observed for repeated transitions from one phase to another only if the temperature was not increased significantly during the course of the experiment. Heating to room temperature re-established the equilibrium dark value of the Curie temperature. An attempt to de-excite the trapping levels with infrared light and to correspondingly quench the "frozen" shift effect did not give a positive result. The absence of a de-excitation effect can be related to the fact that far from the phase transition, the carrier capture cross section at trapping levels is relatively large, while the capture cross section of photons is small. The latter

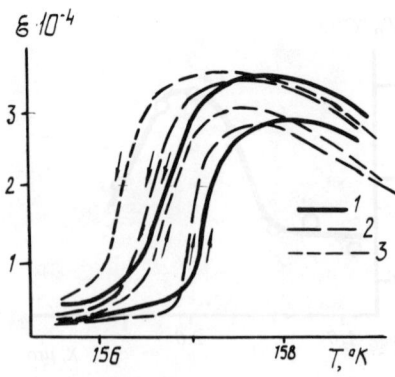

Figure 5.12. Temperature dependence of the dielectric constant for nonilluminated (1) and illuminated (2, 3) crystal of $SbSI_{0.35}Br_{0.65}$. The intensity of illumination is determined by curves 1 and 2, respectively, in Figure 5.13.

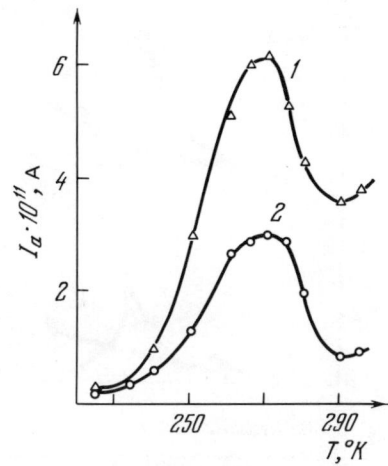

Figure 5.13. Curves of the thermostimulated current measured after various preliminary excitation of the $SbSI_{0.35}Br_{0.65}$ crystal.

assumption agrees with the kinetics of induced impurity photoconductivity in SbSI, described in Chapter 8.

In concluding this section, we discuss the photostimulated phase transition in crystals of HgI_2. It is well known that mercuric iodide experiences a first-order phase transition from the tetragonal phase D_{4h}^{15} to the rhombic phase C_{2v}^{12} at 400°K, where the crystal is evidently a singular ferroelectric in the high-temperature phase [207]. The phase transition in HgI_2 is accompanied by a strong anomaly of the photoconductivity and a large shift of the intrinsic absorption edge (the high-temperature phase is wider-band and less photosensitive) [208]. The effect of nonequilibrium carriers in HgI_2 on the temperature of the phase transition and the temperature hysteresis was investigated in [209]. The investigated crystals were grown from the gaseous phase and had a resistivity of 10^{10}-10^{11} Ω · cm. According to the data in the literature [208], the intrinsic absorption edge and the maximum of the photoconductivity were observed at 580 nm at 87°K.

Two methods for recording the phase transition were used: differential thermal analysis and the ordinary technique of measuring the pyroelectric current. As a result, the effect of illumination on the phase transition was observed and involved the following: (1) the temperature of the phase transition both with heating and with cooling shifted upward by 2-5°; (2) the hysteresis of the phase transition decreased by about 5° (photohysteresis effect). Both effects have a spectral distribution corresponding to the spectral distribution of the photocurrent; they increase with increasing light intensity and naturally depend on the photosensitivity of the crystal. Figure 5.14 presents the results of the pyroelectric measurements, which illustrate both the shift of the transition temperature with illumination as well as the photohysteresis effect. Since the high-

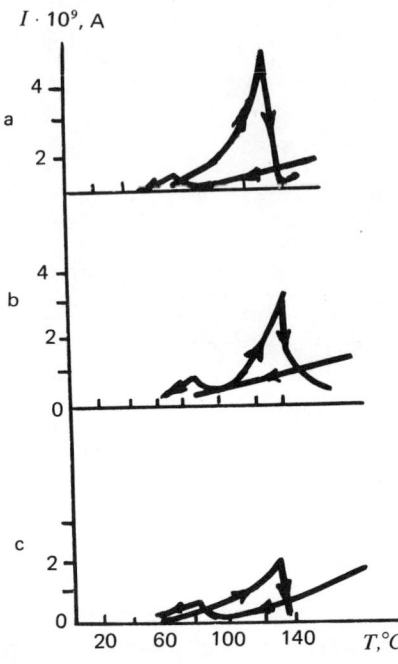

Figure 5.14. Pseudo-pyroelectric current at the phase transition in HgI_2. (a) In darkness; (b) with illumination; (c) recovery in darkness after illumination.

temperature pyroelectric phase is wider-band, the direction of the shift of the transition temperature with illumination is opposite (as compared to SbSI) and coincides with the value predicted theoretically in Section 1.3. However, as in the case of SbSI, the photohysteresis effect and the effect of the shift in the transition point with illumination are close in magnitude, which makes their strict separation difficult. Of course, the effect of nonequilibrium carriers on the phase transition is not an indication of its pyroelectric nature. Moreover, a quantitative comparison of the photoferroelectric effects in HgI_2 with the thermodynamic theory of Section 1.3 is impossible because of the absence in the literature of data on the parameters describing the phase transition in HgI_2. The data of [209] do not give a unique answer to the problem of the nature of the phase transition in HgI_2: dielectric hysteresis loops were not observed near the phase transition, and the anomaly of the dielectric constant at the transition point was insignificant in the pyroelectric phase. Pyroelectric measurements, as is seen in Figure 5.14, showed a current peak of $\sim 10^{-9}$ A at the phase-transition temperature. However, the measured current is evidently not pyroelectric, since the field did not affect the magnitude and direction of the current and, moreover, the direction of the current coincided with heating and cooling. The "pyrocurrent" is somewhat higher in light than in darkness. It was established with visual observation of the phase transition that the "pyrocurrent" peak corresponds to the generation inside one phase of a region of the other phase. Thus, one can assume that the measured current is thermoelectric in nature and is caused by the generation of the boundary of the two coexisting phases. The pseudo-pyrocurrent can obscure the true pyrocurrent, whose magnitude is evidently significantly less.

5.3. Photoferroelectric Phenomena in $BaTiO_3$

The discovery of a significant photosensitivity in barium titanate permitted the observation of the effect of nonequilibrium carriers on the high-temperature transition from the tetragonal phase to the cubic phase [18, 210, 211]. Dielectric, pyroelectric, and optical observations of the effect of illumination on the Curie temperature in $BaTiO_3$ were carried out in [211]. The temperature of the phase transition was fixed with the dielectric measurements from the maximum of the dielectric constant ϵ measured at a frequency of 1 kHz in the [001] direction. Illumination of the crystal with "natural" light was carried out through semitransparent gold electrodes deposited on the (001) surface.

The results of the dielectric measurements for a pure single crystal of $BaTiO_3$, grown by the Remeyka method, are presented in Figure 5.15. The dark temperature of the phase transition with heating and cooling of the crystal can be determined from curves 1 and 2, respectively, in this figure. Curves 3 and 4 were taken with illumination of the crystal during the phase transition. Thus, in accordance with the theory, the Curie temperature of $BaTiO_3$ is shifted downward under the effect of nonequilibrium carriers. The dependence of the magnitude of the shift on the light intensity displays saturation. The magnitude of the shift determined from the branch of $\epsilon = \epsilon(T)$ corresponding to heating is $\Delta T_N \simeq 7°$, and that corresponding to cooling is $\Delta T_N \simeq -3.7°$. Moreover, the illumination decreases the temperature hysteresis of the phase transition, where $\Delta T_{hN} \simeq -3.3°C$.

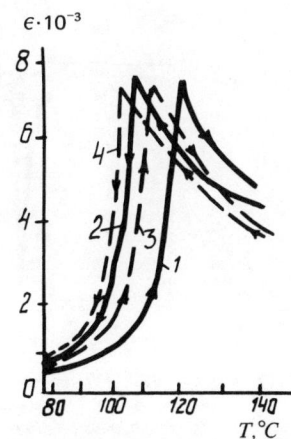

Figure 5.15. Effect of illumination on the phase-transition temperature in $BaTiO_3$. (1, 2) Heating and cooling of the crystal in darkness; (3, 4) heating and cooling of the crystal with illumination.

Analogous results are obtained with the help of pyroelectric measurements. The temperature of the phase transition is fixed in this case from the maximum of the pyroelectric current measured in the [001] direction. Figure 5.16 presents the results of pyroelectric measurements for crystals of BaTiO$_3$ alloyed with iron. In the pyrocurrent measurements, in accordance with the usual technique, the crystal was first polarized by a field above the coercive field, cooled in darkness in the field to room temperature, after which it was brought to the para region in light. The different signs of the pyroelectric change in Figure 5.16a and b correspond to the different directions of the spontaneous polarization (the pyrocurrent with cooling is not presented in the figure). Curves 1 correspond to the dark mode, and curves 2 were taken in the "natural" illumination mode. Thus, the shift $\Delta T_N \simeq -4.5°$ obtained from the pyroelectric measurements agree in sign and magnitude with the result of the dielectric measurements. The effect of illumination on the magnitude of the pyroelectric charge noted in Figure 5.16 will be considered in the following chapter. We will also discuss there the optical observations of the photostimulated shift of the phase-transition temperature in BaTiO$_3$.

We also note that, in contrast to SbSI, the photostimulated shift of the phase transition in BaTiO$_3$ has a significant relaxation time. The equilibrium value of the phase-transition temperature is re-established after a time of the order of several hours. Such a slow relaxation indicates the localization of the nonequilibrium carriers at rather deep levels and the possible role of multiple trapping.

These same measurements display satisfactory agreement with (1.42) by showing that the decrease in temperature hysteresis of the phase transition is accompanied by the photodielectric effect in BaTiO$_3$. For example, it is seen

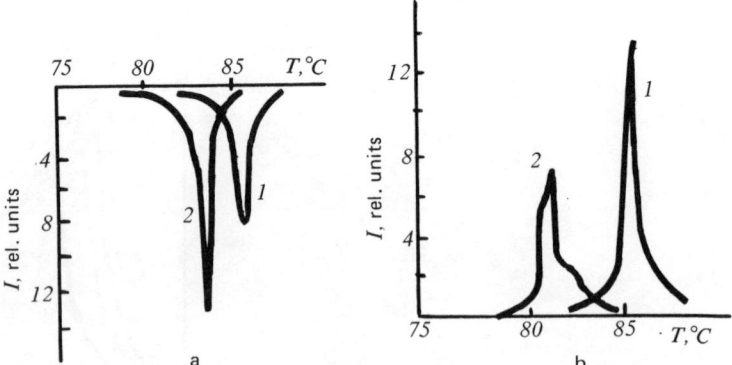

Figure 5.16. Effect of illumination on the phase-transition temperature of BaTiO$_3$ alloyed with 2.5 at. % Fe ($T_1 = 85°$ C). (1) Pyroelectric current in darkness; (2) pyroelectric current with illumination; a and b correspond to two opposite directions of spontaneous polarization.

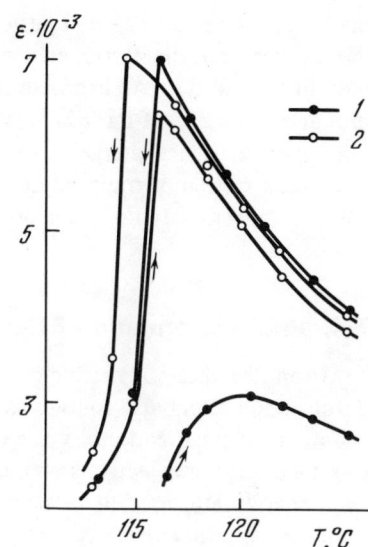

Figure 5.17. Temperature dependence of the dielectric constant of $BaTiO_3$ in darkness (1) and with illumination (2).

from the curves of Figure 5.17 that the "natural" illumination simultaneously shifts the temperature of the phase transition toward lower temperatures, decreases the temperature hysteresis, and, moreover, increases the value of the dielectric constant in the heating mode. The photodielectric effect was also investigated for SbSI [194, 212]. It should be kept in mind that another possible mechanism responsible for the photodielectric effect in $BaTiO_3$ can be the effect of the "natural" illumination on the domain structure (the photo-domain effect, cf. Chapter 6).

The determination of the coefficients a, b, and c in the expansion (1.46) from the form of the intrinsic optical absorption edge in $BaTiO_3$ permits one to compare the magnitude of the photostimulated shift of the phase-transition temperature and the photohysteresis effect with the thermodynamic calculations. Substituting the values $a \simeq 0.8 \times 10^{-23}$, $b \simeq +2.6 \cdot 10^{-33}$ cgs esu, $c \simeq 0$ (Section 4.2), $\beta \simeq -7 \cdot 10^{13}$ cgs esu, and $C \simeq 3 \cdot 10^5$ deg^{-1} [14] for $BaTiO_3$ into (1.36) and (1.39), and taking $\Delta T_N \simeq -5°$ and $\Delta T_{hN}/\Delta T_h \simeq -0.3$ according to the experiment described here, we obtain from both relations the same order of magnitude of the concentration of nonequilibrium carriers $N \simeq 10^{19}$ cm^{-3}. This result permits one to reach two conclusions. First, the photoferroelectric phenomena in $BaTiO_3$, as in other ferroelectrics, are related to carriers localized in traps. Second, the values of ΔT_{hN} and ΔT_N corresponding to saturation evidently correspond to extreme filling of the traps in $BaTiO_3$. As for the concentration of levels $\sim 10^{19}$ cm^{-3}, its value is plausible for such a high-resistance semiconductor as barium titanate.

Unfortunately, direct determination of the concentration of traps by the method of thermostimulated current, as was done for SbSI (cf. Section 5.2), has not yet been successful, evidently because of the low values of carrier mobility in $BaTiO_3$. An estimate of the depth and concentration of traps in unreduced crystals of $BaTiO_3$ was made in [213] by measuring the currents limited by the space charge. A value of $N \simeq 3 \cdot 10^{18}$ cm^{-3} was obtained for the concentration of traps, while $u \simeq 0.35$ eV for the activation energy. As we see, the concentration of traps obtained in [213] is close in order of magnitude to our estimate.

5.4. Photodeformation Effect

From the thermodynamic considerations of Section 1.3, the photodeformation effect also follows as a result of the effect of electrons on the spontaneous polarization. By the photodeformation effect is meant the effect of nonequilibrium electrons on deformation of the ferroelectric and, in particular spontaneous deformation at the phase transition. According to (1.40), the photodeformation effect can be related to the effect of nonequilibrium carriers on the spontaneous polarization $(\Delta P_0/P_0)^2$, on the electrostriction coefficient $\Delta v/v$, and also on the lattice deformation caused by the pressure of nonequilibrium carriers and described by the term $N(\partial \widetilde{E}/\partial \sigma)$, where $\widetilde{E}$ is the energy of the excited carriers and σ is the corresponding component of the mechanical stress tensor.

The photodeformation effect was discovered and investigated for crystals of SbSI [214-217]. The SbSI crystals, having the usual form of needles, were placed in the chamber of a dilatometer, which permitted changing and stabilizing the temperature, illuminating the crystals with monochromatic light (in particular, in the region of its natural photosensitivity), and measuring the deformation of the crystal in the direction of the spontaneous polarization axis (i.e., in the direction of the needle axis).

Curve a in Figure 5.18 demonstrates the temperature dependence of the deformation (elongation) of multidomain single crystals of SbSI (i.e., crystals not subjected to preliminary polarization by an external field) as a result of their illumination with "natural" light. Each point of this curve was taken in the following manner. The crystal was heated to a temperature of +50°C and gradually cooled for several hours in the darkened chamber below the phase-transition temperature, which was about +19°C for the batch of crystals investigated in [216]. On reaching the specified temperature, the light was turned on and the deformation Δl was measured.

We note first of all that the spectral distribution of this effect coincides with the spectral dependence of the photocurrent in SbSI. The photodeformation effect has a positive sign, goes to zero near the Curie point, and is not observed in the paraelectric region. It is also seen from curve a of Figure 5.18

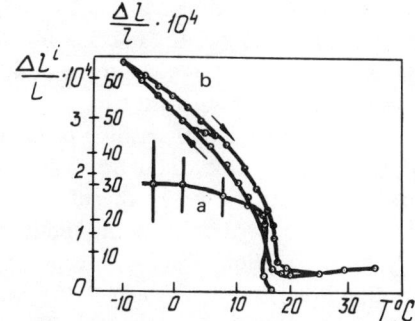

Figure 5.18. Temperature dependence of the deformation of SbSI with illumination (a) and in darkness (b).

that far from the Curie point (in the ferroelectric region) significant scatter of the experimental points occurs. If the crystal was first polarized in an external field greater than the coercive field, then illumination either did not cause additional deformation or the deformation was small. Curve b in Figure 5.18 demonstrates the temperature dependence of the deformation of the crystal near the phase transition in the darkened chamber. The arrows indicate the direction of the change in temperature. To eliminate errors, a continuous change in temperature was not used in [216], but it was stabilized at each point. Curve b shows hysteresis of the dark deformation. The form of the hysteresis curves depends on the degree of cooling of the crystal, which is illustrated in the figure for two values of temperature. These same measurements for the illuminated crystal displayed some decrease in the value of the hysteresis. Figure 5.19 presents the relaxation curves for the photodeformation process for two values of light intensity. The curves are exponential with a time constant τ of the order of several minutes. The time τ is significantly greater than both the

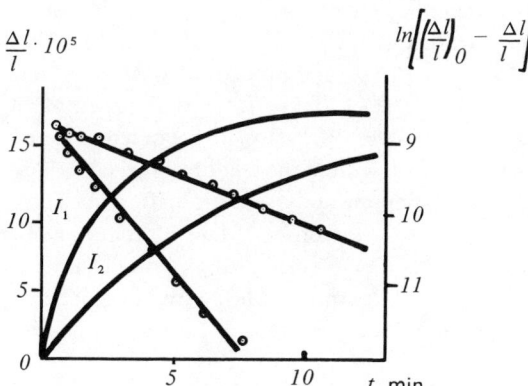

Figure 5.19. Kinetics of the photodeformation effect for two intensities of light.

photocurrent rise time and the Maxwellian relaxation time. Cutoff of the illumination stopped the photodeformation process, after which a very gradual relaxation is observed toward the dark equilibrium value, which was not actually reached in the experiment.

We compare the magnitude of the photodeformation effect with the thermodynamic calculation of Section 1.3. According to the data of Figure 5.18, the experimental value of this effect is $u_N/u = \Delta l/l \simeq 10^{-4}$. According to (1.40), the contribution of the spontaneous polarization to the spontaneous photodeformation effect is determined by the term $(\Delta P_0/P_0)^2$. The following section is devoted to the effect of nonequilibrium carriers on the spontaneous polarization of the ferroelectric. There, in particular, it is shown that from independent measurements the maximum observed value $\Delta P_0/P_0 \simeq 3 \cdot 10^{-3}$ (for LiNbO$_3$) corresponds to the strong filling of traps in the ferroelectric, $N \simeq 10^{18}$ cm^{-3}. Thus $(\Delta P_0/P_0)^2_{max} \simeq 10^{-5}$ and, consequently, the contribution of the change in spontaneous polarization to the photodeformation effect should be neglected. Evaluation of the third term in (1.40) $N(\partial E/\partial \sigma)$ leads to an analogous conclusion.

Investigation of the effect of hydrostatic pressure on the width of the forbidden band E_g and the piezoresistance effect in SbSI [91, 92, 218] leads to the conclusion that over a wide range of temperatures in the ferro- and paraelectric region, $(\partial E_g/\partial p)_T$ and $(\partial E_a/\partial p)_T$ for SbSI change over the range 10^{-5}-10^{-6} eV/atm (E_a is the activation energy of the donor levels). Thus, for a concentration of nonequilibrium carriers in traps $N \simeq 10^{18}$ cm^{-3}, the term $\tilde{E}'N$ for SbSI can vary over the range 10^{-5}-10^{-6}. Thus, the contribution of the third term in (1.40) is also insignificant. Although data on the effect of electrons on the electroconstriction coefficient ν for SbSI are absent in the literature, all other experimental results also refute the possibility of evaluating the photodeformation effect from (1.40). It is sufficient to turn to the photodeformation kinetics (Figure 5.19). It is determined neither by the time for optical recharging of the levels in SbSI nor by the Maxwellian relaxation time $\tau = \epsilon/4\pi\sigma$ (for $\epsilon \simeq 10^3$-10^4 and $\sigma \simeq 10^{-9}$-10^{-10} $\Omega^{-1} \cdot$ cm^{-1}, the time $\tau \approx 10^{-1}$-10 sec). With cutoff of the light, the time for emptying the recharged levels in SbSI is small (at temperatures close to room temperature) and the values of T_1 and ΔT_h accordingly return to the equilibrium dark values practically instantaneously. At the same time, photodeformation after cutoff of the light displays a very slow dark relaxation of the order of tens of hours. The entire set of experimental data leads to the conclusion that photodeformation is caused by a change in the domain structure of SbSI with illumination (photodomain effect). The relaxation time of the effect and the presence of temperature hysteresis also speak in favor of this conclusion.

It was indicated above that both the mechanism of optical recharging of levels in the neutral crystal, as well as the mechanism of screening the spontaneous polarization by nonequilibrium carriers, can be the basis of photoferroelectric phenomena. The thermodynamic approach developed in Section 1.3 naturally does not make it possible to separate these two possible mechanisms. Nonetheless, although the photodomain effect is one of the screening phenomena, the unsuitability of (1.40) for describing photodeformation of SbSI crystals is explained by the fact that this relation is obtained for a single-domain crystal. The change in domain structure with illumination leads to additional deformation of the crystal due to deformation of the domain walls. This mechanism evidently makes a decisive contribution to the photodeformation of SbSI. We will discuss this in somewhat more detail below with the description of the photodomain effect.

5.5. Effect of the Photoinduced Change in Birefringence†

The phenomenological consideration presented in Section 1.3 on the effect of nonequilibrium carriers on the ferroelectric properties of a crystal leads to the conclusion about the effect of illumination of the spontaneous polarization P_0, which in turn must be accompanied by a corresponding change in birefringence (the Pockels effect). Thus the comparatively recently discovered effect [219] of intense illumination on birefringence in ferroelectrics can be related with complete justification to the class of photoferroelectric phenomena.

The effect of the photoinduced change in birefringence has been called in the literature "optical damage" and has found important application in recording three-dimensional holograms. The "optical damage" effect reduces in practice to the fact that with local illumination of the ferroelectric crystal with an intense incident light (a focused laser beam), a reversible change in birefringence occurs inside the light beam in the volume of the crystal, due mainly to a change in the index of refraction of the extraordinary beam n_e. The magnitude of this change for some pyroelectrics reaches 10^{-4}-10^{-3} ($LiNbO_3$, $LiTaO_3$), and the storage time of the effect varies over very wide limits from milliseconds for $BaTiO_3$ to months for $LiNbO_3$. The recording of holograms is achieved by the three-dimensional modulation of Δn corresponding to modulation of the recording beam; the resolution of the recording is very high (10^2-10^3 lines/mm).

Without dwelling in detail on the problems of the practical utilization of the "optical damage" effect, to which extensive investigations have been devoted (cf., for example, [219-231], we turn directly to a survey of works devoted to

†This section was written together with T. R. Volk.

the investigation of the photoferroelectric nature of this phenomenon.† It should be noted that a significant photoinduced change in birefringence in a majority of ferroelectrics (KTN [221], BaTiO$_3$ [222], PZLT [231]) is observed only with the application on the crystal of a constant field of the order of several kV/cm, and only in a limited number of cases (LiNbO$_3$, LiTaO$_3$ [219, 220]) does the effect occur without an external field. In the first case, the existence of spontaneous polarization in the crystal mainly causes a quantitative dependence of the effect on the field, while the phenomenon does not differ qualitatively from that which occurs in nonferroelectric crystals (CdS, PZLT in the paraelectric phase [231]). In the second case, i.e., in the absence of an external field, the existence of the "optical damage" effect itself is caused by the ferroelectric nature of the crystal: the effect arises only because of the effect of nonequilibrium carriers or excited electronic states on the spontaneous polarization. We will discuss this case below.

Like all photoferroelectric phenomena, the "optical damage" effect can be related to two independent mechanisms. This is, first of all, the optical recharging of centers which also causes a change in spontaneous polarization [cf. (1.37)], besides the already described photoferroelectric phenomena. Moreover, a screening of the spontaneous polarization accompanied by a volume separation of the carriers and the formation of internal fields of the space charges can also occur. Investigations of the microscopic nature of the "optical damage" effect (still very far from complete) have been devoted generally to consideration of just these two possible mechanisms.

An attempt to relate the observed effect of photoinduced change of birefringence was first made by Chen with LiTaO$_3$ and LiNbO$_3$ [221]. Lithium niobate still remains the most popular subject of investigation. Chen proposes that with local illumination of the crystal, the carriers photoexcited from certain levels under the effect of the internal field always present in the volume of the ferroelectric drift independently of the screening conditions (Section 3.5) to the periphery of the light spot where they are captured in deep levels.‡ The volume separation of the charges caused by screening of the internal field leads to the formation of a local field $\mathcal{E}_s$ and, consequently, to a local change in birefringence due to the linear electro-optical effect:

---

†It was shown recently that the "optical damage" or photorefractive effect in ferroelectrics without an external field is caused mainly by the anomalous photovoltaic effect in ferroelectrics. See, for example, A. M. Glass and D. von der Linde, *Ferroelectrics*, 10, 163 (1976); P. Günter and F. Micheron, *Ferroelectrics*, 18, 27 (1978); F. Micheron, *Ferroelectrics*, 18, 153 (1978).

‡As has been indicated above, the "optical damage" effect occurs in a majority of cases only in the presence of an external field responsible for the volume separation of nonequilibrium carriers; after removal of the field, the effect quickly disappears due to emptying of the capture levels.

$$\Delta n_s = \frac{1}{2}\left(n_0^3 r_{13} - n_e^3 r_{33}\right)\mathscr{E}_s \sim N, \qquad (5.8)$$

where n_0 and n_e are the indices of refraction of the ordinary and extraordinary rays, respectively, N is the concentration of captured carriers, and r are the electro-optical coefficients.† The experimental results agree with this qualitative model (Figure 5.20). First of all, the effect is anisotropic, and is significant only when the illumination is perpendicular to the ferroelectric axis of the crystal c, i.e., when the direction of the drift of the screening carriers is determined by the direction of the internal field. The spatial distribution of Δn_e agrees with the distribution of the local screening field (Figure 5.21). The calculation made by Chen for the electro-optical equation (5.8) showed that to reach the maximum values observed in the experiment $\Delta n_e \sim 10^{-3}$, it is necessary that the space-charge field $\mathscr{E}_s \sim 7 \cdot 10^4$ V/cm arise in the crystal, which corresponds to a density of captured carriers of the order of 10^{14}-10^{15} cm^{-3}. As we have already seen with BaTiO$_3$ and SbSI as examples, such values of the density are quite reasonable if one takes into account that the density of traps reaches values of 10^{18}-10^{19} cm^{-3} in high-resistance semiconductors similar to LiNbO$_3$ and other ferroelectrics.

Another approach in explaining the microscopic mechanism of optical damage was proposed by Johnston [223]. The author proposed that with

†Calculations have indicated that carrier drift occurs, and not the Dember effect, since the values of Δn in this case would be an order of magnitude lower than those observed.

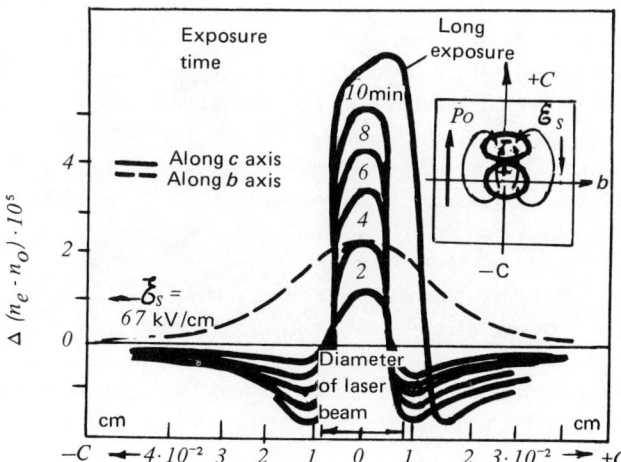

Figure 5.20. "Optical damage" effect in LiTaO$_3$. Distribution of the space-charge field $\mathscr{E}_s$ causing the "optical damage" effect is shown on the upper right corner [221].

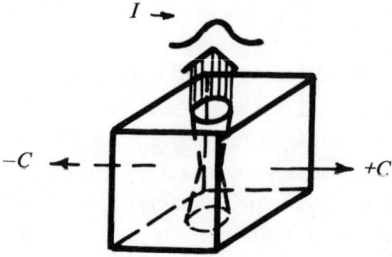

Figure 5.21. Geometry of crystal and light beam.

illumination of the ferroelectric, the change in dipole moments of the impurities due to their optical recharging leads to a change in the macroscopic polarization and a corresponding change in birefringence inside the light spot $\Delta n \sim \Delta P_0$. We add to this the fact that the change in spontaneous polarization ΔP_0 due to nonequilibrium carriers inside the light spot can have a purely Jahn-Teller nature (cf. Chapter 2).

Thus, there arises inside the light spot a field of the spontaneous polarization gradient caused by drift of nonequilibrium carriers beyond the light spot. At the boundary of the light spot, as is also proposed in the Chen model, the carriers are captured by deep levels, where equilibrium is reached when the field created by the carriers captured by traps is completely screened by the field of the polarization gradient. After cutoff of the light, recombination of the excited electrons (localized at shallow trapping levels) with ionized centers occurs for a short time, which leads to $\Delta P_0 \to 0$ inside the light spot. As for the screening charge at the periphery of the light spot, having been formed by carriers captured at deep levels, it is preserved for a long time depending on the energy of these levels and the temperature. After cessation of the illumination, an electric field is thereby formed within the light spot, which leads to a change in the index of refraction n_e, thanks to the electro-optical effect. This field is $\sim 4\pi\Delta P_0/\epsilon$, where ΔP_0 is the initial change in spontaneous polarization inside the light spot. At equilibrium, the steady-state value of Δn_e is determined by the value of ΔP_0. Thus, two basic mechanisms responsible for the photoferroelectric phenomena (optical recharging of levels and screening of the spontaneous polarization) emerge together in this model.

Both in the Johnston model and the Chen model, the change in Δn_e is related to the field of the screening charge formed by carriers captured at deep levels and by the linear electro-optical effect. The Johnston model has only the advantage that the assumption about the existence of an internal field in the ferroelectric is not used, although the latter actually exists, for example, thanks to the surface layers (Section 3.5). Some quantitative estimates were made in [223]. By using the maximum value $\Delta n_e \sim 10^{-3}$ obtained by Chen for $LiNbO_3$, Johnston obtained from the electro-optical equations the value for the change

PHOTOFERROELECTRIC PHENOMENA

in spontaneous polarization corresponding to this change in birefringence, $\Delta P_0/P_0 \simeq 3 \cdot 10^{-3}$ ($P_0 \simeq 70\ \mu C/cm^2$ for $LiNbO_3$). Such a change in spontaneous polarization is possible only if at least 1-2% of the unit cells in the $LiNbO_3$ crystal contain optically recharged defects (under the assumption that the magnitude of the dipole moment of the photoexcited impurity is about 5% of the magnitude of the dipole moment of the unit cell of the ideal polar crystal). However, the most reasonable theoretical estimate for ΔP_0 is the thermodynamic estimate from equation (1.37), which does not require a microscopic model. This estimate will be made below.

Before turning to a discussion of the experimental works devoted to investigation of the "optical damage" effect, it should be noted that there are some general conditions necessary to observe the effect. First of all, to obtain a significant value of Δn_e in the crystals under investigation, the electro-optical coefficients must be sufficiently large. Since the effect is volume in nature, it is necessary that the photoexcitation wavelength be located sufficiently far from the intrinsic absorption edge (which does not exclude theoretically the possibility of producing the effect with illumination by natural light). Thus, in investigating the "optical damage" effect observed, as a rule, in very wide-band materials, (He–Ne) (0.628 μm) or Ar (0.488 μm) lasers are usually used as sources of illumination. Generation of the "optical damage" effect does not require coherence of the exciting light; the basic condition for a significant change in birefringence is sufficiently high energy of the radiation. The value of Δn_e also depends on the duration of the exposure. It was shown in [221] that for short exposures of ~0.01 sec in $LiNbO_3$, the value of Δn_e is proportional to the product of the exposure time and the radiation intensity $\Delta n_e \sim It$ right up to

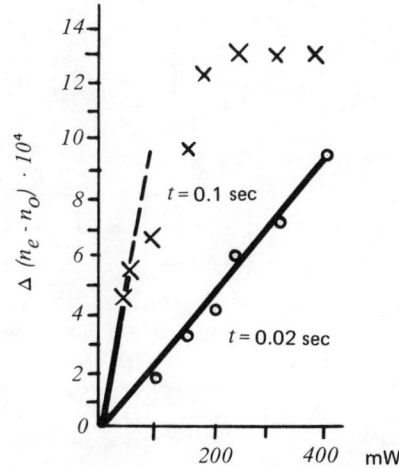

Figure 5.22. Dependence of $\Delta(n_e - n_0)$ in lithium niobate on the intensity of illumination for exposures of 0.02 sec and 0.1 sec. Laser beam diameter 10^{-2} cm.

very high intensities. For long exposures of ca. 1 min, $\Delta n \sim \sqrt{I}$ occurs, where saturation is reached for intensities of 20 mW (Figure 5.22).

There are at present a number of high-precision techniques permitting measurement of the value of Δn_e with an accuracy no worse than 10^{-6} [221]. As an example, Figure 5.23 presents a block diagram for measuring Δn_e by the compensation method. This method was used in the experimental works considered below. It should be stated that the survey of the experimental works on the optical damage effect presented here does not pretend to be complete. There is no detailed description of the crystals used, and it does not touch upon the problem of the role of effective impurity and other aspects of the method. Consideration of the set of all these problems could be the subject of a special monograph. The authors have tried to isolate from the entire mass of experimental material the results related directly to the investigation of the mechanism of the phenomenon within the framework of the ideas about the interaction of nonequilibrium carriers with the spontaneous polarization.

The mechanism of the "optical damage" effect has been studied in a majority of works in crystals of $LiNbO_3$ doped with impurities of transition metals, and recently in the works of the author with co-workers in crystals of barium strontium niobate (BSN) of composition $Ba_{0.25}Sr_{0.75}Nb_2O_6$ [232-234].

The optical damage effect in crystals of $LiNbO_3$ is observed in the temperature interval up to ca. 170°C, where the effect is thermally annealed [221a]. In BSN crystals, the optical damage effect occurs at room temperature only with the application of a significant electric field [230]; a stable change of Δn_e in the

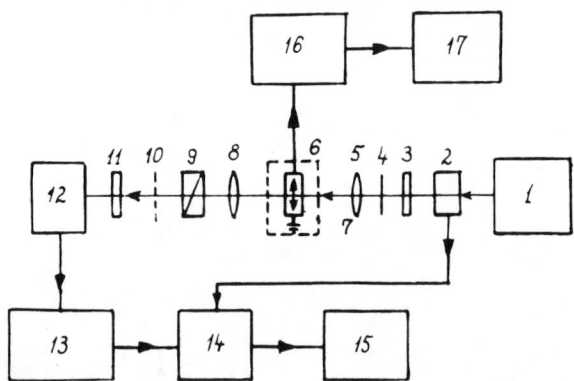

Figure 5.23. Block diagram of installation for investigating the "optical damage" effect and short-circuit photocurrents. (1) He–Ne laser; (2) laser beam modulator; (3) neutral filter; (4) polarizer; (5) lens; (6) cryostat; (7) crystal; (8) lens; (9) quartz wedge; (10) analyzer; (11) neutral filter; (12) photomultiplier; (13) AC amplifier U2-6; (14) synchronous detector V9-2; (15) recorder KSP; (16) electrometric amplifier U5-6; (17) two-coordinate recorder PDS-02.

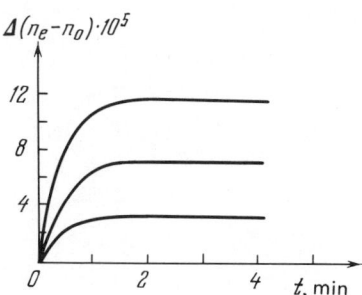

Figure 5.24. Kinetics of the growth of the "optical damage" effect in crystals of BSN for various polarizing fields ($T = 120°K$).

absence of the field is observed only at a temperature below 200°K [234]. (Some consideration with respect to the dependence of the effect on temperature and, consequently, on the depth of the capture levels will be made below.) The BSN crystal is a most convenient object for investigating the relation of the optical damage effect to the ferroelectric properties since it is possible to vary the ferroelectric polarization ($T_1 \simeq 50°C$) by polarization reversal; it is impossible to carry out such an experiment in $LiNbO_3$ because of the high values of the coercive fields [221a]. It was shown using BSN crystals that the value of Δn_e is maximum in the single-domain crystal; the "optical damage" effect does not arise in a previously unpolarized crystal (Figure 5.24).

As has already been indicated above, the mechanism of the "optical damage" is based on the assumption of the generation of a space-charge field in the crystals.

Starting from the assumption of the existence of such a field, Chen calculated its magnitude from the linear electro-optical effect equations, starting from the experimentally observed values of Δn_e. Meanwhile, the space-charge field in ferroelectrics can be investigated by direct photoelectric methods. It is indicated in the works of a number of authors [221a, 233, 235-239] that the short-circuit photocurrents I_{sc} observed with illumination of a photoelectric by photoactive light are related to the existence of internal fields. We will turn once again to this problem in Section 6.7. The short-circuit photocurrent I_{sc} consists of transient and steady-state components. This form of signal is observed in the ferroelectric phase for all photosensitive ferroelectrics ($BaTiO_3$ [235], SbSI [236-238], $LiNbO_3$ [221a]); the signal does not contain the transient component in the paraelectric phase (Figure 5.25).

Investigation of the steady-state signal I_{sc} in crystals of $LiNbO_3$ doped with Fe and also in BSN showed [232, 232a, 239] that the steady-state I_{sc} is a volume photovoltaic signal, a specific property of polar crystals. This is confirmed by the quantitative relation between the magnitude of the photovoltaic effect and the internal field. With increasing Fe^{2+} concentration, the increase in "photosensitivity" of the crystal (i.e., the magnitude of the photovoltaic effect) is

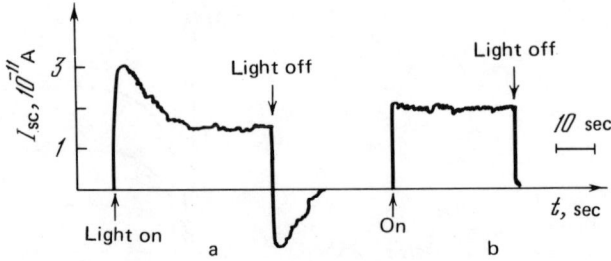

Figure 5.25. Form of the short-circuit photocurrent signal in BSN crystals, $\lambda = 400$ nm. (a) In the ferroelectric phase; (b) in the paraelectric phase.

accompanied by an increase in the "optical damage" effect. The steady-state I_{sc} drops sharply with the transition of BSN to the paraelectric phase.†

The transient signal of the short-circuit photocurrent, as was shown in [233, 239], is a pyroelectric current related to weak heating of the pyroelectric due to absorption of photoactive light. Actually, the transient signal observed with illumination of the strong pyroelectrics $LiNbO_3$ and BSN with a He-Ne laser ($\gamma \simeq 0.5 \cdot 10^{-7}$ and $\simeq 10^{-6}$ C/cm² · deg, respectively [240]) has a magnitude of 10^{-11}-10^{-13} A, which would correspond according to the estimates of the charge to local heating of these crystals by 0.01°. Such heating due to absorption of weakly absorbed light with $\lambda \simeq 0.628$ μm is quite possible. To explain this, Figure 5.26 presents a comparison of the spectral dependences of the transient and steady-state signal of I_{sc} with the spectral dependence of the photoconductivity in $Ba_{0.25}Sr_{0.75}Nb_2O_6$ (BSN) crystals. The weak anomaly of the transient signal of I_{sc} in the natural region ($\lambda \simeq 400$ nm) is evidently related to the stronger heating of the crystal due to increasing absorption.

Since the transient signal of the short-circuit photocurrent is pyroelectric in nature, the magnitude of the local transient signal I_{sc} can clearly serve as an estimate of the spontaneous polarization at the given point of the crystal. Thanks to this, it is possible to probe the local value of P_0 related to the photoferroelectric phenomena occurring in the crystal in parallel with the change in birefringence.

It was established with the simultaneous investigation of the "optical damage" effect and local pyroelectric current in $LiNbO_3$ and BSN that the change in birefringence is always accompanied by an irreversible change in the magnitude

---

†By illuminating some ferroelectrics in the intrinsic or extrinsic optical region the steady-state photovoltaic current and the corresponding photo-emf, which is a few orders of magnitude higher than the energy gap, are observed in the direction of spontaneous polarization (anomalous photovoltaic effect). The mechanism of this effect was studied in a series of investigations. See, for example, A. M. Glass, D. von der Linde, D. H. Auston, and T. I. Negran, *J. Electron. Mater.*, **4**, 915 (1975); V. M. Fridkin and B. N. Popov, *Usp. Fiz. Nauk*, **126**, 657 (1978).

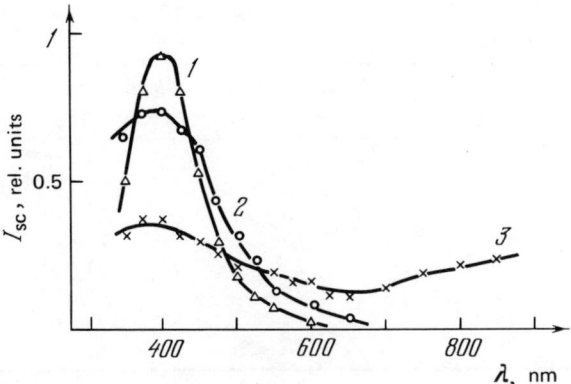

Figure 5.26. Spectral distributions of the photoconductivity (1), steady-state signal of I_{sc} (2), and transient signal of I_{sc} (3) in BSN crystals.

of the local pyroelectric current (Figure 5.27). In accordance with this, under those conditions where the "optical damage" effect somehow does not arise (for example, for small power densities or in BSN at temperatures above 200°K), the local pyroelectric signal also does not change as a result of laser illumination. Thermal obliteration of the "optical damage" effect with heating of the crystal above some characteristic temperature (+170°C for LiNbO$_3$ and 200°K for BSN) is accompanied by thermal annealing of the change in I_{sc}.

The change in the local pyroelectric current can be evaluated by the corresponding charge ΔQ_s. There is a direct quantitative relationship between the

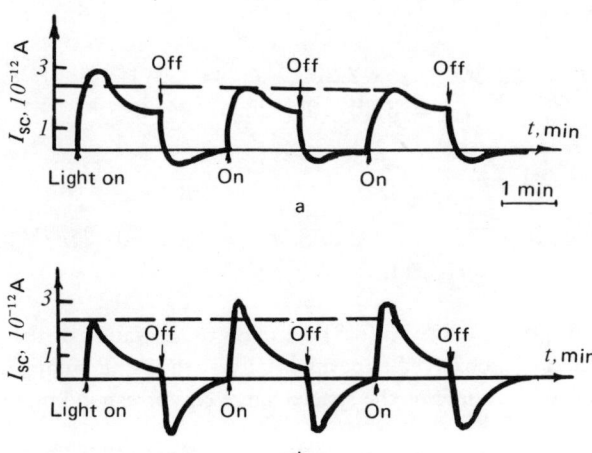

Figure 5.27. Change in I_{sc} as a result of successive illumination of the crystal at one point with a laser beam. (a) In crystals of $Ba_{0.25}Sr_{0.75}Nb_2O_6$, $T = 120°K$; (b) in crystals of LiNbO$_3$, $T = 300°K$.

TABLE 7. Photoinduced Change of Birefringence in LiNbO$_3$ and BSN at Various Points of the Crystal

Crystal	T, °K	Power density, W/cm^2	Δn_e	ΔQ_s, µC
LiNbO$_3$	300	25	$5 \cdot 10^{-5}$	$4 \cdot 10^{-6}$
LiNbO$_3$	300	25	$9 \cdot 10^{-5}$	$9.1 \cdot 10^{-6}$
LiNbO$_3$	300	10	$2.1 \cdot 10^{-5}$	$2.6 \cdot 10^{-6}$
BSN	110	10	$5.7 \cdot 10^{-5}$	10^{-5}
BSN	110	25	$11 \cdot 10^{-5}$	$2 \cdot 10^{-5}$
BSN	160	25	$1.4 \cdot 10^{-5}$	$1.3 \cdot 10^{-5}$
BSN	160	10	$0.95 \cdot 10^{-5}$	10^{-5}
BSN	203	25	0	0

magnitude of the "optical damage" effect Δn_e and the magnitude of the charge ΔQ_s characterizing the change in the local pyroelectric current. Table 7 presents some quantitative values to illustrate this dependence.

Since the magnitude of the local pyroelectric current (the transient I_{sc}) is determined by the spontaneous polarization P_0, the change of this signal indicates the local change in P_0 by the corresponding amount ΔQ_s. An estimate was made for the field $\mathscr{E}_s$† corresponding to the charge ΔQ_s. The linear electro-optical effect equation was then used to calculate the change in birefringence Δn_s, which corresponds to the field $\mathscr{E}_s$:

$$\Delta n_s = \frac{1}{2}\left(n_0^3 r_{13} - n_e^3 r_{33}\right) \mathscr{E}_s, \tag{5.8}$$

where for LiNbO$_3$

$$n_0 \simeq 2.286; \quad n_e \simeq 2.222; \quad r_{13} \simeq 8.6 \cdot 10^{-12} \text{ m/V};$$
$$r_{33} \simeq 30.8 \cdot 10^{-12} \text{ m/V [241]};$$

for BSN

$$n_0 \simeq 2.311; \quad n_e \simeq 2.298; \quad r_{33} \simeq 13.4 \cdot 10^{-10} \text{ m/V};$$
$$r_{13} \simeq 6.7 \cdot 10^{-11} \text{ m/V [242]}).$$

The values of Δn_s obtained from (5.8) are close to the values of the "optical damage" effect Δn_{exp} observed experimentally at the same points of the crystal (cf. Table 8). The local field of the space charge $\mathscr{E}_s$ corresponding to the change

---

†The field $\mathscr{E}_s$ can be evaluated from the formula $\mathscr{E}_s \approx (4\pi/\epsilon)(\Delta Q_s/S)$, were $S = dl$, d is the diameter of the light spot, and l is the thickness of the crystal.

TABLE 8. Values of the Space-Charge Field $\mathcal{E}_s$ in LiNbO$_3$ and BSN

Crystal	T, °K	ΔQ_s, μC	$\mathcal{E}_s$, V/cm	Δn_s	$\Delta n_{\exp}$
LiNbO$_3$	300	$4 \cdot 10^{-6}$	1800	$2.1 \cdot 10^{-5}$	$5 \cdot 10^{-5}$
BSN	110	$2 \cdot 10^{-5}$	600	$4 \cdot 10^{-4}$	$1.1 \cdot 10^{-4}$

in polarization ΔQ_s is responsible for the observed change of birefringence. Thus, the values of the space-charge field $\mathcal{E}_s$ calculated independently from the change in P_0 and the change in birefringence coincide (and are close to the corresponding estimates of Chen [221a]).

Quantitative data can be obtained from investigation of the "optical damage" effect about the effect of nonequilibrium carriers on the spontaneous polarization. As was already indicated in Section 5.4, the other photoferroelectric phenomena, for example, the photodeformation effect, cannot be used for this purpose because of the strongly expressed photodomain effect (cf. Section 6.6). In particular, the concentration of nonequilibrium carriers N responsible for the change in spontaneous polarization ΔP_0 can be evaluated from equation (1.38). Neglecting coefficient c, equation (1.38) can be written in the form

$$\frac{\Delta P_0}{P_0} = \frac{1}{2} \frac{b}{\beta} N. \tag{1.38'}$$

The concentration of nonequilibrium carriers in LiNbO$_3$ and BSN responsible for the independently measured change in polarization ΔP_0 was evaluated with the help of (1.38'). Since the values of the coefficients b and β for LiNbO$_3$ and BSN are unknown, the coefficients for BaTiO$_3$ were used in the calculation (the values of b and β in ferroelectrics with phase transitions of the displacement type do not differ by more than an order of magnitude). Thus by substituting the values $b \simeq +2.6 \cdot 10^{-33}$, $\beta \simeq -7 \cdot 10^{-13}$ cgs esu, and the values of $\Delta P_0/P_0$ corresponding to the data of the preceding table into (1.38'), the concentrations presented in Table 9 were obtained.

If one takes $M \simeq 10^{18}$ cm^{-3} as the concentration of traps in LiNbO$_3$ and

TABLE 9. Evaluating the Concentration N from "Optical Damage" Data

Crystal	ΔQ_s, μC	ΔP_0,[a] $\mu C/cm^2$	P_0, $\mu C/cm^2$	$\Delta P_0/P_0$	N, cm^{-3}
LiNbO$_3$	$4 \cdot 10^{-6}$	$4 \cdot 10^{-3}$	70	$8 \cdot 10^{-5}$	$4 \cdot 10^{16}$
BSN	$2 \cdot 10^{-5}$	10^{-2}	10	10^{-3}	$5 \cdot 10^{17}$

[a] $\Delta P_0 = \Delta Q_s/S$, $S = dl$.

BSN, then the values of $\Delta P_0/P_0$ presented in the table correspond to weak filling. At the same time, if one takes the value $\Delta P_0/P_0 \simeq 3 \cdot 10^{-3}$ corresponding to the saturation of the "optical damage" effect in LiNbO$_3$ [223], then this saturation according to (1.38') corresponds to the concentration $N \simeq 10^{18}$ cm^{-3}, i.e., strong filling of the traps. We recall that the values of the photohysteresis effect and the photostimulated shift of the Curie point at saturation can also be responsible for strong filling of the traps (Section 5.2). The absence in the literature of data on thermostimulated currents, carrier mobility, and the kinetics of photoconductivity in LiNbO$_3$ and BSN do not permit one to evaluate the concentrations N and M directly. However, the values of N corresponding to saturation of various photoferroelectric effects are close and are evidently responsible for strong filling of the traps. Thus, the "optical damage" effect can be quantitatively predicted by the thermodynamics of photosensitive ferroelectrics if one uses the concentrations of carriers N obtained from investigation of other photoferroelectric effects to calculate $\Delta P_0/P_0$ from (1.38').

The experimental results discussed above indicate that the "optical damage" effect is related to the processes of local change in spontaneous polarization with illumination. However, the problem of the nature of the corresponding levels responsible for the processes of excitation and capture can be solved only as a result of parallel investigation of the photoelectric properties. The nature of the levels responsible for the "optical damage" effect is almost uninvestigated. With respect to LiNbO$_3$, one can only say that the effect is impurity in nature [224, 243]. As in BaTiO$_3$, lithium niobate is an oxygen-octahedral ferroelectric, for which the valence band is formed by the niobium ion orbitals. Investigation of the absorption spectra of pure and alloyed LiNbO$_3$ has revealed that LiNbO$_3$:Fe has localized levels Fe^{2+} and Fe^{3+}, where Fe^{2+} are donors, while Fe^{3+} are traps. The generation of the "optical damage" effect is related to photoexcitation of electrons from the Fe^{2+} levels to the conduction band, after which the electrons are captured by the traps Fe^{3+}. In accordance with this, the sensitivity of LiNbO$_3$ to "optical damage" increases with increasing concentration of Fe^{3+} traps and does not depend on the concentration of Fe^{2+} donors [239, 243, 244]. However, there are no quantitative evaluations of the relationship between Δn and the concentration of Fe^{3+}. The thermal annealing activation energy ($T \simeq 170°C$) for LiNbO$_3$ is 1.1 eV [224]. However, attempts to relate this value with the activation energy of the Fe^{3+} capture levels have not been successful. Some authors have tried to relate the long preservation times of the effect (memory) with the large Maxwellian relaxation times [224, 225]. However, one has to assume $\rho \simeq 10^{18}$ $\Omega \cdot$ cm ($\epsilon \simeq 30$) in this case, whereas the experimental values of ρ are several orders smaller. Investigation of photoexcitation processes in LiNbO$_3$ is extremely difficult because of its low photosensitivity, and BSN crystals are a more favorable object in this respect.

The BSN (Ba$_{0.25}$Sr$_{0.75}$Nb$_2$O$_6$) crystal is one of the representatives of a new class of ferroelectric solid solutions $Ba_x Sr_{1-x} Nb_2O_6$ [240]. As is well known, the forbidden band of these compounds contains a set of levels affecting the ferroelectric properties. Thus, in contrast to $LiNbO_3$, the "optical damage" effect in BSN is not related to the introduction of a special impurity. The BSN crystals are very photosensitive, where the photosensitivity increases with decreasing temperature [233]. The "optical damage" effect in BSN (without a field) occurs only at a low temperature, which makes it possible to investigate its temperature dependence and relationship to the activation energy of the corresponding traps.

The generation of the "optical damage" effect, as has already been indicated, is related to processes of optical recharging of levels in the volume of the light spot, screening, and capture of carriers by deep traps at the periphery of the spot. The activation energy of levels responsible for optical recharging and change in spontaneous polarization can be evaluated from the temperature dependence $\Delta n = \Delta n(T)$, which gives $\mu \simeq 0.025$–0.03 eV for BSN. A decrease in the "optical damage" effect with increasing temperature above 200°K evidently corresponds to strong emptying of these shallow levels due to thermal excitation, which leads to small equilibrium filling values $N/M \ll 1$. However, the preservation time for the effect in darkness for BSN does not depend on temperature over the entire range of existence of the effect (and is several hours). Thus, one can confirm that storage is related to capture at deep levels. Investigation of the depth of the capture levels in BSN by the thermostimulated conductivity method gives an energy in the range 0.6–0.7 eV. In accordance with this, the spectral distribution of quenching of the "optical damage" effect displays a maximum in the IR regions for $\lambda \simeq 1800$–2500 nm. One can reach the preliminary conclusion that the change in the value of birefringence (and the corresponding spontaneous polarization) in BSN arises as a result of optical recharging of shallow levels with energy ~ 0.03 eV, after which capture at deep levels of 0.6–0.7 eV causes stable preservation of the effect in darkness.

In connection with the above discussions of the "optical damage" effect, we will briefly discuss the mechanism of the effect of nonequilibrium carriers on the spontaneous polarization. The results obtained permit one to relate the value of the effect of ΔP_0 to the concentration of nonequilibrium carriers N. However, it is still impossible to reach a final conclusion about the microscopic nature of this effect on the basis of these results: it is either related to the optical recharging of centers and the corresponding change in the contribution of the electron–phonon interaction, or the mechanism of screening the internal depolarization field is decisive. We recall that the separation of these two mechanisms as applied to other photoferroelectrics is also a difficult problem.

Useful information about the mechanism of the effect of nonequilibrium carriers on the spontaneous polarization can be obtained with investigation of photoexcitation processes in impurity polar crystals in the mode of illumination with very short light pulses. Relaxation processes of nonradiative excitation of the impurities Cu^{2+} and Cr^{3+} in $LiNbO_3$ and $LiTaO_3$ were investigated in [244-248]. These impurities in the polar lattice have a finite dipole moment μ [245-247]. With illumination of the $LiNbO_3:Cr^{3+}$, $LiNbO_3^-:Cu^{2+}$, $LiTaO_3:Cr^{3+}$, and $LiTaO_3:Cu^{2+}$ crystals with nanosecond pulses from a Nd laser ($\lambda \simeq 1.06\ \mu m$), pyroelectric signals were observed with very short relaxation times τ_p ($\sim 2\ \mu sec$ for $LiNbO_3:Cr^{3+}$ and $LiTaO_3:Cr^{3+}$, and 30 psec for $LiTaO_3:Cu^{2+}$). Such signals cannot be the ordinary pyroelectric response, whose relaxation time is ~ 1 msec, and it is impossible to relate them to the direct phonon excitation characterized by times less than 1 psec [245]. Glass *et al.* [245-247] assume that the generation of such fast pyroelectric responses is related to Cr^{3+} and Cu^{2+} impurity excitation processes. Actually, optical investigations have shown that the lifetime in the excited state is 1 μsec for Cr^{3+} and 30 psec for Cu^{2+} at room temperature, which is close to the obtained values of τ_p. With a decrease in temperature to 10-20°K, the lifetimes of the excited states increase to 11 μsec and 450 psec, respectively. The temperature dependence of τ_p for nonradiative transitions has an analogous nature. The authors reach the conclusion that with illumination of the crystal with a light pulse, the transition of the impurity to the excited state and interaction of the excited impurity with phonons is accompanied by a change in its dipole moment by an amount $\Delta\mu$. Independent measurements have established that $\Delta\mu \simeq 0.6$-0.8 Debye units; $\Delta\mu$ increases with decreasing temperature. The change $\Delta\mu$ makes a contribution in turn to the macroscopic polarization of the crystal. Thus the relaxation time of the fast pyroelectric response τ_p is the vibron (electron–phonon) relaxation time. This effect is not observed in nonpolar crystals where $\mu = 0$. In $LiNbO_3$ and $LiTaO_3$ crystals without impurities, the fast pyroelectric response also does not arise and only the ordinary pyroelectric effect is observed, which is related to phonon absorption. Thus, at least for the indicated crystals, the mechanism for the effect of illumination on the spontaneous polarization is an impurity pseudo-Jahn-Teller effect (Section 2.4).

In concluding this section, we discuss the problem of the relationship between the "optical damage" effect and the shift of the intrinsic absorption edge of a ferroelectric. As follows from (1.46), the change in spontaneous polarization under the effect of nonequilibrium electrons by an amount ΔP_0 leads to a change in the width of the forbidden band by ΔE_g:

$$\Delta E_g = \left(aP_0 + bP_0^3 + cP_0^5\right)\Delta P_0. \tag{5.9}$$

Thus the "optical damage" effect must appear in the shift of the intrinsic absorption edge. Since the "optical damage" is related to the formation of the space-charge field screening the change ΔP_0, the microscopic effect ΔE_g is related to two possible mechanisms.

First, there is the electrical absorption of the ferroelectric (the Kern–Harbeke effect, cf. Section 1.4) for which the sign of the effect is determined by the sign of the electron–phonon interaction constant a. If, for example, for $LiNbO_3$ (by analogy with $BaTiO_3$), $a > 0$, then the "optical damage" leads to a shift of the intrinsic absorption edge toward higher energies. For $LiNbO_3$, an estimate of ΔE_g from (5.9) for $\Delta P_0 \simeq 2 \cdot 10^{-3}$ $C/m^2 > 0$ (the "optical damage" effect is saturated [223]), $P_0 \simeq 0.7$ C/m^2, and $a \simeq 0.5$ eV $\cdot$ $m^4/C^2 > 0$ (the value of a for $LiNbO_3$ is measured independently of the electrical absorption) leads to $\Delta E_g \simeq 1.5 \cdot 10^{-3}$ eV > 0.

Second, the Franz–Keldysh effect can be responsible for the change in width of the forbidden band with "optical damage," for which the field shifts the intrinsic absorption edge toward lower energies and $\Delta E_g < 0$.

A $LiNbO_3$: Fe crystal was illuminated by an argon laser with exposures corresponding to saturation of the "optical damage" effect, which was monitored independently by measuring Δn_e. Spectrophotometry of the crystal after illumination revealed a shift of the intrinsic absorption edge toward higher energies $\Delta E_g \simeq 2.5 \cdot 10^{-3}$ eV > 0, which is electrical absorption of the ferroelectric in sign and magnitude and agrees with (5.9). After annealing the crystal at a temperature of +170°C, both effects (the "optical damage" and the shift of the absorption edge) disappear.

It appears that the Moss rule [249]

$$\Delta E_g \simeq -64.4 \frac{n^3 \Delta n}{\lambda^2}, \qquad (5.10)$$

is satisfied for $LiNbO_3$, where $n = n_e$ is the index of refraction and λ is the wavelength corresponding to the absorption edge in nanometers. Calculation with (5.10) for $\Delta n \simeq -4 \cdot 10^{-3} < 0$ ("optical damage" in saturation) gives $\Delta E_g \simeq 3 \cdot 10^{-3}$ eV > 0, which agrees with the experimental value.

Calculation for the $Ba_{0.25} Sr_{0.75} Nb_2 O_6$ crystal shows that the "optical damage" effect must be accompanied by the shift of the intrinsic absorption edge toward lower energies and that the Franz–Keldysh effect must evidently be responsible for this shift. Actually, the value $a \simeq 0.2$ eV $\cdot$ $m^4/C^2 > 0$ was obtained from data on electrical absorption of BSN, while $\Delta P_0 \simeq -10^{-4}$ $C/m^2 < 0$ was obtained from data on investigation of "optical damage" (cf. Table 9). Calculation from (5.9) for $P_0 \simeq 10 \cdot 10^{-2}$ C/m^2 leads to the implausibly small value $\Delta E_g \simeq -2 \cdot 10^{-6}$ eV. Meanwhile, calculation from the Moss formula for $\Delta n \simeq 10^{-4} > 0$ (cf. Table 8) leads to $\Delta E_g \simeq -0.6 \cdot 10^{-3}$ eV. This value agrees in sign and magnitude with the value $\Delta E_g \simeq -1.5 \times 10^{-3}$ eV calculated from the Franz–Keldysh formula for the space-charge field $\mathcal{E}_s \simeq 10^3$ V/cm corresponding to "optical damage" in BSN (cf. Table 8).

Unfortunately, direct measurement of ΔE_g in BSN has not been carried out, since the "optical damage" effect in BSN is stable only at low temperatures.

5.6. Photostimulated Phase Transitions in Nonferroelectric Materials

Ferroelectrics and pyroelectrics are two of the many materials in which photostimulated phase transitions have been detected and studied. By analogy with ferroelectrics, the role of electrons (and, in particular, nonequilibrium electrons) in the mechanism of phase transitions in other materials can be understood from general thermodynamic considerations.

In the approximation of identical frequencies of atomic vibrations in the crystal lattice, the total internal energy of the crystal has the form

$$E = E' + \hbar \omega n, \quad (5.11)$$

where E' is the internal energy of the crystal at $T = 0$ and n is the concentration of phonons. The entropy S can be expressed as the configuration part of the energy:

$$S = k \ln P, \quad (5.12)$$

where P is the number of possible distributions of n phonons over $3L$ degrees of freedom (L is the number of projections of the wave vector within the first Brillouin zone):

$$P = \frac{(3L + n - 1)!}{(3L - 1)! \, n!}. \quad (5.13)$$

By substituting (5.13), (5.12), and (5.11) into the expression for the free energy $F = E - TS$ and by using the condition of the minimum of the free energy $dF/dn = 0$ and the Stirling formula $\ln n! \simeq n \ln n$, we arrive at the following expression for the concentration of phonons n and the free energy F:

$$n = 3L \frac{1}{e^{\hbar \omega / kT} - 1}, \quad (5.14)$$

$$F = E - TS = E' + 3LkT \ln \left(1 - e^{-\hbar \omega / kT}\right). \quad (5.15)$$

We consider the free energy of two different phases of the crystal α and β, differing by the vibrational frequencies ω_α and ω_β. According to (5.15), the free energies of the α and β phases at a temperature T satisfy the following expression:

$$\begin{aligned} F_\alpha(T) &= E'_\alpha + 3LkT \ln \left[1 - e^{-\hbar \omega_\alpha / kT}\right], \\ F_\beta(T) &= E'_\beta + 3LkT \ln \left[1 - e^{-\hbar \omega_\beta / kT}\right]. \end{aligned} \quad (5.16)$$

Equating the energies F_α and F_β, one can determine the temperature of the phase transition $T = T_1$ from the following equation:

$$\exp\left[-\frac{E'_\alpha - E'_\beta}{3LkT_1}\right] = \frac{1 - e^{-\hbar\omega_\alpha/kT}}{1 - e^{-\hbar\omega_\beta/kT}}. \qquad (5.17)$$

If $E'_\beta > E'_\alpha$, then (5.17) has a solution for $\omega_\alpha > \omega_\beta$, i.e., the phase transition occurs when the β phase is "looser" (with respect to atomic vibrations of the lattice) as compared to the α phase.

An analogous description of the phase transition can be obtained with consideration of the frequency dispersion in both phases $\omega_\alpha = \omega_\alpha(\mathbf{k})$ and $\omega_\beta = \omega_\beta(\mathbf{k})$. The expressions for the free energy of the α and β phases in this case have the form

$$F_\alpha(T) = E'_\alpha + kT \sum_{\mathbf{k},s} \ln\left(1 - e^{-\hbar\omega_\alpha^s(\mathbf{k})/kT}\right),$$

$$F_\beta(T) = E'_\beta + kT \sum_{\mathbf{k},s} \ln\left(1 - e^{-\hbar\omega_\beta^s(\mathbf{k})/kT}\right), \qquad (5.18)$$

where $\omega^s(\mathbf{k})$ is the frequency of the phonon with wave vector $\mathbf{k}$ and polarizations $s = 1, 2, 3$. The summation in (5.18) is carried out over all discrete values of the wave vector $\mathbf{k}$ in the first Brillouin zone and over all vibrational branches s.

Equating the values F_α and F_β at the point of the phase transition $T = T_1$, we have

$$E'_\alpha - E'_\beta = kT \sum_{\mathbf{k},s} \ln \frac{1 - e^{-\hbar\omega_\alpha^s(\mathbf{k})/kT}}{1 - e^{-\hbar\omega_\beta^s(\mathbf{k})/kT}}. \qquad (5.19)$$

The right side of equation (5.19) is a function of temperature, $\tilde{E} = \tilde{E}(T)$. Figure 5.28 presents the function $\tilde{E} = \tilde{E}(T)$ and its intersection with the straight line $\tilde{E} = E'_\alpha - E'_\beta$ which determines the temperature of the phase transition $T = T_1$. For temperatures above the Debye temperature $T > \theta$, the right side of equation (5.19) is a linear function of temperature:

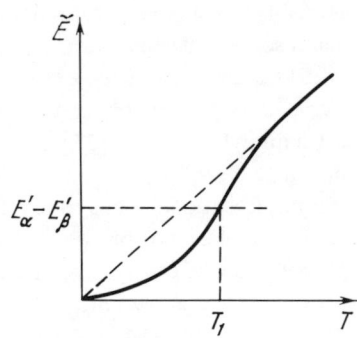

Figure 5.28. Determination of the first-order phase transition temperature T_1.

$$\widetilde{E} = kT \sum_{\mathbf{k},s} \ln \frac{\omega_\alpha^s(\mathbf{k})}{\omega_\beta^s(\mathbf{k})} = kT \ln \frac{\prod_{\mathbf{k},s} \omega_\alpha^s(\mathbf{k})}{\prod_{\mathbf{k},s} \omega_\beta^s(\mathbf{k})}. \quad (5.20)$$

As was shown in Chapter 2, one of the possible mechanisms responsible for the transition from the less "loose" phase to the more "loose" phase (or the appearance of the "soft" mode in terms of the dynamic theory of ferroelectric phase transitions) can be caused by the interaction of electrons with phonons of any of the branches, in the general case, for $\mathbf{k} \neq 0$.

The effect of electrons on first- and second-order phase transitions was shown in the work of Lifshits and Krivoglaz [250–252], starting from the idea of fluctuation stabilization, thanks to localization of electrons and the formation of new quasi-particles—phasons and fluctuons.

For first-order phase transitions, the heterophase fluctuation (nucleation center of the β phase, arising in the equilibrium phase α at the given temperature) increases the free energy of the system due to the difference in densities of the free energies of both phases φ and the surface energy of the phase interface. With the transition of electrons from the conduction band of the α phase to the heterophase fluctuation level and their localization at this level, the electron energy decreases to the activation energy of this level u. If this decrease in free energy of the electron subsystem compensates the increase of free energy related to the formation of the nucleation center of the β phase, then this nucleation center (phason) becomes stable. If the self-consistent radius of the phason R_0 at the specified temperature (and consequently the specified value of the difference in densities of the free energies of both phases φ) and specified value of the surface energy density of the interphase boundary σ is greater than the critical radius of the nucleation center of the β phase, then the phase transition $\alpha \to \beta$ occurs. As was shown in [251], if the free energy of the system depends nonlinearly on the appropriate internal parameter (in the particular case of ferroelectrics, on the spontaneous polarization) and, moreover, if $u/kT \gg 1$, then R_0 is much greater than the lattice constant and the phason is accordingly a macroscopic quasi-particle. It follows from the above discussion that the basis of [251] is the idea of the existence of a contribution of the electron free energy to the free energy of the lattice near the phase-transition temperature (cf. Chapter 1, and also [25]). The phason model permits predicting the existence of two groups of phenomena in the most general case.

First, this model leads to the effect of electrons and, in particular, non-equilibrium electrons on the first-order phase transition. Thus, the formation of a phason is significantly simplified with an increase in the number of localized electrons. It was shown in [251] that $R_0 \sim n^{2/3}$, where n is the number of localized electrons. An increase in concentration of nonequilibrium electrons in the conduction band of the α phase aids the formation of nucleation centers of critical size at larger values of φ, i.e., it leads to a photostimulated shift of the temperature of the phase transition $\alpha \to \beta$. The same mechanism also leads to a decrease of the temperature hysteresis (photohysteresis effect).

Second, the phason model predicts a change of electron properties of the crystal in the region of the phase transition. According to this model, new localized electronic states corresponding to the energy level of the phason in the forbidden band of the α phase arise in the vicinity of the phase-transition temperature in the crystal. New levels can lead to anomalies of the electrical conductivity and photoconductivity and significantly affect the kinetics of the processes of recombination and trapping of nonequilibrium electrons in the region of the phase-transition temperature. In other cases, the phasons can be considered as a nondegenerate gas of moving quasiparticles with an effective mass M giving a contribution to the transport phenomenon. In all cases, the temperature region of existence of phasons near the phase-transition temperature is wider the smaller the latent heat of transition.

Following [251], we evaluate the concentration of phasons near the first-order nonferroelectric phase transition. We consider the region of the α phase containing a spherical nucleation center of the β phase, having captured one electron. The free energy ΔF of this region has the form

$$\Delta F(R) = \frac{4\pi}{3} R^3 \varphi + 4\pi R^2 \sigma + E(R), \tag{5.21}$$

where $E(R)$ is the energy of the ground state of the electron in the spherical potential well (m is the mass of the free electron):

$$E(R) = -u + \frac{\hbar^2 k^2}{2m}, \quad k \cot kR = -\sqrt{\frac{2mu}{\hbar^2} - k^2}. \tag{5.22}$$

The equilibrium self-consistent values of the phason radius $R = R_0$ and activation energy u_0 can be determined from (5.21) and (5.22) and the conditions $\Delta F(R_0) = 0$ and $\partial \Delta F / \partial R = 0$. The following values were obtained in [251]:

$$u_0 = 11.5 \hbar (\sigma/m)^{1/2}, \quad R_0 = 0.6 (\hbar^2/m\sigma)^{1/4} \quad \text{for} \quad \varphi R_0 \ll \sigma,$$

$$u_0 = 5.81 (\hbar^2/m\varphi)^{3/5} \varphi, \quad R_0 = 0.73 (\hbar^2/m\varphi)^{1/5} \quad \text{for} \quad \varphi R_0 \gg \sigma \tag{5.23}$$

and numerical evaluations were carried out. According to (5.23), $u \simeq 1$ eV and $2R_0 \simeq 10$ Å for $\sigma \simeq 20$ ergs/cm^2 and $\varphi = 0$. If $\sigma \ll 10^2$ ergs/cm^2, then the order of the magnitude of the values u_0 and $2R_0$ remains the same far from the phase-transition temperature (for example, for $\varphi \simeq 10^9$ erg/cm^3). Thus the estimates of R_0 show that the phason is a macroscopic quasi-particle. Corrections to the effective mass of the electron leads to an increase in the equilibrium values of R_0.

To evaluate the concentration of phasons, Krivoglaz [251] starts with the model of localized and mobile quasi-particles. The free energy N_{ph} of the mobile phasons in the nondegenerate-gas approximation has the form

$$F = N_{ph} \Delta F(R) - N_{ph} kT \ln\left[e \frac{N}{N_{ph}} v \left(\frac{MkT}{2\pi\hbar^2}\right)^{3/2}\right]. \quad (5.24)$$

Here, v is the volume of the unit cell of the α phase, and N is the number of cells. The free energy N_{ph} of localized phasons takes into account the configuration contribution related to the permutation of N_{ph} phasons over N cells of the α phase:

$$F = N_{ph} \Delta F(R) - kT \ln\left(e \frac{N}{N_{ph}}\right). \quad (5.25)$$

We have for the nondegenerate electron gas in the α phase, according to (1.70),

$$F = -N_e kT \ln\left[e \frac{N}{N_e} v \left(\frac{mkT}{2\pi\hbar^2}\right)^{3/2}\right]. \quad (5.26)$$

Here, N_e is the concentration of free electrons in the α phase. Since the formation of phasons must not change the free energy at equilibrium, then, by equation (5.26) with (5.24) or (5.25), we obtain the following expression for the concentration of phasons:

$$N_{ph} = N_e \delta \exp(-\Delta F/kT), \quad (5.27)$$

where $\delta = (M/m)^{3/2}$ for the quasi-particles and $\delta = (2\pi\hbar^2/v^{2/3} mkT)^{3/2}$ for the localized phasons ($\delta \gg 1$). It is shown in [251] that for the numerical parameters presented above, $\Delta F(R_0) \lesssim 10\,kT$. Thus, according to (5.27) in the region of the $\alpha \to \beta$ phase transition, a large number of electrons transfer to the phason level or to the mobile phason state. Thus, the phason model assumed a significant effect of the first-order phase transition on the electrical conductivity and even a change in the nature of the electrical conductivity in the region of the phase transition.

According to the definition contained in [251], the fluctuon is a macroscopic quasi-particle formed in the region of phase transitions of the order–disorder type and caused by fluctuations of the order parameter (for example, the magnetization) and its stabilization with capture of electrons. The fluctuon model also leads to the photostimulated phase transition and the corresponding effect of the phase transition on the electronic properties of the material.

The possibility of the existence of photostimulated phase transitions was predicted in 1966 in independent works [25, 253], as a result of which Fridkin and Galashin [25, 253] not only established the significant role played by the electron subsystem in the thermodynamics of phase transitions but also proposed possible microscopic mechanisms for the photostimulated phase transitions for specific systems (in [25], in particular, for ferroelectric phase transitions). The phason and fluctuon models [251], as was shown above, make it possible to consider from a unified point of view the effect of light on the phase equilibrium, including photosensitive phase transitions in a solid, the phenomena of photoassociation and photocondensation, photostimulated nucleation formation, and certain other related phenomena. Thus, in concluding this chapter, we present a brief survey of these phenomena—photostimulated phase transitions in nonferroelectric materials.

The phenomenon of vapor condensation in a beam of light was described by Tyndall in 1869 [254]. The author expressed doubt about the possibility of explaining it on the basis of an assumption about some sort of irreversible photochemical reactions occurring in the system. The possibility of formation of neutral condensation centers of water vapor with illumination of the vapor in the near ultraviolet was discovered later [255]. In this connection, Wilson [255] proposed that the formation of fog drops in a beam of light occurs, thanks to a decrease in surface tension of water as a result of photochemical reaction forming traces of hydrogen peroxide. This point of view of the mechanism of photocondensation of water vapor was developed in [256]. In later investigations of the photocondensation phenomenon, the decisive effect of traces of contaminants (ammonia, hydrogen sulfide, sulfur dioxide) was confirmed and the possibility of optical sensitization of the process was thereby shown. The sensitization of the effect by ammonia molecules led a number of authors to the assumption that the photocondensation of water vapor is related to electron excitation of the ammonia molecules. The increase in the dipole moment with excitation leads to association of the water molecules with the ammonia molecules. The fog drops are formed as a result of collision of the associates. The possibility of forming neutral condensation centers in vapors of various liquids under the effect of photoactive illumination was confirmed in the works of Galashin [253, 257], who observed the formation of fog drops in vapors of benzene, bromobenzene, carbon tetrachloride, and other substances. No traces of any sort of photochemical reactions were observed in the system, which was the basis in turn for the author to reach the conclusion that the general cause of photocondensation is an increase in the energy of the intermolecular interaction with electron excitation of the molecules [258]. It was thereby reliably established that condensation centers playing the role of nucleation centers of critical size, on which begins the condensation and growth of drops of the liquid phase, arise with illumination of vapors of many substances in the region of their natural absorption or in the region of absorption of a sensitizer. The works of Galashin stimulated observation of analogous effects in other systems, for example, with illumination of cooled alcohol solutions of dyes. A change in the fluorescence spectrum was observed in solutions of pyrene with their excitation in the near ultraviolet [259]. Förster and Kasper [259] were led to the conclusion that the change in the fluorescence spectrum is caused by the formation of dimers of unexcited and excited molecules of pyrene. The formation of excited associates was also observed in a number of other investigations.

The effect of increasing the melting temperature of the photoisomer of anthracene under the effect of electron excitations was first noted in [260]. The assumption was expressed that the generation of excitations and the formation of excited associates leads to an increase in the energy of the crystal lattice and, as a consequence, to an increase in the melting temperature of the crystals. The shift in the melting temperature of anthracene and its derivatives depends, according to [260], on the intensity of the "natural" illumination and can be tens of degrees.

A number of works reported on the phenomena of photocrystallization or an increase in the crystal growth rate under irradiation by photons. It was noted that the growth rate of certain crystals, for example, selenium, increases with illumination [261]. The contraction of the induction period of ice formation under the effect of ultraviolet radiation was observed [262]. The phenomenon of photocrystallization of amorphous selenium was described in [263]. The effect of photocrystallization can be particularly clearly observed with illumination of solutions and melts of certain aromatic hydrocarbons with photoactive light [257]. Galashin, in particular, showed that for a supercooled melt of pyrene-anthracene eutectic, the effect of photocrystallization is not related to the formation of fused crystals of dianthracene and that the mechanism of this phenomenon is somehow or other caused by electron excitation of part of the molecules [258]. It was established in [264] that the change in the nature of oriented crystallization of anthraquinone on the surface of single crystals of germanium occurs under the effect of light. Both the density of crystals and their orientation change with illumination of the semiconductor in the spectral region of its natural photosensitivity. The authors called this phenomenon photoepitaxy. The experimental data lead to the conclusion that the basis of photoepitaxy, as well as photocrystallization in general, is the mechanism of electron excitations and, in particular, the generation of nonequilibrium current carriers by light.

Photosensitive phase transitions, besides those of ferroelectric semiconductors described in this chapter, are observed in a large number of crystals and solids. One can recall here the ferromagnet $Fe_{1-x}S$ [265], ferromagnetic semiconductors of the type of chalcogenides and europium oxides [266], certain vanadium oxides, and lead oxides. However, the transition from the orthorhombic to the tetragonal phase of PbO can be considered pyroelectric according to a series of data [267]. The photostimulated nature of this phase transition (shift of the phase-transition temperature with illumination) was established in [267].

6

Screening Phenomena

A number of photoferroelectric phenomena are caused by screening of spontaneous polarization by nonequilibrium carriers. Moreover, as has been emphasized in Chapter 3, screening of spontaneous polarization by equilibrium or nonequilibrium carriers determines the basic ferroelectric and semiconductor properties of the crystal. We will discuss some of these phenomena in this chapter. The group of screening phenomena related to the formation of stable electret polarization in ferroelectrics (photoelectret polarization) will be treated separately in Chapter 7.

6.1. Experimental Observation of the "Intrinsic" Field Effect

Both for the field effect at the free surface of a ferroelectric and for the field effect for contact of a ferroelectric with a semiconductor, the theory predicts the formation of degenerate electron–hole layers only if the surface levels do not significantly affect the screening (the "intrinsic" field effect, cf. Section 3.3). The field effect in ferroelectrics was investigated in a number of works [268-271]. Here we will discuss mainly two works [268, 269].

The "intrinsic" field effect for single crystals of $BaTiO_3$ was investigated in [269]. The crystals were reduced by annealing in a hydrogen atmosphere at 650°C, which somewhat decreased its surface resistance. Polarization of the crystal in two opposite directions led to a change in resistance along the free surface (001) by five times at most. From a comparison of the surface resistance with the direction of spontaneous polarization, Würfel and Ruppel [269] concluded that the change in direction of P_0 does not change the n-type conductivity and there is no degeneracy in the surface layer. Würfel and Ruppel [269] arrived at an analogous conclusion by investigating the "intrinsic" field effect for contact of the $BaTiO_3$ crystals with deposited layers of CdS. The ferroelectric polarization reversal of $BaTiO_3$ in the c-direction did not change the n-type conductivity of the semiconductor. The concentration of surface levels $N \simeq 10^{13}\text{-}10^{14}$ cm$^{-2} \cdot$ eV^{-1} in CdS was determined from the change in concentration of carriers in CdS corresponding to the two opposite directions of P_0 in $BaTiO_3$. An analogous result was obtained in [270], where the electrical conductivity of a thin layer of n-type germanium on a $BaTiO_3$ crystal was measured depending on the direction of spontaneous polarization in the ferroelectric. Since the change in electrical conductivity was no more than 10-20% and the n-type electrical conductivity did not change, Bogatko and Kovtonyuk [270] concluded that the spontaneous polarization is screened mainly by surface levels. Such an insignificant field effect was observed in [271] for tellurium films deposited on crystals of triglycine sulfate.

A significant field effect was evidently first observed in [268] for contact of the ferroelectric ceramic PZT (lead and zirconium titanate) with a tin oxide film having n-type conductivity in the range of values 10-100 $(\Omega \cdot cm)^{-1}$. The change in the direction of the spontaneous polarization in the ferroelectric ceramic led to the formation in the tin oxide of screening layers, respectively enriched and depleted in electrons, where the enriched layer displayed degeneracy. Figure 6.1 shows the surface current in the semiconductor as a function of the ferroelectric polarization at various temperatures. The direction of polarization was taken so that its increase leads to the conversion of the layer enriched with electrons to a depleted layer (in the interval of polarizations 70-80 $\mu C/cm^2$). In [268], the change in electrical conductivity was recorded during observation of the ferroelectric hysteresis loop. Figure 6.2 presents the temperature dependences of the surface current I in the semiconductor and the corresponding electron concentrations n for the electron-enriched layer and three states of the depleted layer corresponding to the different values of ferroelectric polarization. The electron concentration was determined for the mobility $\mu \simeq 2$ cm$^2 \cdot$ (V $\cdot$ sec)$^{-1}$.

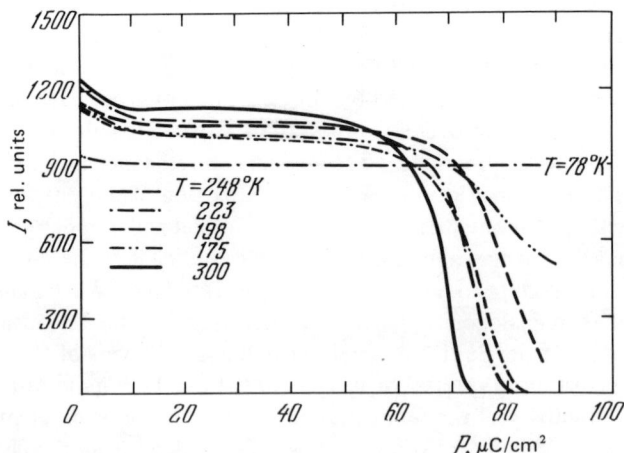

Figure 6.1. Effect of the field in the tin oxide–PZT ceramic system for various temperatures and points of the dielectric hysteresis loop [268].

SCREENING PHENOMENA

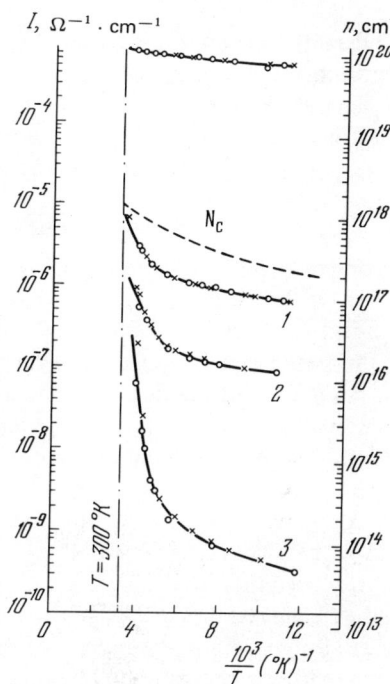

Figure 6.2. Temperature dependence of the field effect in the tin oxide-PZT ceramic system for the enriched layer and three states of the depleted layer [268].

The data of Figure 6.2 show that the transition from the enriched to the depleted layer at room temperature is related to the removal of $\sim 1.2 \cdot 10^{20}$ electrons per cubic centimeter, which corresponds to a screening charge density ~ 38 $\mu C/cm^2$. If one keeps in mind that $P_0 \simeq 74$ $\mu C/cm^2$, then 51% of the charge screening the spontaneous polarization is localized in the volume of the semiconductor. Of course, this calculation does not take into account the possible effect of the field on the carrier mobility in the tin oxide. One can determine from the temperature dependences of Figure 6.2 the position of the Fermi level in the semiconductor corresponding to the different sections of the ferroelectric hysteresis loop. If the electron-enriched layer is in the degenerate state, where the Fermi level at $(2-3)kT$ is above the bottom of the conduction band of the semiconductor, then, according to (3.49), the electron concentration n depends on the energy of the Fermi level E_F in the following manner:

$$n = \frac{8\pi}{\hbar^3}[2m^*(E_F - E_c)]^{3/2}. \quad (6.1)$$

Accordingly, if the depleted layer is in the nondegenerate state, then

$$n = N_c \exp\left[-\frac{E_c - E_F}{kT}\right], \quad N_c = 2\left(\frac{2\pi m^* kT}{\hbar^2}\right)^{3/2}. \quad (6.2)$$

The value of the effective mass $m^* = 0.15 m_0$ was used in [268] to calculate the density of states $N_c = 2.8 \cdot 10^{14} T^{3/2}$. [The dependence $N_c = N_c(T)$ is also shown in Figure 6.2.]

These results lead to the following conclusions. First, the concentration of electrons 1.2×10^{20} cm^{-3} corresponding to the enriched state of the layer corresponds to a degenerate electron gas since $N_c \simeq 1.5 \cdot 10^{18}$ cm^{-3} at room temperature. The Fermi level in the electron-enriched layer of the semiconductor lies 0.6 eV above the bottom of the conduction band. Second, the temperature dependences $n = n(T)$ corresponding to the enriched and depleted layers and presented in Figure 6.2 satisfy (6.1) and (6.2). The Fermi level in the depleted nondegenerate layer is located at 0.057, 0.09, and 0.28 eV below the bottom of the conduction band (for the three curves of Figure 6.2, respectively). Thus, the data of [268] confirm the conclusion of the theory (cf. Section 3.3) that the ferroelectric field effect can be related to the formation of a degenerate screening layer in the semiconductor if the surface levels do not play a significant role.

A photoelectric probing of the screening layer arising in the semiconductor CdS with its contact with a BaTiO$_3$ crystal was carried out in [269]. As has already been indicated above, a screening space charge much less than P_0 arises in the CdS layers on the (001) face of the barium titanate crystal. This screening charge leads to surface curvature of the bands in CdS, whose sign is determined by the direction of the spontaneous polarization in the ferroelectric. With illumination of the BaTiO$_3$-CdS contact, nonequilibrium carriers drifting in the field of the screening space charge arise in the CdS semiconductor. The transient signal of the photocurrent then arises, whose sign is determined by the direction of the field in the screening layer. Thus, the signs of the photoresponses are opposite for the electron-enriched and depleted screening layers. The measurement circuit is presented in Figure 6.3. Illumination of the contact was carried out through a semitransparent gold electrode and the BaTiO$_3$ crystal, which is transparent in the region of the natural photosensitivity of the CdS. Figure 6.4 presents the spectral distribution of the photoresponse for the electron-depleted (a) and enriched (b) layers. The signs of the photoresponses are opposite in the cases a and b, while their spectral distribution corresponds

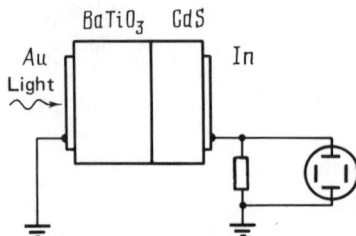

Figure 6.3. Schematic for measurement of the photoresponse with illumination of the BaTiO$_3$-CdS contact [269].

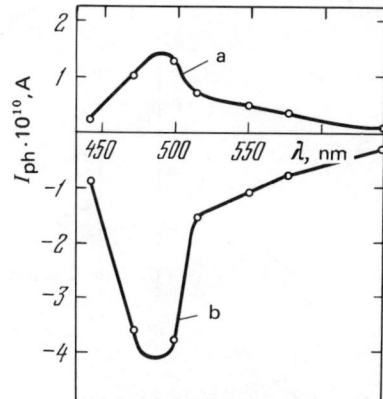

Figure 6.4. Spectral distribution of the photoresponse for the electron-depleted (a) and enriched (b) layers [269].

to the spectral distribution of the photocurrent in CdS. As is noted in [269], the pyroelectric currents caused by heating of the ferroelectric and by the photodomain effect (cf. Section 6.6) affect the kinetics of the photoresponse.

6.2. Electroluminescence of Ferroelectrics

As was shown by Guro, Ivanchik, and Kovtonyuk [272], electroluminescence is another possible method for observing screening layers and charges. The idea of this method involves the fact that polarization reversal of a ferroelectric in a variable electric field must be accompanied by recombination of the screening charges. If this recombination is radiative in nature, then the polarization reversal of a ferroelectric must be accompanied by electroluminescence. Thus, this phenomenon is analogous to the previously observed luminescence of electrets [273]. In the case of an electret, the external depolarizing field also leads to recombination of the electrons and holes forming the space charge.

The electroluminescence of $BaTiO_3$ in a variable electric field was first observed by Harman [274]. The author observed emission of $BaTiO_3$ crystals in a field with intensity ~ 1000 V/cm and frequency ~ 500 kHz over a wide temperature interval, including the paraelectric region (up to 300°C). The intensity of the electroluminescence had a maximum at the Curie point, and Harman [274] proposed that its temperature dependence is determined by the Curie-Weiss law (1.28). This fact led in turn to the proposal that the emission of $BaTiO_3$ is caused by electroluminescence of the surface layers (dielectric "gaps" [15]). Actually, if one assumes that the dielectric constant of the surface layer ϵ_l is small ($\epsilon_l \ll \epsilon_s$) and does not depend on temperature, while the dielectric constant of the volume of the ferroelectric ϵ_s satisfies the Curie-

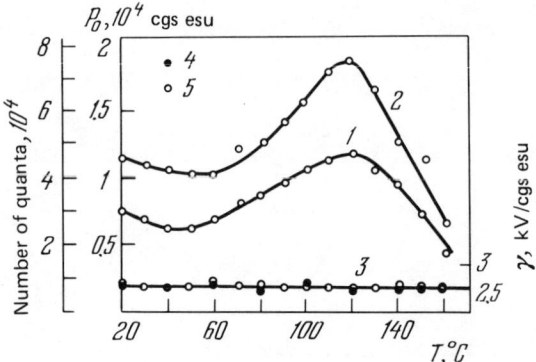

Figure 6.5. Temperature dependence of the radiation energy (1) and spontaneous polarization (2). (3) Ratio of radiation energy to spontaneous polarization; (4) measured with decreasing temperature; (5) measured with increasing temperature [272].

Weiss law, then near the Curie point a large fraction of the external high-frequency voltage falls at the surface layer, enhancing its electroluminescence.

The results obtained in [272] speak in favor of the mechanism based on the recombination of screening charges with polarization reversal. Single crystals of $BaTiO_3$ with Cd and Ca impurity with contacts of InGa were investigated. Electroluminescence was observed with delivery to the crystal of bipolar rectangular pulses of amplitude 250 V (the thickness of the crystal was ~ 0.1-1.4 mm). Polarization reversal of the crystal, monitored by observing the polarization reversal current, led to emission for sufficiently steep pulse fronts (front width less than 10^{-4} sec). For a polarization reversal time of approximately 10^{-4} sec, the emission time was 10^{-3} sec. A parallel investigation of the polarization-reversal kinetics and the electroluminescence showed that the beginning of the emission pulse corresponds to the decay of the depolarization current to 0.7 of the maximum value. The temperature dependence of the electroluminescence correlates with the temperature dependence of the spontaneous polarization, where the ratio γ of the emission energy to the spontaneous polarization remains constant for all temperatures (cf. Figure 6.5). The electro-

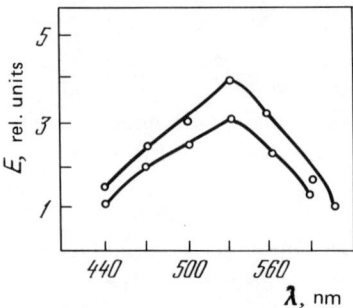

Figure 6.6. Emission spectrum of $BaTiO_3$ [272] for two values of the field.

luminescence is absent in the paraelectric region. Although the values of spontaneous polarization and emission energy display temperature hysteresis near the first-order phase transition in $BaTiO_3$, the hysteresis does not affect γ. Figure 6.6 presents the spectral distribution of the electroluminescence energy of $BaTiO_3$, showing that the emission is related to band-level transitions. The electroluminescence mechanism proposed in [272] assumes that the concentration of recombining pairs is proportional to the screening charge and, consequently, to the spontaneous polarization, $N \simeq P_0/q$, while the intensity of the electroluminescence is proportional to N. We are thereby led to the conclusion of the temperature independence of γ and the absence of electroluminescence in the paraelectric region, which was observed experimentally by Guro et al. [272].

These results were confirmed to a significant degree by Godefroy and his colleagues [275]. However, it was shown that the temperature dependence of the electroluminescence depends significantly on the form of the electric signals, and accordingly on the kinetics of ferroelectric polarization reversal of $BaTiO_3$. With polarization reversal under the effect of a sinusoidal alternating field with a frequency of 2 kHz, a "slow" luminescence is observed with a duration of the light pulses of $\sim 10^{-3}$ sec, as in [272]. In this mode, the temperature dependence of the electroluminescence coincides with the temperature dependence of the spontaneous polarization P_0, where γ does not depend on temperature, which confirms the conclusions of [272]. Short light pulses $\sim 10^{-4}$ sec were observed with excitation of barium titanate by bipolar rectangular electric pulses with frequency of 50 Hz [275]. The temperature dependence of the electroluminescence did not correlate in this case with $P_0 = P_0(T)$, and a maximum was detected at the Curie point and observed in the paraelectric region. Accordingly, the kinetics of the ferroelectric polarization reversal differed in these two cases. Analogous results were obtained [275] for tryglycine sulfate.

The aggregate of all these data permits one to reach at least two conclusions. First, the electroluminescence mechanism of ferroelectrics is related to recombination of electron–hole pairs screening the spontaneous polarization, which does not exclude the appearance of other mechanisms, for example, a mechanism related to electroluminescence of the surface layers. Second, the relation between the electroluminescence and the kinetics of polarization reversal of a ferroelectric indicates the essential effect which the screening charges have on the ferroelectric polarization reversal. Section 6.5 of this chapter will be devoted to this last problem.

6.3. Opposing Domains in SbSI

As was already shown in Section 3.4, opposing ferroelectric domains can arise in a ferroelectric semiconductor under the effect of an injecting contact field.

Their formation must lead to certain peculiarities involving the nonmonotonic distribution of the potential and discontinuity of the electrical conductivity at the boundary between the opposing domains. The latter is related to the formation of a screening layer along the boundary of the opposing domains, in which degeneracy of the electron gas occurs. Opposing domains have been observed experimentally in PbTiO$_3$ [276] and SbSI [227]. We will discuss here the results obtained in [277] for SbSI.

Measurements of the potential distribution with the help of electric probes performed in [277], as well as by other authors [278, 279], on single crystals of SbSI did not exhibit any sort of peculiarities. At the same time, it follows from scanning electron microscope measurements of the potential distribution for SbSI [87] and dielectric hysteresis loops for solid solutions SbSI$_x$Br$_{1-x}$ (Chapter 7) that a nonequilibrium distribution of the spontaneous polarization can be realized in a uniaxial ferroelectric semiconductor.

An attempt was undertaken in [277] to form opposing domains in a single crystal of SbSI under conditions where the latter was abruptly converted from the paraelectric phase to the ferroelectric by rapid cooling from the ends. In this case, the boundaries of the phases moved from the ends of the sample toward the middle, where they opposed each other. The state thus obtained in the case of the absence of an external electric field will be called the state of natural polarization. The formation of opposing domains in the natural polarization mode can be explained by the existence of the surface fields oppositely directed in the paraelectric region near the end faces of the SbSI. These fields can be related both to the transfer of carriers from the electrodes into the crystal (the mechanism considered in Section 3.4), as well as to the presence of charged surface states.

Along with the observation of opposing domains in the natural polarization mode in [277], the formation of opposing domains was realized by the opposing field method. The idea of this method is understood from the diagram presented in Figure 6.7. Two pairs of electrodes are applied to the crystal. One pair of electrodes (1 and 2) is applied at the ends of the crystal, while the other pair of point electrodes (3 and 4) is applied to the (110) faces at the interface, i.e., perpendicular to the boundary of the opposing domains. Aquadag was used as the electrode material. The contacts (3 and 4) were used to measure the electrical conductivity perpendicular to the ferroelectric axis

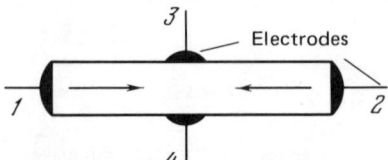

Figure 6.7. Method of forming opposing domains in SbSI.

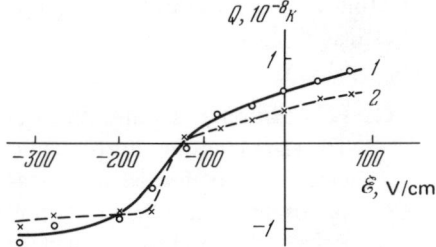

Figure 6.8. Dependence of the pyroelectric charge on the electric field intensity. (1, 2) Left and right halves of the crystal, respectively.

in the assumed ferroelectric polarization screening layer. The installation also permitted the recording of the pyrocurrent between any pair of electrodes. It was established that the pyroelectric charges of both halves of the naturally polarized sample are of one sign (Figure 6.8). An external electric field applied under these conditions to the entire sample in the ferroelectric phase increases the polarization of one half of the crystal, while decreasing the polarization of the other half. The opposing field method involved short-circuiting electrodes 3 and 4 and then applying oppositely adjusted fields (opposing fields) to the electrodes 1 and 3 and to 2 and 3. The dependence of the pyroelectric charge on the external opposing field is presented in Figure 6.8. If the direction of the external field coincided with the direction of natural polarization, then the pyroelectric charge increased, whereas if it was opposite, then the latter decreased, changing sign for a field intensity of ~ 120 V/cm, which corresponds to the value of the coercive field for SbSI. Of course, this also changes the sign of the ferroelectric polarization screening charges.

The application of a measuring field to the electrodes 3 and 4 permitted the measurement of the electrical conductivity in the screening layer $\sigma_\perp$. The first measurements by the opposing field method have already shown that the transverse electrical conductivity $\sigma_\perp$ increases by one to two orders of magnitude with the formation of opposing domains (external field intensity of ~ 400 V/cm). The increase of $\sigma_\perp$ was greater in the case where the direction of the opposing field coincided with the direction of polarization of the naturally polarized crystal. We also note that failure of the crystals near the electrodes 3 and 4 was observed in a number of cases after a single measurement. The latter is possible both because of the high current densities and because of the significant mechanical stresses in the region of the screening layer.

We will discuss in somewhat more detail the results of measurement of $\sigma_\perp$ depending on the conditions under which the opposing domains were produced in SbSI. Before polarization, the sample was annealed at the temperature of 50-55°C for 10-15 h. An opposing field of ~ 1.5 kV/cm was used, which was applied to the short-circuited pair of middle electrodes and end electrodes. The polarization was removed discontinuously with the lapse of time.

The following conditions for the formation of opposing domains were investigated:

1. Polarization in darkness in an opposing field in the ferroelectric phase. The sample in the short-circuited state was converted from the paraelectric to the ferroelectric phase and cooled to 10°C, after which the opposing electric field was applied in darkness.
2. Polarization in darkness in an opposing field near the phase transition. The sample was cooled to temperatures exceeding the phase-transition temperature by 1-2°, the opposing electric field was applied at this temperature, and further cooling was carried out to 10°C.
3. Polarization with illumination in an opposing field in the ferroelectric phase. The sample was cooled to 10°C in the short-circuited state in darkness, the opposing electric field was applied at this temperature, and the crystal was simultaneously illuminated in the region of natural photosensitivity. The illumination was cut off after a specified time and the sample was short circuited.
4. Polarization in darkness in an opposing field in the paraelectric phase. The opposing electric field without illumination of the sample was applied at 50-60°C.

Formation of opposing domains occurred in the first three cases, i.e., with the exception of polarization in the paraelectric region. A screening layer, whose electrical conductivity $\sigma_\perp$ increased by one to two orders of magnitude (as compared to the electrical conductivity of the same layer before polarization of the crystal in the opposing field), was generated along the boundary of the opposing domains simultaneously in all three cases. The effect of the phase transition on the electrical conductivity $\sigma_\perp$ was investigated corresponding to the first three conditions for formation of opposing domains. Figure 6.9 presents the temperature dependences of the electrical conductivity $\sigma_\perp$ plotted during continuous heating at a rate of ~ 1 deg/min. These dependences are identical in nature to polarization without illumination: a broad maximum preceded by somewhat of a drop is observed at a temperature of 30-35°C. After holding at a temperature of 50-55°C for 30 min, the value of $\sigma_\perp$ returns to the initial value before polarization. A sharp maximum with an abrupt drop at the phase transition temperature is seen in the temperature dependence of $\sigma_\perp$ for polarization by the third method. The presence of the broad maximum in the temperature dependence of $\sigma_\perp$ at temperatures above the phase-transition temperature (Figure 6.9a and b) is related to the liberation of carriers, which form the screening space charge, from the traps. It is possible that in the case presented in Figure 6.9c, these carriers are localized in the volume at less deep levels or even at

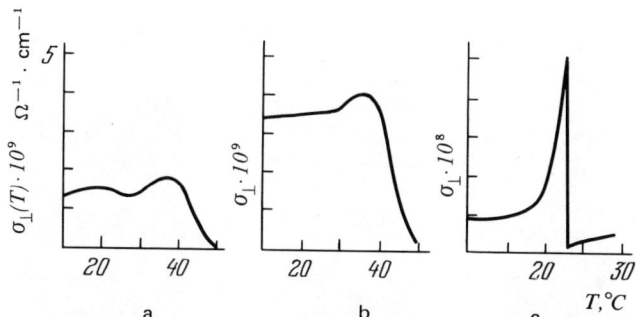

Figure 6.9. Temperature dependence of the electrical conductivity $\sigma_\perp$ along the boundary of the opposing domains for the three polarization modes (a, b, c correspond to methods 1, 2, 3, respectively).

surface levels coupled to the boundary between the opposing domains. This leads to the sharp maximum in the temperature dependence of $\sigma_\perp$ at the phase-transition temperature in SbSI.

A photoelectric probe of the screening layer at the boundary of the opposing domains was also carried out. The region of the boundary of the opposing domains was illuminated with "natural" light, and the photoresponse related to the drift of nonequilibrium carriers in the screening layer field was measured. The photoresponse signal is presented in Figure 6.10. Separate measurements of the photoresponse for an unpolarized crystal (without opposing domains) show that the photoresponse in Figure 6.10 is a superposition of the ordinary photo-emf (short-circuit photocurrent) on the photoresponse caused by the field of the opposing domains.

The formation of opposing domains and screening charge at their boundary can be observed with the help of the hysteresis loop. In the absense of opposing domains, SbSI crystals display saturated hysteresis loops of the usual form. With the formation of opposing domains, the hysteresis loops acquire the form of double loops, where the intensity of the variable field in which the hysteresis loops are observed increases significantly (Figure 6.11a). If a variable electric field sufficient for polarization reversal is applied to a crystal in which the

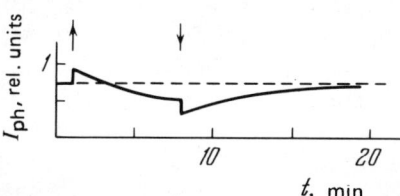

Figure 6.10. Photoresponse with illumination of the boundary of the opposing domains in SbSI. Arrow denotes time of light on and off.

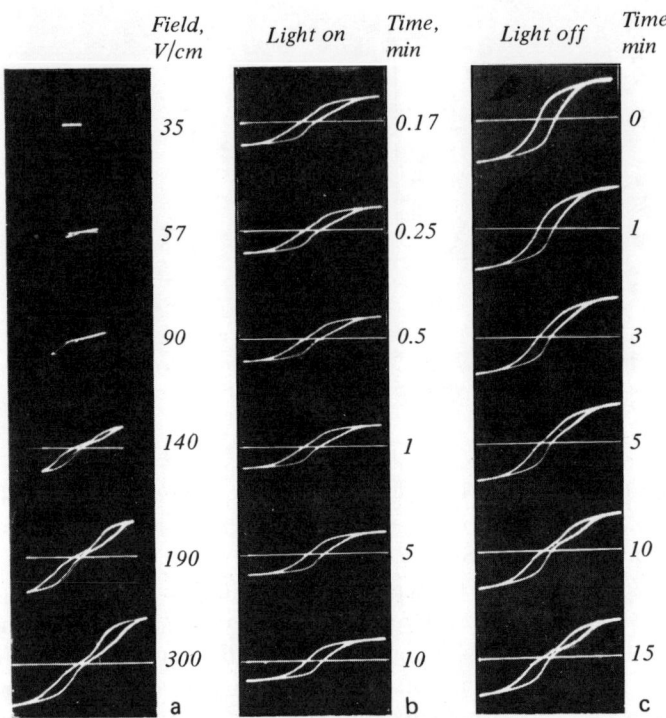

Figure 6.11. Structure of the opposing domains and the effect of "natural" illumination and variable field according to the dielectric hysteresis data.

opposing domains are formed and the crystal is simultaneously illuminated in the spectral region of its natural photosensitivity, then the double loops will transform in the course of some time to loops of the usual form, which indicates the disappearance of the opposing domain structure (Figure 6.11b). The time necessary for this is much greater than the corresponding Maxwellian time, since this process is determined not only by the liberation and dissipation of the screening charge with illumination, but also by the simultaneously occurring restructuring of the domain structure. With cutoff of the light after a time much greater than the Maxwellian time, the opposing domains are again generated, which is accompanied by the appearance of the double hysteresis loops (the variable field is applied only for the short time necessary for photographing the hysteresis loops). The process of the recovery of the opposing domain structure in darkness is presented in Figure 6.11c.

These observations permitted obtaining yet another mode for the formation of the opposing domain structure in SbSI. A variable electric field sufficient for polarization reversal of the ferroelectric was applied to the crystal in the ferroelectric phase and the crystal was simultaneously illuminated by "natural" light. The variable field and light were cut off after some time. The ferroelectric in this state has a mobile domain structure for some time, as a result of which the opposing domains are formed in an opposing field of lower intensity. In this mode, a number of crystals after polarization in an opposing field of ~ 400 V/cm displayed a sharp increase of $\sigma_\perp$, where the dependence of $\sigma_\perp$ on the direction of the opposing field was first successfully detected. Thus, for the direction of the opposing field coinciding with the direction of the natural polarization, the electrical conductivity in the screening layer $\sigma_\perp$ increased by three orders of magnitude [$\sim 10^{-6}$ ($\Omega \cdot$ cm)$^{-1}$, while for the opposite direction of the opposing field—by one order of magnitude [$\sim 10^{-8}$ ($\Omega \cdot$ cm)$^{-1}$]. One should keep in mind that this dependence of $\sigma_\perp$ on the direction of the opposing field in SbSI is most likely related not to the transition from the electron boundary to the hole (cf. Section 3.4), but rather to the effect of the direction of the opposing field on the ferroelectric polarization near the boundary between the opposing domains and on the structure of this boundary. Imperfection of this structure can also explain the absence of degeneracy in the screening layer at the boundary between the opposing domains, which is predicted by the theory. This explanation does not exclude the possible role of surface levels at the boundary between the opposing domains, which were not taken into account in Section 3.4.

In concluding this section, we recall again one method for observing opposing domains in SbSI and PbTiO$_3$.† The strong field in the screening layer is distinguished in the form of a light band on the dark background of the crystal. The effective thickness of the screening layer in SbSI estimated by this method is $\sim 10^{-4}$ cm.

6.4. Effect of Nonequilibrium Carriers on the Screening of Interphase Boundaries in SbSI

In ferroelectric semiconductors displaying a first-order phase transition, such as BaTiO$_3$ and SbSI, the coexistence of both phases is observed in some temperature interval near the Curie point, where the sections of the ferroelectric phase are single-domain [204, 280-282]. The spontaneous polarization is then screened near the phase interface by free carriers or carriers bound at trapping levels. The problem about the screening of spontaneous polarization at the phase interface in a ferroelectric semiconductor and about the formation of a periodic structure of the interphase boundaries was considered in Section 3.9. We will present experimental data in this section about the effect of nonequilibrium carriers on the formation of such a structure in SbSI crystals.

†This method was first proposed for PbTiO$_3$ crystals by M. A. Malitskaya.

According to [204], under conditions where strong screening occurs and the screening energy W_2 is less than the elastic energy of interphase boundary W_1, plane boundaries are formed between single-domain sections of the ferroelectric and paraelectric phases. The elastic energy density of the interphase boundary for perovskite ferroelectric semiconductors is $W_1 \simeq 30$ ergs/cm^2 [204]. Under conditions where the screening energy density is much less than this value, plane phase interfaces oriented along (110) are formed in the semiconductor ferroelectrics BaTiO$_3$ and KTa$_{0.65}$Nb$_{0.35}$O$_3$ in the region of the transition from the cubic phase to the tetragonal. Orientation of the boundaries is determined by the anisotropy of the elastic stresses and, according to [204], can be calculated on the basis of the theory of Wechsler, Lieberman, and Read [283]. Since the screening energy decreases with increasing carrier concentration according to (5.5), it becomes clear why plane interphase boundaries were observed only for low-resistance perovskite ferroelectrics ($\rho < 10^3$ $\Omega \cdot$cm, $n > 10^{16}$ cm^{-3}) [204]. This also explains why these boundaries were not observed in the region of the phase transition for high-resistance dielectric crystals of BaTiO$_3$.

Investigation of the ferroelectric SbSI is of particular interest in connection with the study of the effect of the screening energy on the formation of interphase boundaries. This is related to the presence in SbSI of nonequilibrium current carriers and to the possibility of varying the concentration of free carriers over wide limits with illumination of the crystal in the region of the natural photosensitivity. A number of works [203, 284, 285] have been devoted to investigation of the effect of nonequilibrium carriers on the formation of interphase boundaries in SbSI. It is possible, for single crystals of SbSI in particular, to investigate the conditions for generation of interphase boundaries near the Curie temperature depending on the carrier concentration at the trapping levels N. For low illumination intensities, N is small according to (5.7), the screening energy W_2 is correspondingly large, and plane interphase boundaries do not arise. With increasing light intensity, filling of the trapping levels increases, and for some values of N in accordance with (5.5), the screening energy becomes less than the elastic energy of the phase interface $W_2 \lesssim W_1$. The latter inequality, in accordance with [204], also should be considered as the condition for the generation of plane boundaries in the crystal between single-domain sections of the ferroelectric phase and sections of the paraelectric phase. A further increase in illumination intensity must not lead to a change in the observed system of interphase boundaries.

The transition to the paraelectric phase in SbSI is accompanied by an abrupt decrease in the width of the forbidden band and a correspondingly long-wavelength shift of the intrinsic absorption edge (cf. Chapter 4). The difference in absorption of both phases caused by this anomaly permits visual observation of the coexisting phases in transmitted light with a wavelength near the intrinsic absorption edge: light regions correspond to the ferroelectric phase, and dark regions correspond to the paraelectric phase [280]. The possibility of optical observation of coexisting phases in the region far from the intrinsic absorption edge is due to another mechanism. The observation of the structure here is related to the conditions of refraction and reflection at the interphase boundaries and crystal faces [286, 287].

Optical observation of the interphase boundaries was carried out in [203] with the help of the MIK-1 infrared microscope provided with an electro-optical converter. A silicon filter was used to observe the crystals in transmitted infrared light with wavelength $\lambda > 1.1$ μm. The temperature of the SbSI crystal was preselected so that the crystal displayed the ordinary structure of the interphase boundaries with additional "natural" illumination. The crystal, held for a long time in darkness, was thermally stabilized near the Curie temperature and was simultaneously observed in the microscope in transmitted infrared light. If the ordinary interphase boundaries did not arise, then additional "natural" illumination was implemented. Measurement of the photocurrent was carried out simultaneously at constant voltage applied in the [001] direction and the concentration of free carriers n was evaluated for an electrical mobility of $\mu \simeq 50$ cm^2/V · sec). This method was used to measure the value of n corresponding to the light intensity for which plane interphase boundaries appeared, oriented along the (101) planes. Measurements of the thermosimulated current were carried out to evaluate the corresponding value of N, with the help of which the values of the activation energy of the trapping levels u, the reduced density of states N_{cm}, and the concentration of trapping levels M were calculated according to the known technique discussed in Section 5.2. Evaluation of the value of N was carried out with equation (5.7). Single crystals of SbSI grown from the gaseous phase were investigated, for which the ordinary structure of the interphase boundaries was observed in transmitted visible light by the ordinary technique.

The investigated crystals can be divided into two groups according to the results obtained. The first group includes crystals not displaying the ordinary interphase boundary structure with observation in transmitted infrared light. The interphase boundaries for these crystals appear with additional "natural" illumination. The interphase boundaries for crystals of the second group were also observed without additional illumination. Parallel measurement of the electrical conductivity showed that these crystals are low-resistance crystals and, as will be shown below, the condition $W_2 \lesssim W_1$ is satisfied for them due to the high concentration of equilibrium carriers.

Figure 6.12 presents microphotographs of a crystal of the first group, taken for various light intensities and correspondingly various concentrations of nonequilibrium carriers. The microphotograph 6.12a, taken in transmitted infrared light without additional illumination, and the corresponding equilibrium concentration $n_0 \simeq 2 \cdot 10^8$ cm^{-3} indicate the absence of interphase boundaries. Plane interphase boundaries with the ordinary orientation for SbSI arise with addi-

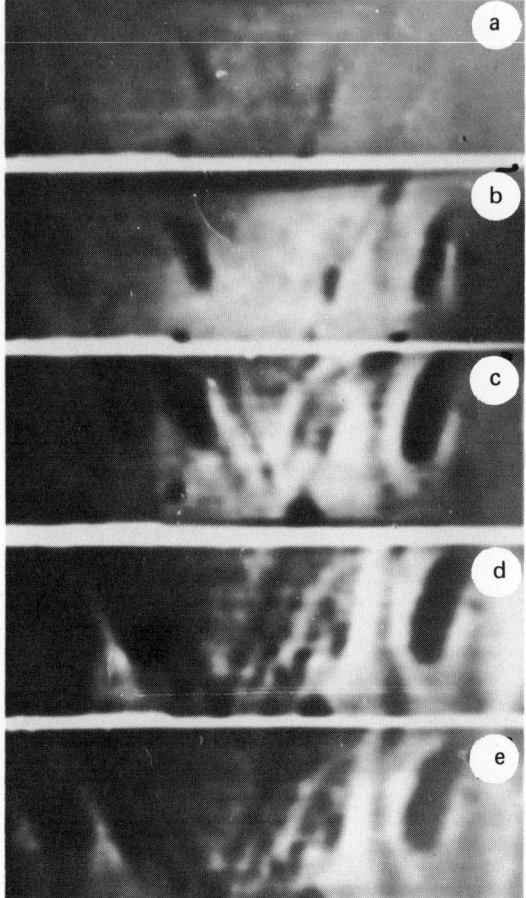

Figure 6.12. Microphotographs of a crystal of the first group taken for various light intensities. (a) Without additional illumination ($n_0 = 2 \cdot 10^8$ cm^{-3}); (b, c, d, e) with additional natural illumination ($n_1 = 4 \cdot 10^8$ cm^{-3}, $n_2 = 6 \cdot 10^8$ cm^{-3}, $n_3 = 8 \cdot 10^8$ cm^{-3}, $n_4 = 10^9$ cm^{-3}, respectively).

tional "natural" illumination (Figure 6.12c), which corresponds to a concentration of nonequilibrium carriers $n_2 \simeq 6 \cdot 10^8$ cm^{-3}. As is seen from Figure 6.12, a further increase in light intensity does not lead to a change in the observed structure. On the contrary, for a long holding time (for an hour) of the crystal in darkness, the interphase boundaries disappear and the initial pattern presented on the microphotograph 6.12a is re-established.

The thermostimulated current method was used for this crystal to measure $M \simeq 1.4 \cdot 10^{18}$ cm^{-3} and $u \simeq 0.53$ eV (corresponding to $N_{cm} \simeq 8 \cdot 10^9$ cm^{-3}).

The concentration $N \simeq 10^{17}$ cm^{-3} corresponding to the appearance of interphase boundaries was determined by substituting these values and also $n \simeq 6 \cdot 10^8$ cm^{-3} into (5.7). To evaluate the screening energy W_2 from equation (5.5), one should measure independently the spontaneous polarization P_0 and the dielectric constant ϵ. The quantities P_0 and ϵ should be understood as parameters corresponding to the separate sections of the ferroelectric phase, and not the averaged macroscopic parameters of the crystal. Actually, the crystal is significantly inhomogeneous in the temperature region of phase coexistence. In this case, the temperature dependences of P_0 and ϵ measured for the crystal in the temperature region where it is inhomogenous does not characterize the behavior of one of the phases, but rather the change in the relative distribution of these phases in the crystal volume. Substituting the values $\epsilon \simeq 6 \cdot 10^3$, $P_0 \simeq 14$ μC/cm^2, and $\rho = Nq$ into (5.5) and using the condition $W_2 \lesssim W_1$, Groshik and Fridkin [203] found $W_1 \gtrsim W_2 \simeq 10$ ergs/cm^2 for SbSI. This calculated value of the elastic energy of the boundary W_1 agrees well in order of magnitude with the value of W_1 presented above for perovskite ferroelectrics [204]. It should be emphasized once again that the indeterminacy in evaluating the parameters P_0 and ϵ appearing in (5.5) is fundamental and the value of W_1 presented above is no more than an estimate of the order of magnitude. Evaluation of the screening length $l_D = P_0/\rho$ leads to the value $l_D \simeq 6 \cdot 10^{-4}$ cm. This value agrees satisfactorily with the screening length independently measured by the optical method in [204] for perovskite ferroelectric semiconductors and in [277] for SbSI at the boundary of opposing domains (cf. Section 6.3).

Crystals of the second group, for which the interphase boundaries were observed without additional "natural" illumination, were low-resistance crystals. The concentration of equilibrium carriers $n \simeq 3.6 \cdot 10^9$ cm^{-3} was measured for one of these crystals. The values $u \simeq 0.55$ eV, $N_{cm} \simeq 3.5 \cdot 10^9$ cm^{-3}, and $M \simeq 3 \cdot 10^{17}$ cm^{-3} were determined from thermostimulated current data for this same crystal, and the value $N \simeq 1.5 \cdot 10^{17}$ cm^{-3} was calculated from equation (5.7). Hence, it is seen that these crystals had a high carrier concentration at trapping levels in the absence of illumination, which fulfills the condition $W_2 \lesssim W_1$.

As was noted in Section 3.9, the kinetics of the formation of the periodic structure of interphase boundaries is determined by the time for establishing diffusion-drift equilibrium. This means that the possibility of experimental observation of the interphase boundaries in SbSI is determined not only by the screening conditions, but also by the rate of heating or cooling of the crystal at the transition from the ferroelectric to the paraelectric phase or with the reverse transition. The higher the heating (cooling) rate, the higher must be the concentration of nonequilibrium carriers (and, accordingly, the less the diffusion-drift time) required for the appearance of the structure of the interphase boundaries. This conclusion was experimentally verified for SbSI in [285].

Figure 6.13 presents microphotographs of a crystal heated at varying rates in the region of the phase transition. It is seen that the higher the heating rate, the higher the nonequilibrium conductivity required for the appearance of the periodic structure of the interphase boundaries.

The experimental data obtained in [203, 285] permit one to make some quantitative estimates and compare them with the theoretical values obtained in Section 3.9. Thus, according to (3.181), the period of the structure of the interphase boundaries equals the diffusion length $L_d = \sqrt{D\tau}$, which is assumed much greater than the Debye length l_D. If one assumes that for SbSI the lifetime of nonequilibrium electrons is $\tau \simeq 10^{-3}$ sec and their mobility is $\mu \simeq 50$ cm^2/ (V-sec) (cf. Chapter 8), then $L_d \simeq 3 \cdot 10^{-2}$ cm. If the diffusion length for holes is close to this same value, then the theory predicts a value of $\sim 10^{-2}$ cm for the period of the interphase structure, which agrees well with the experimental value. Moreover, the experimental value of the period of the structure is much greater than the screening length in SbSI. Substituting the value $\sigma = W_1 \simeq 10^2$ ergs/cm^2 and also the corresponding experimental values of ΔE_g and L_d into (3.195), we estimate the critical concentration $n_k \simeq 10^{17}$ cm^{-3} below which the periodic structure does not arise. As we see, the estimate from (3.195) leads to a value of the critical concentration agreeing in order of magnitude with the estimates presented above.

Such high values of the critical concentration confirm the assumption that the effect of illumination of the crystal in the region of its natural photosensi-

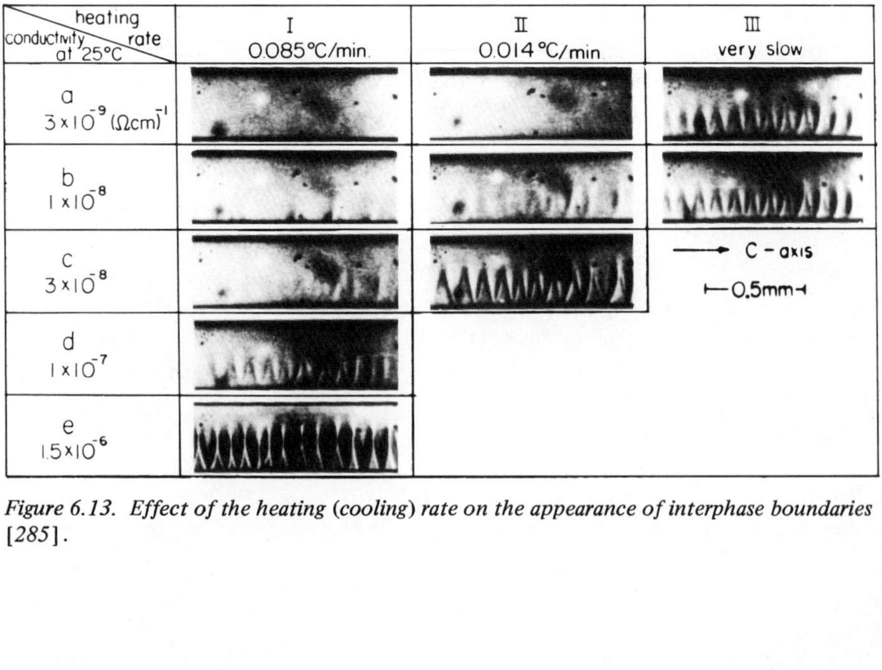

Figure 6.13. Effect of the heating (cooling) rate on the appearance of interphase boundaries [285].

tivity on the formation of interphase boundaries is related to screening of single-domain sections of the ferroelectric phase due to carriers localized at trapping levels. The slow dark relaxation of the structure of coexisting phases and the disappearance of the interphase boundaries after a long holding time of the crystal in darkness is explained by the slow emptying of the trapping levels, for which multiple capture evidently plays a significant role. The presence in SbSI of the slow relaxation of the nonequilibrium conductivity also speaks in favor of this mechanism.

In conclusion, let us dwell once again on the problem of the temperature interval of the coexistence of the phases. According to the theory discussed in Section 3.9, for a specified value of carrier concentration $n_0 > n_k$, the temperature interval $T_1 - T_1'$ in which the periodic structure of the interphase boundaries is maintained is directly proportional to the discontinuity in the width of the forbidden band at the phase transition ΔE_g [cf. (3.187) for $\Delta E_g \ll kT$]. This theoretical conclusion has been verified experimentally in [162]. The state diagram of SbSI shows that the line of first-order phase transitions approaches the line of second-order phase transitions near the triple point with coordinates ($p \simeq 1500$ atm, $T = -40°C$) [160]. The discontinuity ΔE_g accordingly decreases with the approach toward this point and disappears at the point itself. In this connection, it was of interest to investigate the structure of the interphase boundaries in SbSI along the state diagram, i.e., in the region of pressures and temperatures where the nature of the phase transition in SbSI changes near the triple point. Thus, the theory predicts that the temperature interval for the existence of the periodic structure of the interphase boundaries must decrease with approach toward the triple point, and the structure itself must disappear near this point.

The corresponding measurements were performed in [162] in a thermostated high-pressure chamber. The chamber permits one to carry out microphotography of the crystal in transmitted monochromatic light. The discontinuity in the width of the forbidden band at the phase transition ΔE_g was determined from the temperature dependence of the intrinsic absorption edge. Figure 6.14 presents microphotographs of the crystal taken in polarized light near the phase transition, which for SbSI demonstrate the ordinary structure of the interphase boundaries oriented along the (101) plane. The six microphotographs of Figure 6.14 correspond to six different pressures in the range from 1 to 1500 atm. These values of pressure on the state diagram correspond to the values of the Curie temperature in the range from +23 to −40°C. As should be expected, the structure disappears near the triple point. As is seen from Figure 6.14, the orientation of the interphase boundaries does not change with pressure. The individual observations indicated that the temperature interval for the coexistence of the phases $T_1 - T_1'$ continuously decreases with increasing pressure and goes to zero near the triple point. The values of ΔE_g were measured in this

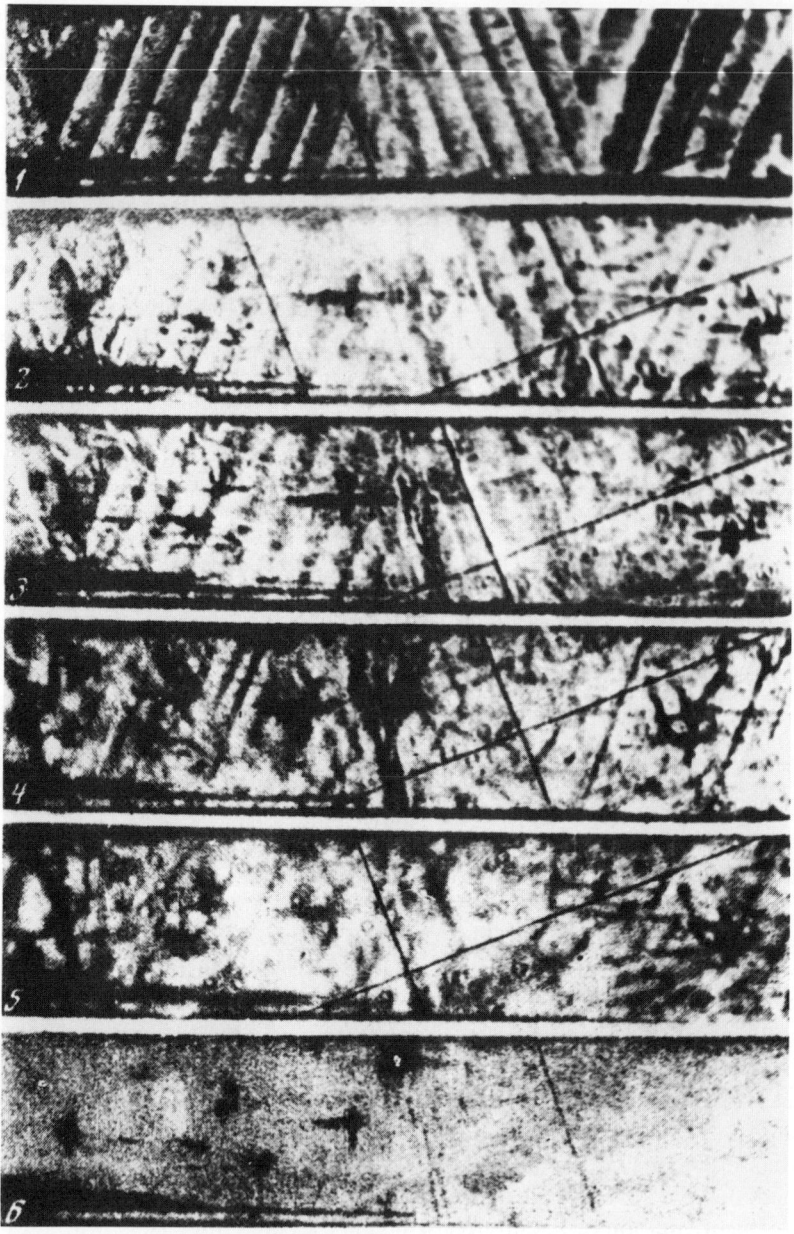

Figure 6.14. Microphotographs of a single crystal of SbSI for various values of hydrostatic pressure near the phase-transition temperature.

SCREENING PHENOMENA

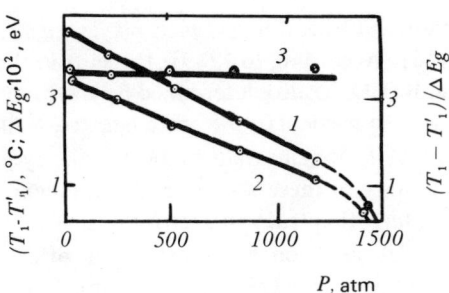

Figure 6.15. (1) Dependence of the discontinuity in the width of the forbidden band on pressure $\Delta E_g(p)$; (2) temperature range of coexisting phases T_1-T_1'; (3) ratio $(T_1-T_1')/\Delta E_g$ for single crystals of SbSI.

same range of pressures. Figure 6.15 presents the dependences of ΔE_g, $T_1 - T_1'$, and also the ratio $(T_1 - T_1')/\Delta E_g$ on pressure p. It is seen from these data that $(T_1 - T_1')/\Delta E_g$ is almost unchanged with pressure, i.e., in accordance with (3.187), there is a linear dependence between $T_1 - T_1'$ and ΔE_g, where the coefficient of proportionality does not depend on pressure over some interval.

6.5. Effect of Screening on Polarization Reversal Processes

A large number of works [15] have been devoted to investigation of ferroelectric polarization reversal processes. These works have considered these processes without regard to screening. Nevertheless, as we have already seen in the example of electroluminescence (Section 6.2), ferroelectric polarization reversal is accompanied by a redistribution of the screening charge. Thus, it is obvious that the screening charges play a significant role in polarization reversal processes. It was shown in Section 3.5 that the presence of surface layers is an inherent property of the ferroelectric crystal. The presence of these layers is somehow or other related to the existence of a space charge screening the spontaneous polarization. According to the various models, this charge can be localized in the surface layer [15, 72] or distributed at the boundary between this layer and the crystal [15]. The space charge at the surface of BaTiO$_3$ crystals above the Curie point was detected with the help of pyroelectric measurements by Chynoweth [235] and from birefringence measurements by Meyerhoffer and Triebwasser [288]. The thickness of the space-charge layer depends on the intensity of the field applied to the crystal and the duration of the polarization. However, in view of the special role of the surface, nucleation of antiparallel domains with polarization reversal is accomplished near it [15]. Thus, the formation of the screening space charge and the corresponding increase in the surface layer must significantly affect the polarization reversal mechanism.

The polarization reversal mechanism can be characterized by the magnitude of the polarization reversal field. Translational motion of the domain walls predominates in strong fields [15]. On the contrary, in weak fields, lateral motion of the domain walls plays a significant role in the switching process [15]. According to [289], the shift in the polarization reversal mechanisms in BaTiO$_3$ is also determined by the process of screening space-charge relaxation. In particular, the space charges, while not hindering the lateral motion of the 180° domain walls in low-frequency polarization reversal fields, reduce the mobility of these walls in a high-frequency field. The latter leads to a decrease in the high-frequency dielectric constant.

All these conclusions about the effect of the space charge on the polarization reversal process are primarily indirect in nature. In photoconducting ferroelectrics, such as SbSI, the change in concentration of the nonequilibrium carriers with illumination permits one to observe directly the relation between the ferroelectric polarization reversal and the screening. The effect of nonequilibrium carriers on the polarization reversal process in SbSI was studied by various methods in a number of works [290–296]. We will discuss here some of these results.

The experimental methods for studying the polarization reversal process in ferroelectrics can be divided into two main groups: (1) direct observation of the motion of the domain walls with polarization reversal and (2) the change in electrical parameters characterizing the value of the polarization.

Only the 180° domain structure is possible in SbSI. Thus, observation of the domains in polarized light is impossible. Study of the etch figures on the surface of SbSI is very difficult. Thus, the polarization reversal phenomena in SbSI and their relation to screening are naturally studied by observing the interphase boundaries arising near the phase transition. As has already been stated in the preceding section, the charge screening the spontaneous polarization of the single-domain section of the ferroelectric phase is concentrated near the interphase boundary. The presence of the screening space charge can be shown by direct optical observation. If the crystal is cooled and transforms to the ferroelectric phase, then the interphase boundaries naturally disappear, but light bands evidently caused by the shift in the absorption edge under the effect of the space-charge field remain for a long time at their position.

The polarization reversal of sections of the ferroelectric phase must be accompanied by a redistribution of the screening charge. This redistribution in turn leads to a change in the form of the boundary. Thus, observation of the relaxation of interphase boundaries can serve as one of the methods for studying the polarization reversal processes. This method was used in [294].

Optically homogeneous crystals of SbSI with aquadag electrodes applied on their ends were placed in a thermostated chamber having a window. Two ultrathermostats E-142 maintained a constant temperature with an accuracy of ±0.02°C over the entire chamber volume or produced a specified temperature gradient along the [001] axis of the crystal. Light from a monochromator was transmitted through the parallel (100) faces of the crystal, which could be

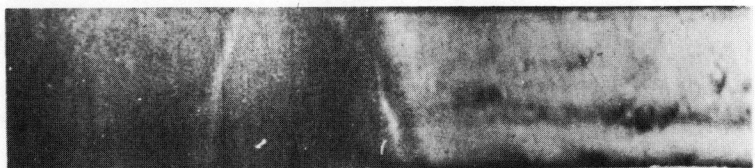

Figure 6.16. Boundary between regions of ferroelectric and paraelectric phases. A temperature gradient is produced along the ferroelectric [001] axis of the crystal.

microphotographed. The usual periodic structure of the interphase boundaries for SbSI were observed near the phase-transition temperature. If the ends of the crystal were maintained at different temperatures T_2 and T_3 with $T_2 > T_1 > T_3$, where T_1 is the Curie temperature, then the ferro- and paraelectric regions, into which the crystal is divided because of a temperature gradient, could be clearly seen with observation in transmitted monochromatic light at the absorption edge of the crystal (Figure 6.16). Only the boundary between the phases was noted with observation of the crystal in the long-wavelength region in accordance with the mechanism considered above.

The results of investigation of the processes of interphase boundary relaxation with pulsed switching of the crystal are presented in this section. The boundary between the phases was produced by the temperature gradient along the ferroelectric c axis of the crystal. The form of the boundaries did not change significantly in constant fields up to $\mathcal{E} \simeq 800$ V/cm. The boundary broadens with a further increase of the field, which is obviously related to the broadening of the phase transitions in the electric field. We note at once that the values of the field presented here, $\mathcal{E} = V/d$ (V is the voltage applied to the crystal, d is the distance between electrodes), are average values. Actually, screening of the spontaneous polarization of the ferroelectric region leads to the formation of a section near the interphase boundary with a high field intensity. The measured values of the electrical conductivity of the crystal σ are also average values for the same reason. It is most likely that the field intensity in the ferroelectric region is lower, while the electrical conductivity is greater than the average values presented.

The optical method was used to study the behavior of the interphase boundary with a change in polarity of the voltage applied to the crystal. If the field is small, the interphase boundary does not change with its switching. On reaching some critical value $\mathcal{E}_{cr}$, switching of the field leads to broadening of the boundary with subsequent re-establishment of the initial pattern after a time τ. The value of the critical field increases with increasing intensity of the illumination (and the corresponding increase of electrical conductivity of the sample) (Figure 6.17), while the boundary formation time τ decreases from several tens of seconds to seconds (Figure 6.18). The time τ also depends on the direction of

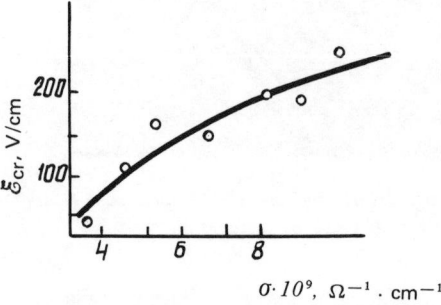

Figure 6.17. Dependence of the magnitude of the critical field causing relaxation of the boundaries on the sample photoconductivity.

switching of the field, while this dependence is not related to unipolarity of the crystal.

The boundary relaxation mechanism can be understood in the following manner. Polarization reversal of the ferroelectric region of the crystal occurs on reaching the critical field. The switching time is small; it does not exceed milliseconds in SbSI in fields of ~200 V/cm [3, 296]. Illumination does not affect the switching time [297]. However, the polarization reversal processes do not conclude with the reorientation of the domains. Redistribution of the screening charge in the contact regions of the crystal and in the interphase boundary region occurs after this. This leads to redistribution of the field in the crystal, which can obviously cause in turn the occurrence of additional switching processes. Thus, the polarization reversal time as a whole is determined essentially by the time for screening of the spontaneous polarization and the external polarization reversal field. The boundary relaxation time measured experimentally obviously equals the time of formation of the screening space charge at the boundary of the phases. Only after formation of the screening layer does the boundary become sharp. Actually, the boundary relaxation time coincides

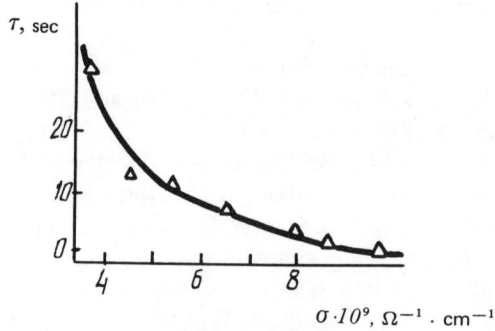

Figure 6.18. Dependence of the relaxation time of the interphase boundary after switching of the electric field on the sample photoconductivity.

in order of magnitude with the Maxwellian relaxation time of the crystal $\tau = \epsilon/4\pi\sigma$. This coincidence could have been worse if it is taken into account that the electrical conductivity near the boundary differs from the measured average value. The effective internal field causing the polarization reversal decreases with increasing photoconductivity because of screening, which also leads to an increase of $\mathcal{E}_{cr}$ (Figure 6.17). Since the time of formation of the boundary corresponds to the Maxwellian time, then it is inversely proportional to the photoconductivity (Figure 6.18). Of course, the dependence of τ on the direction of switching of the field can be related to the dependence of the electrical conductivity of the screening layer near the interphase boundary on the direction of spontaneous polarization ("intrinsic" field effect). The presence of a layer near the boundary with enhanced concentration of the principal carriers leads, on the one hand, to an increase in τ and, on the other, to a decrease in the effective field intensity in the remaining volume of the crystal. The latter leads to some shift of the boundary and to a decrease in the size of the ferroelectric phase previously induced by the field, which is observed experimentally. We note that the shift of the interphase boundary with a change in the direction of the external field was also noted in [298].

A study of the relaxation process of the dielectric constant ϵ with pulsed switching of the crystal consisting of two regions of differing phases has also been carried out in parallel with the optical observations of the formation of boundaries. The oscillograms obtained are presented in Figure 6.19. An increase in the dielectric constant is observed at the time of switching. The curves have a break or minimum, after which follows a second lower maximum. The break appears more clearly with increasing photoconductivity. The increase is evidently caused by an increase in the number of domain boundaries and their contribution to the dielectric constant. This problem was considered phenome-

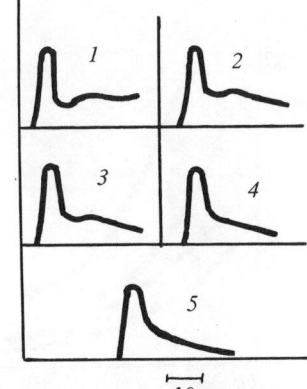

Figure 6.19. Relaxation of the dielectric constant of the SbSI crystal with an interphase boundary with switching of the electric field. (1) 560 V/cm; (2) 600 V/cm; (3) 640 V/cm; (4) 680 V/cm; (5) 720 V/cm.

nologically in [299]. The appearance of the second maximum in the relaxation curves of ϵ is naturally related to screening. The redistribution of the screening charges at the boundary and at the contact regions does not occur at first with delivery of the opposing pulse of external field on the crystal. The field of these charges has the same direction as the external field and aids the switching. The redistribution of the screening charges occurs after the Maxwellian time, the direction of the field of these charges changes to the opposite, and the effective field intensity decreases in the crystal. The formation of the interphase boundary also affects this process. All of this leads to partial depolarization of the sample (reverse switching). According to [291], an analogous mechanism leads to inversion of the sign of the Barkhausen discontinuities in SbSI in switching a previously illuminated crystal.

Thus, the second maximum in the relaxation curve of ϵ is related to partial reverse switching of the crystal. For large external fields, the effective field intensity in the crystal becomes sufficiently large even after screening, and reverse switching is less notable (the second maximum is less). On the contrary, with increasing photoconductivity, the internal field decreases and the corresponding break in the relaxation curve of ϵ becomes distinctive.

The position of the minimum (or break) must correspond to the time of formation of the boundary, which is confirmed by direct optical observations. Thus, the time of formation of the boundary can be evaluated from the time τ_b corresponding to the break in the relaxation curve of ϵ. Figure 6.20 presents the dependence of τ_b on the intensity of the switching electric field for various values of photoconductivity of the sample. In accordance with the results of optical observations, the time of formation of the boundary decreases with increasing photoconductivity. The value of τ_b decreases with increasing field intensity, tending toward saturation. This decrease can be related to the decrease

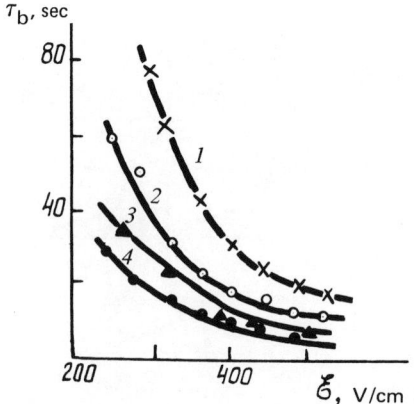

Figure 6.20. Dependence of the time of boundary formation between phases on the magnitude of the switched electric field for various values of the sample photoconductivity σ $(\Omega \cdot cm)^{-1}$. (1) $-3.1 \cdot 10^{-9}$; (2) $-4.2 \cdot 10^{-9}$; (3) $-1.2 \cdot 10^{-8}$; (4) $-2.1 \cdot 10^{-8}$.

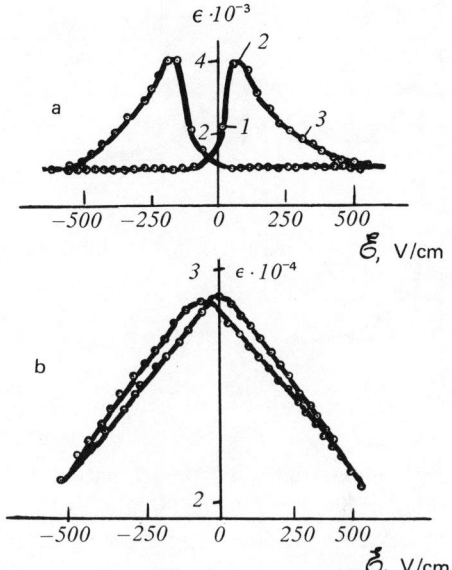

Figure 6.21. Dependence of the dielectric constant on the magnitude of the constant bias electric field. (a) $T = 18°C$, ferroelectric phase; (b) $T = 24°C$, paraelectric phase. Curie temperature $T_1 = 21.5°C$.

in the dielectric constant and the corresponding decrease in the Maxwellian relaxation time. Since the dielectric constant of the ferroelectric phase is less than the paraelectric near the Curie temperature, the screening charge at the interphase boundary is localized predominantly in the ferroelectric phase. In large fields, ϵ of the ferroelectric phase ceases to depend on the value of the field (dielectric saturation), which also leads to saturation of τ_b.

The independently measured dependence of the reverse dielectric constant on the intensity of the constant bias field (Figure 6.21a) actually displays a maximum in the region ~100 V/cm and saturation in a field of ~500 V/cm. The value of ϵ on the saturation section remains fixed with subsequent decrease of the field to zero (hysteresis). The dependence $\epsilon = \epsilon(\mathcal{E})$ in the paraelectric phase at a temperature somewhat exceeding the Curie temperature is presented in Figure 6.21b. The dielectric constant increases in small fields and then drops with increasing field. This behavior, as is well known, is usually related to the formation of contact layers of space charge and a decrease in the effective capacitance of the sample [72].

The effect of nonequilibrium carriers on the polarization reversal processes in SbSI is also expressed in the change in nature of the dielectric constant relaxation with illumination of the sample. An increase in ϵ occurs with a gradual increase of the bias field, and then a drop to a steady-state value (Figure 6.22). The characteristic time for establishment of the process is of the order of 30 sec. The relaxation time of ϵ in the paraelectric phase is even less and is ob-

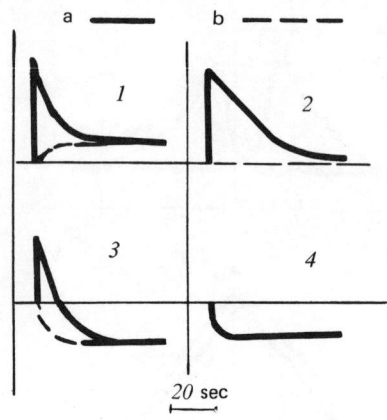

Figure 6.22. Relaxation of the dielectric constant with a change in the magnitude of the electric field. Oscillograms 1–3 were obtained at the corresponding points noted in Figure 6.21. Oscillogram 4 was obtained for a crystal in the paraelectric phase. (a) In darkness; (b) with illumination.

viously determined by the formation time of the contact layers of space charge. Illumination has practically no effect on the relaxation process of ϵ in the paraelectric phase and significantly changes the nature of the relaxation in the ferroelectric phase. The initial discontinuity of the dielectric constant caused by the increase of the field is "quenched" by light. The relaxation time of the dielectric constant decreases abruptly, where this decrease is maintained for a long time after cutoff of the illumination. The steady-state values of ϵ are somewhat lower than the corresponding values obtained with measurements in darkness. These phenomena are naturally related to the change in nature of the relaxation of the domain structure with an increase in concentration of nonequilibrium carriers. In particular, the disappearance of the discontinuity of ϵ can be caused by a decrease in mobility of the domain boundaries because of the accumulation of space charges at them, and also by the increased role in the polarization reversal process of the lateral motion of the 180° domain walls because of a decrease in the depolarization field.

Since the polarization reversal of sections of the ferroelectric phase is accompanied by a redistribution of the screening space charge, the time characteristics of the polarization reversal depend on the Maxwellian relaxation time. This conclusion was confirmed additionally by investigation of polarization reversal of SbSI in extralow-frequency sinusoidal fields [294]. The SbSI crystals were observed in transmitted light; the ordinary structure of the interphase boundaries was observed near the phase transition. The variable low-frequency field was applied in the direction of the ferroelectric axis of the crystal. Polarization reversal of a section of the ferroelectric phase under the effect of the variable electric field is accompanied by motion of the boundaries between the phases. The minimum value of the applied field intensity $\mathcal{E}_{cr}$ at which motion of the boundaries was observed was measured in these experiments.

The frequency characteristics of the critical fields are presented in Figure

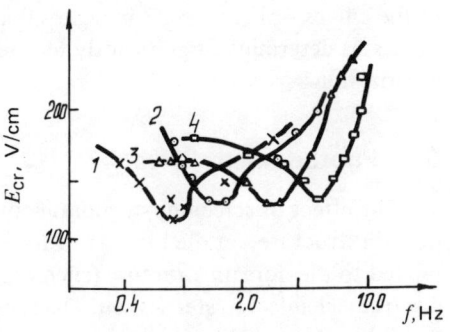

Figure 6.23. Frequency dependence of the critical fields obtained for various levels of illumination of the crystal. The electrical conductivity of the sample is $(\Omega \cdot cm)^{-1}$, respectively, (1) $-1.1 \cdot 10^{-8}$; (2) $-2.7 \cdot 10^{-8}$; (3) $-3.9 \cdot 10^{-8}$; (4) $-4.5 \cdot 10^{-8}$.

6.23. It follows, first of all, from these data that the times corresponding to the minima of the curves 1–4 are close to the corresponding values of the Maxwellian relaxation time. Actually, the values of $1/\tau$ lie in the range 1–10 Hz for $\epsilon \simeq 10^4$ and $\sigma \simeq (10^{-9}\text{–}10^{-8})\,(\Omega \cdot cm)^{-1}$. It is well known that deep trapping levels in SbSI are responsible for the slow kinetics of the dark decay of the photocurrent. Corresponding to this, the minimum of the frequency characteristic must be shifted toward lower frequencies after cutoff of intense light, where the kinetics of this shift must correlate with the kinetics of the dark decay of the electrical conductivity. This phenomenon was actually observed in [294].

The four frequency characteristics in Figure 6.23 correspond to different intensities of light used in observing the crystal and to the correspondingly different concentrations of nonequilibrium carriers. The latter can be determined from measurements of the photocurrent corresponding to these same values of light intensity. It follows from the data of this figure that the minima of the curves are shifted toward higher frequencies with increasing light intensity and corresponding electrical conductivity. Figure 6.24 presents the dependence of the frequency f_{min} corresponding to the minima of the curves of Figure 6.23 on the electrical conductivity of the crystal.

The results obtained indicate that the frequencies f_{min} can be identified with the frequencies corresponding to the Maxwellian relaxation and that the nature

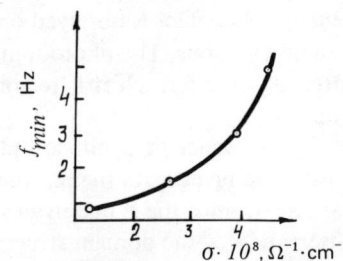

Figure 6.24. Increase in frequency f_{min} with increasing photoconductivity of the sample.

of the curves of Figure 6.23 indicates that the ferroelectric polarization reversal process is determined significantly by the time for screening of the spontaneous polarization.

6.6. Photodomain Effect

The effect of screening of spontaneous polarization on the equilibrium domain structure, on the kinetics of its formation, and on the properties directly related to the domain structure (such as, for example, the pyroelectric charge, electromechanical hysteresis, etc.) has been called the photodomain effect. The possible nature of the photodomain effect and its relation to surface layers, volume screening, and recharging of the surface levels with illumination have been discussed in Sections 3.5 and 3.8. We will discuss here some results of experimental investigation of the photodomain effect in barium titanate and antimony sulfoiodide. We will first discuss the effect of nonequilibrium carriers on the domain structure, i.e., the direct observation of the effect, and then the number of properties related to the photodomain effect.

6.6.1. The Photodomain Effect in SbSI and $BaTiO_3$. The photodomain effect was first observed in SbSI crystals [300]. Investigation of the photodomain effect in SbSI was carried out by indirect methods since the method of direct observation of the domain structure of this crystal did not exist until recently. The effect of illumination on the magnitude of the pyroelectric current in crystals of SbSI and SbSeI was noted in [301, 302]. The photodomain effect was subsequently observed in SbSI by investigating the effect of illumination on the Barkhausen discontinuity [290]. Direct observation of the photodomain effect in SbSI was performed in [303], where etching of a single-domain crystal of SbSI was carried out before and after illumination. Etching in SbSI reveals the 180° domain boundaries parallel to the ferroelectric axis of SbSI. It was shown in [303] that illumination of a short-circuited single-domain crystal sharply increases the number of domain boundaries.

The principal results of experimental investigation of the photodomain effect in SbSI crystals are summarized as follows. The spectral distribution of the photodomain effect coincides with the spectral distribution of the photoconductivity. The magnitude of the effect increases with increasing light intensity. The effect is observed only for sufficiently high-resistance and photosensitive crystals. The photodomain effect is observed over a wide temperature interval in the ferroelectric region and increases with approach toward the Curie point.

The presence of significant photosensitivity in $BaTiO_3$ crystals makes it possible to investigate the photodomain effect in detail and its kinetics in particular, since the symmetry of $BaTiO_3$ crystals permits direct optical observation of the domain structure. The photodomain effect was first observed in $BaTiO_3$ in [211, 304]. Investigation of the effect was first carried out for unreduced single crystals of $BaTiO_3$ grown by the Remeyka method which had a resistivity of $\sim 10^9$–10^{10} Ω · cm. Observation of the domain structure was carried out in the MIN-8 polarization microscope in the [001] direction through semitransparent gold electrodes applied to the (001) faces, which permitted applying the electric field in the direction of the c axis. The observation was carried out in transmitted monochromatic light λ = 550 nm. The illumination of

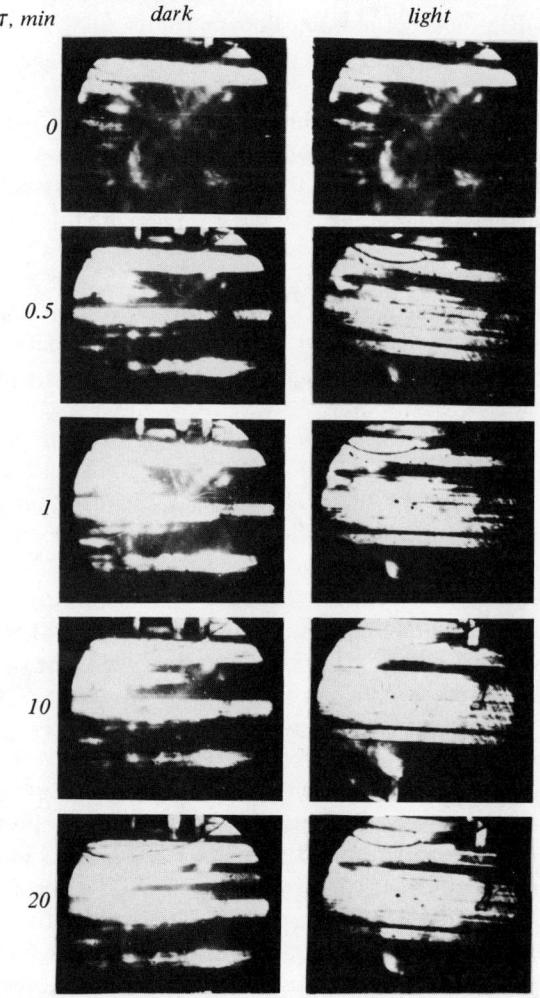

Figure 6.25. Photodomain effect in BaTiO with anode illuminated ($T = 70°C$). (a) Multiple-domain formation of short-circuited crystal in darkness after removal of the field; (b) multiple-domain formation with illumination.

the crystal was carried out in the natural region $350 < \lambda < 450$ nm, where the photoconductivity of $BaTiO_3$ occurs. All measurements carried out with illumination by long wavelength nonphotoactive light will be called "dark." An investigation of the effect of illumination on the kinetics of the decay of the c-domain state in a crystal of $BaTiO_3$ was carried out in [211, 304]. Figure 6.25 illustrates the kinetics of the multiple domain formation of a short-circuited crystal previously polarized in the c direction by a field of 5 kV/cm. In the course of time, the single-domain crystal changed into multiple-domain crystals consisting of a and c domains. It is seen from Figure 6.25 that illumination accelerates the process of multiple-domain formation in the crystal. This effect is enhanced with increasing temperature (with approach toward the Curie point) and with an increase in intensity of photoactive illumination. The effect is polar in a number of cases: with illumination of the electrode used as the anode in polarization, the rate of multiple-domain formation is greater than with illumination of the cathode. We recall that acceleration of the multiple-domain formation with illumination of a single-domain crystal of SbSI was also noted in [300, 303]. Enhancement of the effect with approach toward the Curie point is noted by comparing the microphotographs of Figure 6.25 with the microphotographs of Figure 6.26, on which are presented results of the same observations at temperatures close to the Curie point. It is seen that the effect of "natural" illumination on the rate of multiple-domain formation increases significantly with approach toward the Curie point.

Figures 6.25 and 6.26 illustrate the effect of "natural" illumination on the kinetics of the transition of the nonequilibrium domain structure of $BaTiO_3$ to the equilibrium. As was shown in Section 3.8, the theory predicts the effect of illumination on the equilibrium domain structure. Some results relating to this case were obtained in [304].

It is well known that for the transition of a free short-circuited $BaTiO_3$ crystal from the ferroelectric to the paraelectric phase, the so-called square-lattice structure of Forsberg [305], occurs near the phase-transition temperature (this is seen, in particular, in Figure 6.26). It was established that the Forsberg structure is not formed with illumination of the crystal by photoactive light. Figure 6.27 presents the successive changes in the "equilibrium" domain structure of the $BaTiO_3$ crystal with increasing temperature in darkness and with illumination, respectively. The region of existence of the Forsberg structure in darkness corresponds to 123–127.5°C. The Forsberg structure does not arise with illumination of the crystals by photoactive light. Further investigations showed that the effect of illumination on the Forsberg structure occurs only when the illumination is produced at the time corresponding to the formation or dissipation of this structure, whereas illumination of the crystal at a temperature corresponding to the already-formed stable Forsberg structure does not lead to a change in the domain structure. The mechanism of the effect of illumination

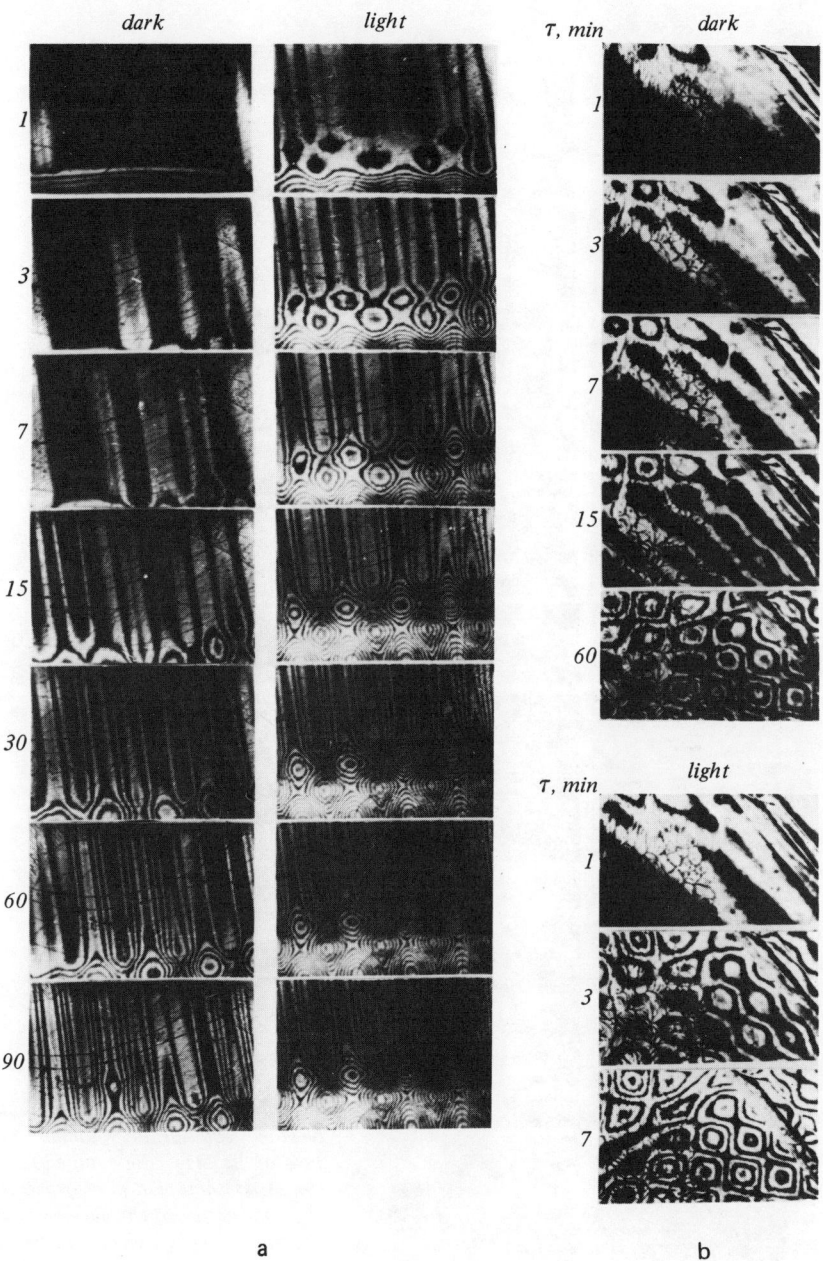

Figure 6.26. Effect of temperature on the photodomain effect in $BaTiO_3$. (a) Multiple-domain formation of a c-single-domain crystal in darkness and with illumination. $T = 110\,°C$; (b) same process at $T = 125\,°C$ ($T_1 = 127.5\,°$). Multiple-domain formation is accompanied by the formation of typical Forsberg structures.

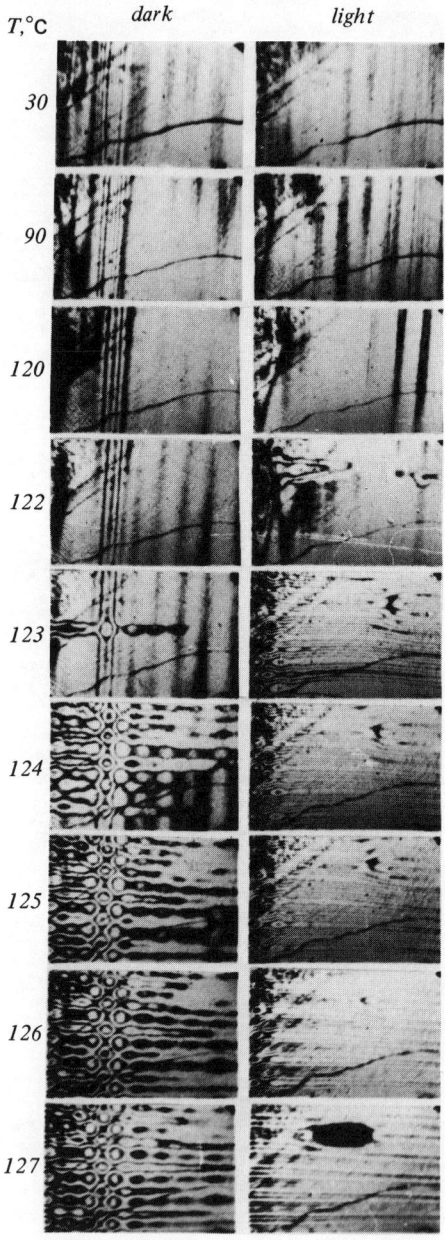

Figure 6.27. Effect of illumination on the "equilibrium" domain structure of $BaTiO_3$ (the temperature of the phase transition is recorded from the appearance of the dark section of the paraelectric phase inside the ferroelectric phase).

on the square-lattice structure is unclear. It is not clear, for example, whether it is related to the screening processes or to the direct effect of nonequilibrium carriers on elastic stresses arising because of inhomogeneity of the impurity distribution. As is seen from Figure 6.27, illumination does not affect the "equilibrium" domain structure of $BaTiO_3$ far from the Curie point. Figure 6.27 also illustrates the shift of the Curie point toward lower temperatures with "natural" illumination. The frame corresponding to a temperature of 127°C shows that a section of the paraelectric phase arises with illumination.

We will make some remarks with respect to the possible mechanism of acceleration of the multiple-domain formation of the crystal with illumination [306]. As has already been emphasized in Section 3.5, the single-domain state is energetically favored in the ideal short-circuited crystal. However, the existence in ferroelectrics, including $BaTiO_3$, of surface layers (which is equivalent to the presence of a dielectric gap) leads to the generation of some depolarization field in the short-circuited single-domain crystal. The surface maxima of the depolarization field causes switching processes (multiple-domain formation) in the short-circuited single-domain crystal. The dark processes of multiple-domain formation occur rather slowly, since the magnitude of the surface field is insignificant. The screening length decreases with illumination, which causes a redistribution of the depolarization field: the field increases at the surface and decreases in the volume. The increase of the surface field accelerates the process of multiple-domain formation, which is illustrated by Figures 6.25 and 6.26. It was also shown in Section 3.5 that the effectiveness of the effect of the nonequilibrium carriers on the domain structure is determined by the ratio of the thickness of the surface layer and the screening length. The polarity of the photodomain effect indicated above can be explained if one takes into account the asymmetry of the surface layer with respect to the direction of the polarization.

Data on the effect of illumination on the rate of polarization of the crystal in an external field are also evidence in favor of the mechanism under consideration. According to present ideas about the mechanism of polarization of $BaTiO_3$, the process of c-domain formation under the effect of an external field in the region of small fields is determined mainly by the surface field, whereas the process of a-domain formation depends on the field in the volume of the crystal. The nonequilibrium carriers, in redistributing the field applied to the crystal, must affect the rate of its polarization. Optical observations indicated that the process of polarization of the crystal in a small field ($\sim$100 V/cm) in the c direction is accelerated with illumination. On the contrary, polarization of the crystal in the [100] direction is not accelerated, and is slowed in some cases with illumination. One can assume that the effect of illumination on the polarization kinetics is also caused by redistribution of the field in the crystal, with which the surface field increases while the field in the volume decreases. Of course, these effects are expressed more notably with increasing light intensity.

The results obtained for $BaTiO_3$ permitted investigating for the first time the kinetics of the photodomain effect. The latter is determined by the larger of two characteristic times: the Maxwellian relaxation time and the switching time. Calculation of the Maxwellian relaxation time gives a value of 10^{-1} to 1 sec for $BaTiO_3$ in the temperature interval under investigation.

Investigations of the kinetics of the photodomain effect and polarization with illumination show that these processes are determined by times of the order of tens of seconds (cf., for example, Figure 6.25b), i.e., significantly greater than the Maxwellian time. Thus, in accordance with the proposed mechanism, the kinetics of the photodomain effect is limited by the switching time.

Results, whose explanation requires including the mechanism of optical recharging of surface levels (Section 3.8), were obtained in [20] with investigation of the photodomain effect in reduced crystals of $BaTiO_3$. Barium titanate crystals grown by the Remeyka method, whose reduction in a hydrogen atmosphere at a temperature of 500°C increased the concentration of equilibrium carriers by three orders of magnitude, were investigated. The transition from the c-single-domain state to the (c, a)-multiple domain state and the effect of reduction and "natural" illumination on this transition was investigated for these crystals by the technique described above. Figure 6.28 illustrates the effect of reduction of the crystal on the kinetics of multiple-domain formation in darkness. The absence of an effect of the reduction on the kinetics of the multiple-domain formation draws attention, whereas the "natural" illumination significantly accelerates this process. The photodomain effect is then expressed more sharply for reduced crystals. As in the cases described above, the effect of photoactive illumination becomes less significant with separation from the Curie point, where the relaxation time increases both with illumination and in darkness. Nonetheless, the results presented in Figure 6.28 cannot be directly related to the change of volume concentration of carriers and explained on the basis of the mechanism of space screening considered above. Actually, reduction of the crystal, which increases the concentration of equilibrium carriers by $4 \cdot 10^3$ times, does not significantly affect the relaxation time of the domain structure. At the same time, the increase in concentration of nonequilibrium carriers by 10^2 times decreases the relaxation time of the domain structure for the unreduced crystal by 30 times. The increase of concentration of nonequilibrium carriers by three times in the reduced crystal decreases the relaxation time by 180 times.

These results lead to the conclusion of the necessity of including the mechanism of optical recharging of the surface levels (Section 3.8), for which condition (3.140) ensuring the appearance of the a domains can be satisfied.

It is significant that this mechanism explains the temperature dependence of the photodomain effect, since condition (3.140) is more easily satisfied near the Curie point, where ϵ_a has a maximum. Besides, this same mechanism takes into

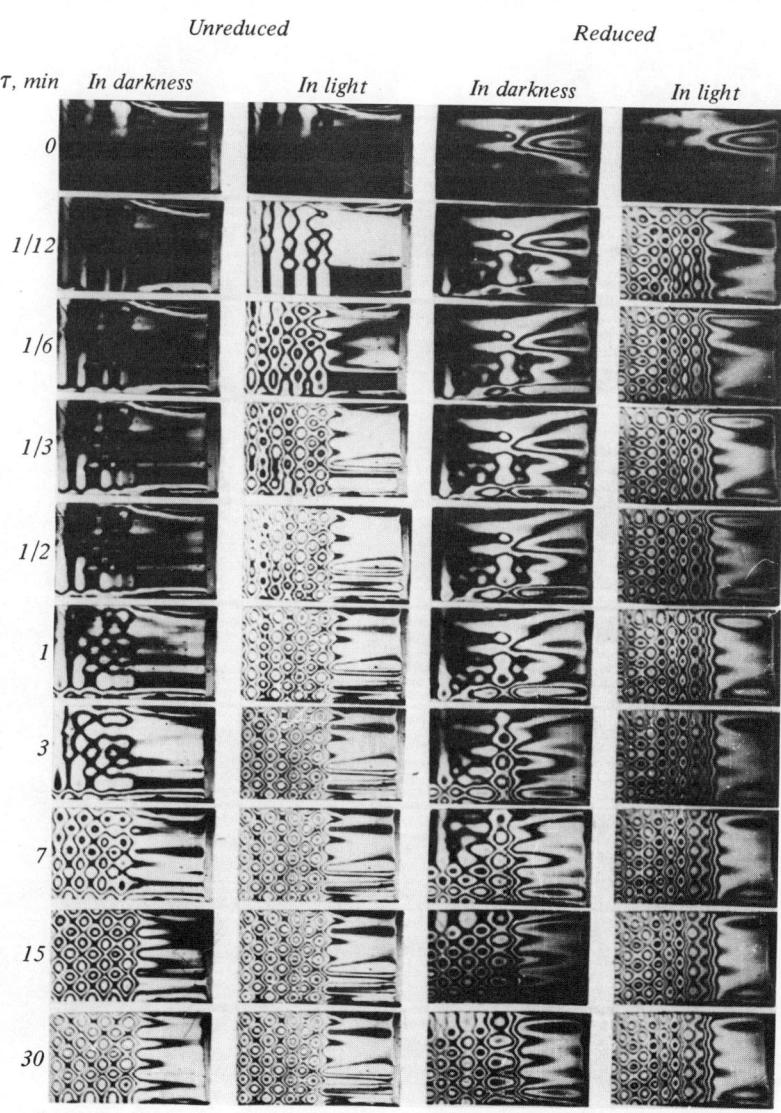

Figure 6.28. Kinetics of the transition of the c-domain structure to the (c, a)-domain structure for unreduced and reduced crystals in darkness and with illumination ($T = +125°C$).

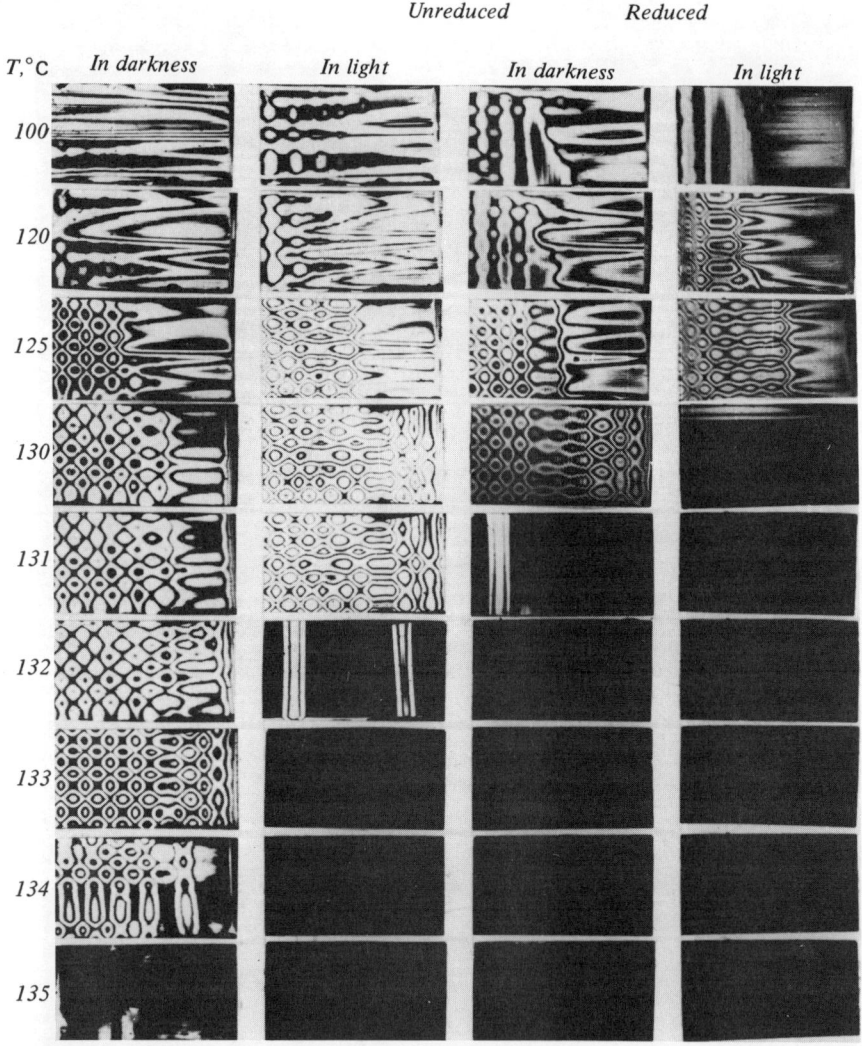

Figure 6.29. Effect of illumination on the "equilibrium" domain structure near the phase transition.

account the space screening. Thus, condition (3.140) cannot be satisfied for any temperature if the screening space-charge density ρ_{0a} is below a definite value.

Figure 6.29 shows the temperature dependence of the "equilibrium" domain structure for crystals in the initial and reduced states. Reduction of the crystal has an insignificant effect on the domain structure, while "natural" illumination has a slight effect on the Forsberg structure near the Curie point. The frames of Figure 6.29 illustrate well the shift of the Curie point toward lower temperatures, which is caused by reduction and illumination.

In all the cases considered above, the nonequilibrium carriers aid the transition of the single-domain state to the multiple-domain, i.e., they decrease the unipolarity of the crystal. Investigation of the photodomain effect with natural polarization, i.e., for the multiple-domain ferroelectric, and also with the formation of the domain structure in the process of the transition of the crystal from the paraelectric to the ferroelectric region gives the opposite effect in a number of cases. Illumination increases the unipolarity of the ferroelectric in these cases. Since these observations were carried out with measurement of the pyroelectric current, we will discuss them in the following section.

6.6.2. Effect of the "Natural" Illumination on the Pyroelectric Currents in SbSI and BaTiO$_3$. In SbSI crystals, for which optical observation of the domains is impossible, investigation of the photodomain effect was carried out by measuring the pyroelectric current at the phase transition.

The transition of the crystal from the single-domain to the multiple-domain state with illumination must be accompanied by a significant decrease in the pyroelectric charge. To observe this effect, the crystal after preliminary illumination in the ferroelectric phase is heated and the temperature dependence of the pyroelectric current is simultaneously measured with the transition from the ferroelectric to the paraelectric phase. The residual polarization of the crystal is determined by integrating the pyroelectric current over time. It was found that illumination of the single-domain SbSI crystal leads to a sharp decrease in the magnitude of the residual polarization as compared to the spontaneous. A typical case is presented in Figure 5.3. Investigation of the spectral distribution of this effect showed that the maximum decrease in the residual polarization occurs in the region of the maximum of the photosensitivity of the crystal. This shows that the effect of the decrease in the pyroelectric charge, or the photodomain effect, is photoelectric in nature and is caused by the nonequilibrium conductivity. The photodomain effect recorded by the pyroelectric current should not be confused with the effect on the pyroelectric current of the interphase boundaries formed with "natural" illumination of the crystal near the Curie temperature [307].

While the second effect is observed only with illumination of the crystal in the region of coexistence of the phases, the photodomain effect also exists at lower temperatures, although, as has already been indicated, it decreases with separation from the Curie point.

With the transition of the SbSI crystal from the paraelectric to the ferroelectric region, the "natural" illumination can both decrease the pyroelectric charge as well as increase it [308-310]. Individual probe measurements have shown that the sign of the effect depends on which of the contact regions is illuminated. Figure 6.30 presents typical pyroelectric current curves for the same crystal, taken with heating of the crystal. Curve a is obtained for the crystal cooled in darkness. Curve b corresponds to the case where the negative end of the crystal was illuminated in the process of transition of the crystal from the paraelectric to the ferroelectric phase; curve c was obtained under the same conditions, but the positive end was illuminated. These results show that illumination of the negative ends decreases the pyroelectric charge and, consequently, the unipolarity of the crystal; conversely, illumination of the positive end increases it. Investigation of the spectral distribution of these effects showed the presence of two maxima: an intrinsic at $\lambda \simeq 620$ nm ($E_g \approx 2$ eV for SbSI) and impurity at $\lambda \simeq 720$ nm.

These results do not contradict the fact of the decrease in the pyroelectric charge of the single-domain crystal with its illumination in the ferroelectric region, and, as was indicated in Section 3.8, they can be explained starting from

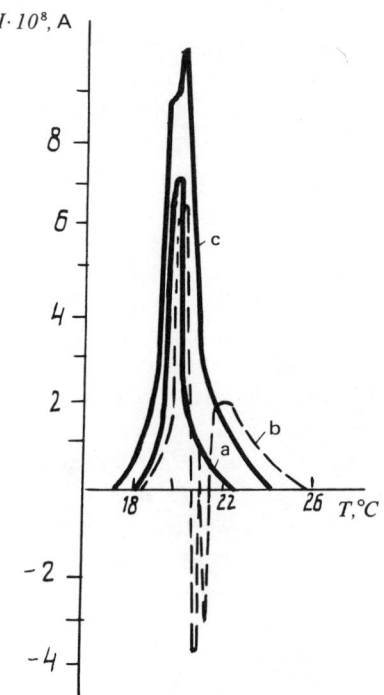

Figure 6.30. Temperature dependence of the pyroelectric current with "natural" polarization.

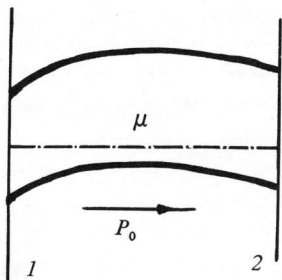

Figure 6.31. Curvature of the bands at the ends of the crystal in the paraelectric phase.

the mechanism of optical recharging of the surface levels in the paraelectric region. We assume that the natural unipolarity of the crystal is caused by the unequal charges of the surface levels at the opposite ends of the crystal in the paraelectric region, i.e., $\Delta\mu = \mu_+ - \mu_- \neq 0$. To be specific, let a positive charge be localized at both end surfaces, where the charge at end 1 is larger than at end 2 (Figure 6.31). Since the field at surface 1 is greater than at surface 2, while the nucleation formation and growth of domains are determined by the surface field and begin from the surface [15], then the ferroelectric polarization, whose direction P_0 in Figure 6.31 is indicated by the arrow (natural unipolarity or natural polarization), arises with the transition of the crystal from the paraelectric to the ferroelectric phase. Thus, the greater $\Delta\mu$, the greater the unipolarity of the crystal. Of course, the curvature of the bands and distribution of the field in the crystal in the ferroelectric phase do not correspond to Figure 6.31, since they are also determined by screening of the resulting natural polarization.

In order to explain the effect of light on the unipolarity, we assume that with illumination of the crystal in the paraelectric region (or with the transition through the Curie point), the positively charged surface levels of both ends capture nonequilibrium electrons and their charge decreases. Nonequilibrium holes are accordingly captured by levels in the volume and decrease the space charge. In this case, illumination of surface 2 increases the difference in charges at the opposite ends and, consequently, increases the unipolarity and natural polarization. On the contrary, with illumination of surface 1, the difference in charges and, consequently, the measured pyroelectric charge decrease. With the illumination of a naturally polarized crystal in the ferroelectric region, the pattern has a more complex nature since it is also determined by the effect of nonequilibrium carriers on the screening of the polarization. As was shown in [308-310], the photodomain effect can be accompanied under these conditions both by an increase as well as a decrease of the unipolarity. If the impurity maximum at $\lambda = 0.72$ μm corresponds to the transition of electrons from the valence band to the surface levels in SbSI, as was proposed in [308-310], then the presence of the two maxima in the spectral distribution of the photodomain effect can be understood. Thus, probe investigation of the photodomain effect with natural polarization permits reaching conclusions about the distribution of the field in the paraelectric region. The nature of the field distribution in the short-circuited ferroelectric can also be investigated with the help of the short-circuit photocurrent measured in the probe mode, which we will discuss in the following section.

6.6.3. Photodomain Effect and Electromechanical Hysteresis in SbSI. It was shown in Section 5.4 that the effect of "natural" illumination on the spontaneous deformation of SbSI is caused by the photodomain effect. This conclusion finds further confirmation with the study of electromechanical hysteresis in SbSI and the effect of "natural" illumination on it.

Electromechanical hysteresis in SbSI under the effect of a linearly changing sawtooth voltage of low frequency 10^{-1}-10^{-3} Hz was investigated in [216-217]. To obtain the deformation loops, the linearly changing voltage was applied in the direction of the c axis of the crystal, and the deformation in the same direction was recorded simultaneously with the help of a dilatometer. After measuring the electromechanical hysteresis, a prolonged annealing of the crystal was carried out in darkness at a temperature of 50°C. After this, the electromechanical hysteresis loop presented in Figure 6.32 was measured at +7°C. The differential piezoelectric modulus d_{33} calculated from the slope of the linear section of the loop M-K is $\sim 2 \cdot 10^{-9}$ C/N, which agrees with independent measurements [311]. After annealing, the unipolarity of the crystal is insignificant, which corresponds to the almost symmetric loop in Figure 6.32a. Illumination of the crystal with natural polarization leads to an increase in unipolarity and the corresponding asymmetric electromechanical hysteresis loops (Figure 6.32b and c). The asymmetric loops in Figure 6.32b and c were obtained after illumination at points K and K', respectively, i.e., with the change

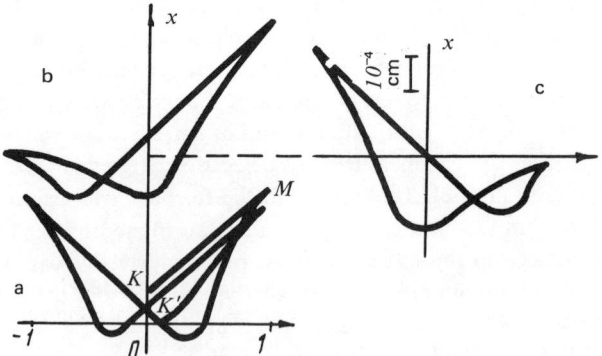

Figure 6.32. Effect of illumination on the electromechanical hysteresis. (a) Hysteresis obtained in darkness with a previously annealed crystal; (b, c) after illumination (light on at points K and K', respectively).

SCREENING PHENOMENA   251

in field in the two opposite directions. The change in the domain structure after illumination is stable; the asymmetric electromechanical hysteresis loops are stable according to this. Annealing of the crystal or its illumination in a variable field above the coercive field makes the form of the loops symmetric again (Figure 6.32a).

6.7. Screening and Short-Circuit Photocurrents

The photodomain effect indicates the presence of an internal electric field in the short-circuited ferroelectric, both multiple domain and single domain. The existence of an internal field in a short-circuited photoconducting ferroelectric can be verified in addition by observing the photoemf or short-circuit photocurrent I_{sc} related to the drift of nonequilibrium carriers in the internal field of the crystal. Moreover, with light probing of the crystal by this method, one can investigate the distribution of the internal field along the axis of spontaneous polarization, which can be compared with that obtained theoretically for the single-domain crystal with a free surface (cf., for example, Figure 3.1). As is clear from section 5.5,† the distribution of the short-circuit photocurrent I_{sc} and the pyroelectric current related to heating of the crystal with illumination is not a simple problem in the general case. However, if the photoconductivity of the crystal is large (for example, in the case of SbSI), these effects are separable.

The distribution of the internal field in SbSI crystals from measurements of the photocurrent I_{sc} with light probing was obtained in [236-238]. The crystal was cooled and transformed without preliminary polarization to the ferroelectric phase, where light probing was carried out and the photocurrent I_{sc} was recorded with motion of the light probe between the electrodes in the direction of the c axis. The results of these measurements showed that two types of distribution of the internal field, a and b in Figure 6.33, occur in SbSI. The less the unipolarity of the crystal, the more frequently it belongs to group b and, conversely, the greater the natural polarization, the more frequently the crystal belongs to group a. The internal field distribution of type a is detected, as a rule, after polarization of a type b crystal in an external field above the coercive

---

†The short-circuit photocurrent I_{sc}, which is due to screening, should not be confused with the photovoltaic current I, which was mentioned in Section 5.5. The different origin of these currents was also mentioned there. The photovoltaic current I is stationary and is observed in the absence of an internal macroscopic field $\tilde{\mathscr{E}}$ in a ferroelectric. The short-circuit photocurrent is a transitional current and is due to the screening of the internal field $\tilde{\mathscr{E}}$ by nonequilibrium carriers. When the light is switched on, this short-circuit photocurrent reaches a maximum and then falls off to zero or to a stationary value corresponding to the photovoltaic current. The value of I_{sc} is usually estimated by the maximum.

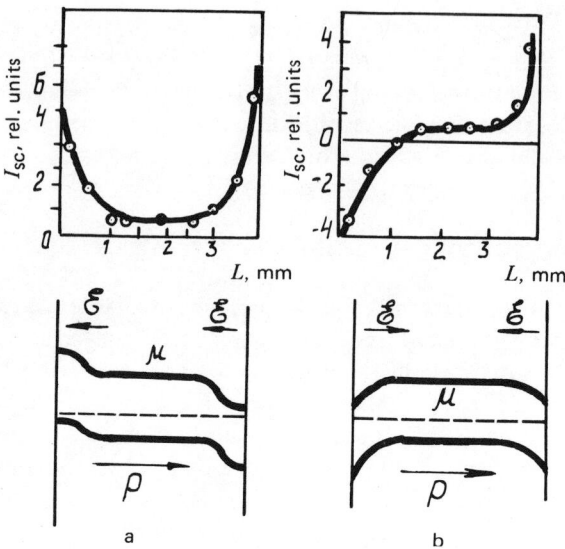

Figure 6.33. Two types of distributions of I_{sc} of a naturally polarized crystal of SbSI with light probing along the c axis. (a) Distribution of I_{sc} and curvature of the bands for group a crystals; (b) distribution of I_{sc} and curvature of the bands of group b crystals.

field. If this crystal is heated to a temperature above the Curie point and then cooled again, it again displays the type b distribution in the ferroelectric phase.

These and other results [236–238, 312] show that the nature of the internal field distribution in SbSI is determined by two mechanisms. The first mechanism leads to curvature of the bands and the internal field distribution presented in Figure 6.33a (type a). The type a distribution is undoubtedly related to the

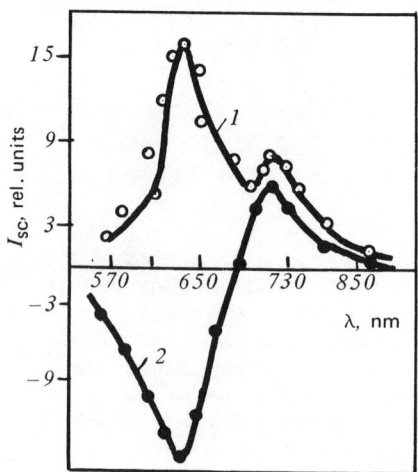

Figure 6.34. Spectral dependence of I_{sc} for crystals of group a (1) and b (2).

participation of the surface levels in screening since there is no degeneracy in SbSI. The second mechanism is related to charge of the surface levels and causes curvature of the bands and the distribution of the surface field in the paraelectric region. In the absence of ferroelectric polarization, this mechanism always leads to the type b distribution. This same mechanism, as was stated in Section 6.6, is responsible for the unipolarity of the crystal. These conclusions are confirmed by the spectral distribution of the photocurrent I_{sc}, taken for opposite ends of the crystal (cf. Figure 6.34). Curve 1 is the spectral distribution of I_{sc} for type a crystals; curve 2 is that for type b crystals. The type a spectral distribution is identical for both ends (curve 1) and displays both an intrinsic maximum at $\lambda \simeq 620$ nm, as well as an impurity maximum at $\lambda \simeq 720$ nm, which Bezdetnyi et al. [312] relate to transitions of electrons from the valence band to surface levels. The spectral distributions of I_{sc} for the opposite ends are different for crystals of group b. The spectral distribution at the positive end is the same as for a type a crystal (curve 1). There occurs at the negative end a spectral distribution with a change in sign of I_{sc} with the transition from the long-wavelength maximum to the short-wavelength maximum. Bezdetnyi et al. [3/2] relate this to the decrease in charge of the surface levels with illumination in the band $\lambda \simeq 720$ nm and, accordingly, with the change in sign of the surface curvature of the bands.

7

Ferroelectric Photoelectrets

One of the phenomena related to the screening of spontaneous polarization by nonequilibrium carriers is the formation of photoelectret polarization in a ferroelectric. The possibility of the formation of photoelectret polarization in ferroelectrics is based on the existence of photoconductivity and deep local levels in the forbidden band. The stability of photoelectret polarization, as is well known [273], is determined by a large value of u/kT (u is the activation energy of the local levels responsible for the photoelectret polarization, T is the temperature of the crystal). In SbSI, for example, photoelectret polarization is stable only at low temperatures, far from the Curie point. Meanwhile, the separation of ferroelectric and photoelectret (i.e., purely electronic) polarization is possible only in the region of the phase transition. In this region, the mutual effect of the two forms of polarization—ferroelectric and photoelectret—leads to a number of interesting phenomena. Thus, the photoelectret effect was investigated for photoconducting solid solutions of $SbSi_xBr_{1-x}$ with low-temperature phase transitions [313-315] and crystals of $BaTiO_3$, for which the photoelectret polarization is stable near the temperature of the transition from the tetragonal phase to the cubic phase [211]. The photoelectret effect can certainly be observed for a large number of wide-band ferroelectric photoconductors, for example, in other perovskite ferroelectrics of the $SbNbO_4$ type, niobates of lithium, barium, and strontium, etc. Thus, investigation of the photoelectret effect in ferroelectrics is of general interest for ferroelectric semiconductors.

7.1. Formation of Photoelectrets with Screening of Ferroelectric Polarization by Nonequilibrium Carriers

It has been established that illumination of a crystal in the ferroelectric phase by light from the region of intrinsic photosensitivity leads to the formation of photoelectret polarization. Since an external field is not applied to the crystal, formation of the photoelectret is naturally related to screening of the spontaneous polarization by nonequilibrium carriers. The coexistence of the two forms of polarization (ferroelectric and purely electronic) causes a number of interesting peculiarities in the behavior of the crystal in the region of the phase transition.

This effect has been investigated principally in SbSI crystals grown from the gaseous phase. The crystals had the needle form usual for $SbSI_{0.35}Br_{0.65}$. The following were determined from the temperature dependence of the reciprocal dielectric constant for these crystals: phase-transition temperature $T_1 = 170°K$, Curie-Weiss temperature $T_0 = 165°K$, Curie Weiss constant $C \simeq 2 \cdot 10^5$ deg, and maximum value of the dielectric constant $\epsilon_{max} \simeq 10^4$. The spontaneous polarization $P_0 \simeq 21 \cdot 10^{-6}$ C/cm^2 was determined from measurements of the hysteresis loop and pyroelectric current. The conductivity of the crystals at room temperature is $\sigma \simeq 10^{-9}$ $(\Omega \cdot cm)^{-1}$. The maximum of the photoconductivity at room temperature was found in the region 570-590 nm.

The sequence of observations and measurements was the following. A constant electric field was applied to the crystal in the paraelectric region in the direction of the ferroelectric axis. The crystal was cooled in the field to a temperature of 113°K, after which the external field was removed and the electrodes were short-circuited. The crystal was then heated at a constant rate. The pyroelectric current was recorded in the region of the phase transition. Its maximum (without preliminary illumination) occurred at $T = 159°K$. In other experiments the sequence of observation was the same, with the only difference being that after short-circuiting of the electrodes the crystal was illuminated through the wide slit of a monochromator in the region of the maximum of the photosensitivity. The short-circuit photocurrent caused by screening of the spontaneous polarization field by nonequilibrium carriers was observed at the time of illumination. The crystal was heated under the same conditions after cessation of the illumination, and the temperature dependence of the pyroelectric current was recorded. It was found that preliminary illumination shifts the maximum of the pyroelectric current by 1-2° toward higher temperatures, while the magnitude of the shift depends on the intensity of the light, the illumination time, and the magnitude of the pyroelectric current curve over time. The pyroelectric charge, as is usual, could be varied by varying the intensity of the external field applied to the ferroelectric in the process of preliminary polarization.

The effect of the temperature shift of the pyroelectric current maximum can be explained by assuming that illumination of the crystal in the ferroelectric phase leads to screening of the ferroelectric polarization by nonequilibrium carriers, while the screening charge, although partial, is formed by carriers captured at deep trapping levels. In this case, the decrease in the ferroelectric polarization at the phase transition is accompanied by an increase of the internal field of the photoelectret, which leads to a shift in the first-order phase-transition temperature to higher temperatures. This assumption was confirmed by direct measurement in the paraelectric region of the photoelectret charge in the crystal by its photodepolarization. After the crystal was illuminated at a temperature below the Curie point, it was converted to the paraelectric region. When the pyroelectric current was no longer observed, the crystal was again illuminated

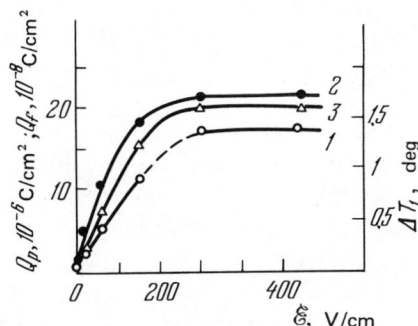

Figure 7.1. Dependence of the photoelectric charge density Q_f (1), pyroelectric charge density Q_p (2), and temperature shift of the maximum ΔT_1 (3) on the intensity of the external polarizing field.

at the sensivity maximum at a temperature of 183°K (in the paraelectric phase). The photodepolarization current, whose direction coincided with the direction of the pyroelectric current, was then observed. By integrating the photodepolarization current over time, the photoelectret charge was determined. The magnitude of the photoelectret charge thus determined correlates with the magnitude of the temperature shift of the pyroelectric current maximum: the shift increases with increasing charge. The results of these measurements are presented in Figure 7.1 in the form of the dependence of the pyroelectric charge density Q_p, photoelectret charge density Q_f, and temperature shift of the pyroelectric current maximum ΔT_1 on the external field intensity $\mathscr{E}$ used in polarizing the ferroelectric. We especially note the following results. All three dependences $Q_p = Q_p(\mathscr{E})$, $Q_f = Q_f(\mathscr{E})$, and $\Delta T_1 = \Delta T_1(\mathscr{E})$ correlate with each other and display a saturation section occurring in the same region of field. The values of Q_f are then two orders of magnitude smaller than the corresponding values of Q_p; this indicates that the measured photoelectret charge Q_f is only a small fraction of the total screening charge. Figure 7.2 presents the dependence of Q_f on the light intensity I (for fixed values of $\mathscr{E}$ and Q_p).

Parallel measurement of the photocurrent I_f displayed a linear intensity-current characteristic and showed that $Q_f \simeq I_f^{1/2}$ for the curve in Figure 7.2. The dependence of Q_f on the time of illumination of the crystal in the ferro-

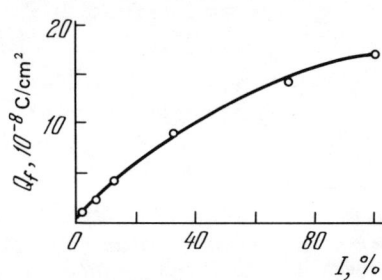

Figure 7.2. Dependence of Q_f on light intensity I.

electric region was measured in an analogous manner. Other measurements, the results of which are presented in Figure 7.3a, also confirm the assumptions made above. The sequence of these measurements was similar to that described above, with the only difference being that the temperature dependence of the dielectric constant $\epsilon = \epsilon(T)$ was measured instead of measuring the pyroelectric current. It is seen from Figure 7.3a that the maximum of $\epsilon = \epsilon(T)$ decreases with increasing Q_f and shifts toward higher temperatures. Illumination of the crystal in the paraelectric region and the corresponding depolarization of the photoelectret returns the maximum of $\epsilon = \epsilon(T)$ to the initial position corresponding to dark conditions (curves 1 and 1'). If illumination in the paraelectric region (and the corresponding photodepolarization) was not carried out, then the shift in the phase-transition temperature caused by the initial illumination of the crystal in the ferroelectric region was observed for an indefinitely long time. Corresponding to this, the photoelectret charge Q_f does not display a significant decrease with time (the time of observation in [313] was limited to several hours).

Further heating of the crystal in the paraelectric region leads to a decrease in both Q_f and ΔT_1. All this also indicates that the observed shift in the phase-transition temperature is related to the internal field of the photoelectret.

Three groups of independent measurements were carried out for additional confirmation of this mechanism.

In the first group of measurements, the dependence of the shift of the

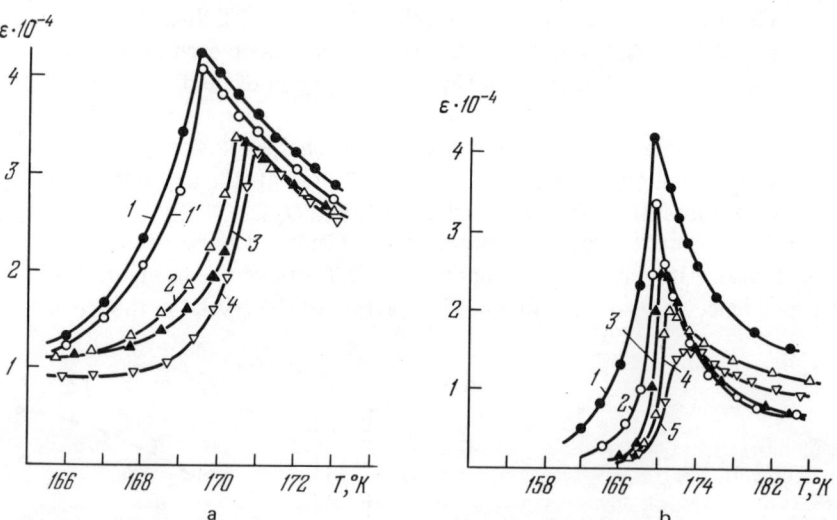

Figure 7.3. Temperature dependences of the dielectric constant. (a) For various photoelectret charge densities Q_f: (1) 0; (2) $5 \cdot 10^{-8}$; (3) $11 \cdot 10^{-8}$; (4) $17 \cdot 10^{-8}$ C/cm^2; curve 1' was obtained after photodepolarization. (b) Reversible dielectric constant for various values of external field.

FERROELECTRIC PHOTOELECTRETS

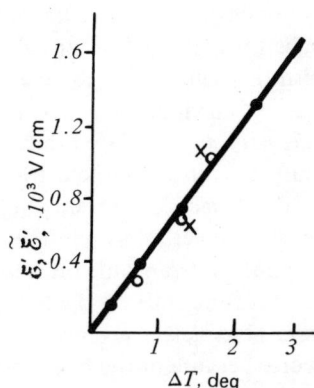

Figure 7.4. Shift of the phase-transition temperature with field, determined from three groups of independent measurements.

transition temperature T_1 on the external field intensity $\mathscr{E}$ and, consequently, the coefficient of the shift $dT_1/d\mathscr{E}$ was determined directly by investigating the temperature dependence of the reversible dielectric constant of the crystal. Figure 7.3b presents the curves of the temperature dependence of the reversible dielectric constant plotted by the usual technique. The curves in Figure 7.3b were used to determine the dependence of the shift ΔT_1 on the field intensity $\mathscr{E}$, which is presented in Figure 7.4, and the coefficient of the shift $dT_1/d\mathscr{E} \simeq 1.7 \cdot 10^{-3}$ deg/(V/cm) was determined.

The dependence of the temperature shift of the maximum of the pyroelectric current (from Figure 7.1) and also the dependence of the maximum of the dielectric constant (from Figure 7.3a) on the photoelectret charge density Q_f were determined in the second group of measurements. In order to compare the data of this group of measurements with the data of the preceding group of measurements, it is necessary to relate the charge density Q_f to the intensity of the internal field of the photoelectret $\tilde{\mathscr{E}}$. As is well known from the theory of photoelectrets [273], the intensity of the internal field $\tilde{\mathscr{E}}$ in the photoelectret with short-circuited electrodes is related to the charge density Q_f in the following manner:

$$\tilde{\mathscr{E}}_1 = \frac{4\pi Q_f}{\varepsilon} \frac{L-2d}{L}, \quad \tilde{\mathscr{E}}_2 = -\frac{4\pi Q_f}{\varepsilon} \frac{2d}{L}, \qquad (7.1)$$

where $\tilde{\mathscr{E}}_1$ and $\tilde{\mathscr{E}}_2$ are the field intensity in the electrode sections and in the volume of the crystal, respectively, d is the effective thickness of the electrode section in which the charge is concentrated, and L is the distance between the electrodes. Setting $d \ll L$ (a barrier type of distribution of the heterocharge [273]), one should assume that the field

$$\tilde{\mathscr{E}}_1 \simeq \frac{4\pi Q_f}{\varepsilon} \qquad (7.2)$$

is responsible for the observed shift in the phase-transition temperature. If one substitutes into (7.2) the values of Q_f corresponding to curve 1 in Figure 7.1 and also the value $\epsilon \simeq 2 \cdot 10^3$ corresponding to the temperature of the pyroelectric current maximum $T = 150°K$ (cf. curve 1 in Figure 7.3b), the values of the field intensity $\tilde{\mathcal{E}}_1$ calculated thus can be compared to the values of the temperature shift ΔT_1 of the pyroelectric maximum (Figure 7.1, curve 3) and the maximum of the dielectric constant (Figure 7.3a). The points corresponding to the calculated values of $\tilde{\mathcal{E}}_1$ and the corresponding values of ΔT_1 are presented in Figure 7.4, from which it is seen that they fit well a linear dependence of the temperature shift on the field, which is obtained from the measurements of the reversible dielectric constant. We note that this correspondence could have been worse because of the heterogeneity of the crystal in the region of the phase transition.

Finally, the values of the spontaneous polarization P_0 and the inverse dielectric constant $\chi_0 = 4\pi/\epsilon_0$ at the Curie point were determined in the third group of measurements. Hence, the coefficient of the shift in the phase-transition temperature with external field $dT_1/d\mathcal{E}$,

$$\frac{dT_1}{d\mathcal{E}} = \frac{C}{4\pi}\left(-\frac{4}{3}\frac{\gamma}{\beta}\right)^{1/2}, \tag{7.3}$$

$$\beta = -4\frac{\chi_0}{P_0^2}, \quad \gamma = 3\frac{\chi_0}{P_0^4}, \tag{7.4}$$

was determined from the Devonshire equations [15].

Table 10 presents values of the Curie-Weiss constant C, the Devonshire coefficients β and γ, and the coefficient of the shift in the transition temperature with field $dT_1/d\mathcal{E}$ for SbSI [126] and the solid solution $SbSI_{0.35}Br_{0.65}$ [313]. Hence it follows that the calculated values of $dT_1/d\mathcal{E}$ for both crystals are close in order of magnitude to the results of the preceding measurements presented in Figure 7.4. In conclusion, attention should also be drawn to the qualitatively identical nature of the dependences of the dielectric constant on the external field $\mathcal{E}$ (Figure 7.3b) and on the heterocharge density Q_f (Figure 7.3a). This, in particular, confirms the assumption expressed in the literature [15] that the dependence of the reversible dielectric constant on the field, similar to that presented in Figure 7.3b, is caused by the formation of electrode space charges.

The data of Figures 7.1 and 7.2 refer to the case of low light intensity I,

TABLE 10. Calculation of $dT_1/d\mathcal{E}$ for SbSI and Solid Solution

Crystal	$C, 10^5$ °K	$\beta, 10^{-13}$ cgs esu	$\gamma, 10^{-23}$ cgs esu	$\dfrac{dT_1}{d\mathcal{E}}, 10^{-3}$ deg/(V/cm)
$SbSI_{0.35}Br_{0.65}$	2	−2.9	5.6	0.8
SbSI	3.8	−3.2	7.4	1.8

when the illumination of the crystal was carried out through the monochromator. The increase in photoelectret charge density Q_f with light intensity (Figure 7.2) was not accompanied by a significant change in the pyroelectric charge density Q_p. Meanwhile, the photodomain effect was considered above (Section 6.6) for SbSI, which involves a decrease of Q_p with illumination of the previously polarized crystal in the ferroelectric region. In connection with this, the dependence of Q_f and Q_p on I in the region of high intensities was measured for the solid solution under investigation. The crystal, which was polarized in a field above the coercive field, was illuminated by "natural" light in the ferroelectric region. The sequence of all the measurements was as before. If was found that in the region of high light intensity I, the increase in the photoelectret charge density Q_f with light intensity I is accompanied by a decrease in the pyroelectric charge density $\Delta Q_p < 0$. This is illustrated by Figure 7.5, in which the dependences of Q_f, $|\Delta Q_p|$, and ΔT_1 on I are presented. The saturation observed at high light intensities I corresponds to a high photoelectret charge density $Q_f \simeq 2 \cdot 10^{-6}$ C/cm². However, evidently $Q_f \ll Q_p$. The existence in a ferroelectric of such large values of the photoelectret charge (as compared with other ferroelectrets [273]) can be understood if one takes into account the high dielectric constant of the ferroelectric, which reduces the internal field in the crystal to values typical for photoelectrets [cf. (7.1)]. It is interesting that although the values of Q_f increased by almost an order of magnitude (as compared to the corresponding values in Figure 7.1), the shift of the phase-transition temperature increased insignificantly. This can be caused by two factors. The values of Q_f along curve 1 of Figure 7.5 correspond to large internal fields $\tilde{\mathcal{E}}$, which can correspond to saturation in the dependence of ΔT_1 on $\tilde{\mathcal{E}}$. Thus, for example, the saturation charge $Q_f = 1.9 \cdot 10^{-6}$ C/cm² and $\epsilon = 2 \cdot 10^3$ correspond according to (7.2) to the field $\tilde{\mathcal{E}} \simeq 10.5 \cdot 10^3$ V/cm. At the same time, the data of Figure 7.4 illustrating the linear dependence of ΔT_1 on $\mathcal{E}$ were obtained for fields not exceeding $1.3 \cdot 10^3$ V/cm. The complex nature of the internal field distribution in the photoelectret can be another possible reason.

Screening of spontaneous polarization in BaTiO₃ also leads to the formation of photoelectret polarization [211]. A previously polarized and short-circuited crystal was illuminated for a long time in the ferroelectric phase ($T = 25°C$),

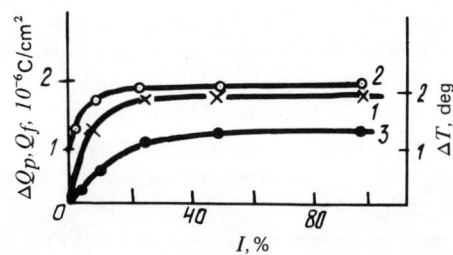

Figure 7.5. Dependence of ΔQ_f (1), ΔQ_p (3), and ΔT_1 (2) on light intensity (region of high intensities).

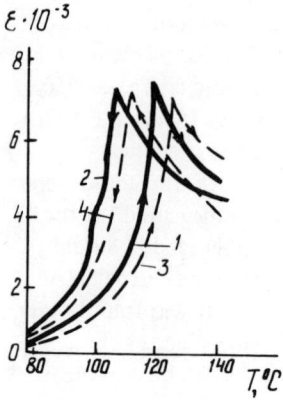

Figure 7.6. Effect of illumination in the ferroelectric phase at the Curie temperature of $BaTiO_3$. (1, 2) in darkness; (3, 4) after illumination by photoactive light in the ferroelectric phase.

after which the Curie temperature was measured in darkness by measuring the temperature dependence of ϵ (Figure 7.6, curves 3, 4). Comparison with the initial equilibrium values of the phase-transition temperature determined from curves 1 and 2 reveals an upward shift of the Curie point, where $\Delta T_1 = +4.8°C$.

The transition point returns to the initial equilibrium value in darkness after a time of the order of an hour. Repeated illumination of the crystal in the paraelectric region depolarizes the crystal and accordingly returns the Curie temperature to its equilibrium value. An estimate of the internal field responsible for the shift of the phase-transition point upward was made by determining the screening space charge in $BaTiO_3$. For this, the crystal was repeatedly illuminated in the paraelectric region, and the photodepolarization current was simultaneously measured. The screening charge density Q_f was determined by integrating the photodepolarization current over time, and the internal field of the space charge responsible for the observed shift of the phase-transition point toward higher temperatures was determined from (7.2). For the case presented in Figure 7.6, $\tilde{\mathcal{E}}_1 \simeq 4 \cdot 10^3$ V/cm, from which $dT_1/d\mathcal{E} \simeq +1.2 \cdot 10^{-3}$ deg/(V/cm), which is close to the coefficient of the phase-transition-point shift in $BaTiO_3$ under the effect of an external electric field $dT_1/d\mathcal{E} \simeq +(1-1.2) \cdot 10^{-3}$ deg/(V/cm) [15]. The magnitude of the screening charge and the corresponding internal field and shift of the Curie temperature increases with increasing light intensity. The shift also increases with increasing preliminary polarization field, reaching saturation for fields corresponding to saturation polarization and pyroelectric charge.

Both shifts of the Curie temperature with illumination described above display a slow dark relaxation caused by localization of the nonequilibrium carriers at relatively deep trapping levels.

We note in this regard that the shift of the Curie temperature toward lower temperatures presented in Figure 5.15 is actually a sum with respect to the above-described effects of opposite sign. To isolate the "pure" effect of the shift of the Curie point downward in $BaTiO_3$, the observation should be made under conditions where the photoelectret polarization does not arise. For this, the crystal with short-circuited electrodes was illuminated only in the paraelectric region. The Curie point determined after this from the maximum of $\epsilon(T)$ was shifted downward with respect to the dark point by 7–8°C.

The dependence of the photoelectret charge on the concentration of nonequilibrium carriers can be obtained from (3.71). Under the assumption Q_p = const, (3.71) gives a clear dependence of the photoelectret charge density Q_f on the screening length l_D and, consequently, on the concentration of nonequilibrium carriers. For low light intensities, when $l_D/\epsilon_s \gg l/\epsilon_l$, $Q_f \sim l_D^{-1}$. This case evidently corresponds well to the data of Figure 7.2. In the region of high light intensities, when $l_D/\epsilon_s \ll l/\epsilon_l$, saturation occurs: $4\pi Q_f \simeq D_0 - 4\pi Q_p$. However, the saturation section in the experimental curve of Figure 7.5 is evidently caused by another mechanism corresponding simultaneously to the decrease in the pyroelectric charge Q_p, whereas relation (3.71) is valid only for Q_p = const.

As for the decrease in the pyroelectric charge in the region of high intensities (curve 3 in Figure 7.5), two possible mechanisms can cause this, The first is related to optical recharging of levels, when carriers transfer from surface levels (or from shallow trapping levels in the volume) to deep levels in the volume of the crystal. The decrease in the pyroelectric charge Q_p is compensated by the increase in the photoelectret charge Q_f. The photodomain effect (Section 6.6) can be another possible cause.†

7.2. Polarization of a Ferroelectric under the Effect of the Internal Field of the Photoelectret

It is clear from the discussion in the preceding section that there must exist an inverse effect involving the polarization of the ferroelectric under the effect of the internal field of the photoelectret. This effect was observed and investigated for solid solutions of $SbSI_{0.35}Br_{0.65}$ in [314].

The crystal was illuminated in the region of the photosensitivity maximum with the application of an external electric field in the paraelectric region at a temperature of 173°K. Thus, the photoelectret state was produced in the crystal, where the photoelectret charge could be measured by photodepolarization in the paraelectric region. The photoelectret with short-circuited plates was then

---

†The last investigations have shown that the above-described phenomenon for $SbSI_{0.35}Br_{0.65}$ crystals is strongly influenced by the anomalous photovoltaic effect [239] and caused mainly by the creation of the high photovoltaic field $\widetilde{\mathscr{E}}$. For $BaTiO_3$ crystals the anomalous photovoltaic effect is excluded due to the high conductivity and, correspondingly, to the low photovoltaic field $\widetilde{\mathscr{E}}$ in the region of the phase transition.

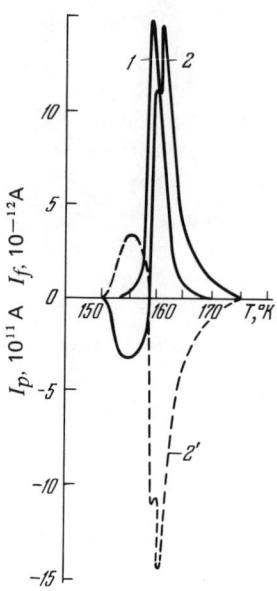

Figure 7.7. Temperature maxima of the pyroelectric current for a ferroelectric polarized in an external field (I_p–curve 1) and internal field of the photoelectret (I_f–curves 2, 2′).

cooled below the phase-transition temperature and held for some time in the ferroelectric region. The crystal was again heated after this and transformed to the paraelectric phase, while the pyroelectric current, whose temperature dependence is presented in Figure 7.7, was recorded in the region of the phase transition. Curve 1 in this figure corresponds to the ordinary pyroelectric current obtained after polarization of the crystal in an external electric field of ∼200 V/cm (in darkness). Curve 2 is the temperature dependence of the pyroelectric current corresponding to the photoelectret obtained in the paraelectric region in a polarizing field of the same intensity and the same direction. Curve 2′ is the pyroelectric current obtained for the photoelectret formed in a field of the same intensity, but in the opposite direction. It is seen from Figure 7.7 that the pyroelectric currents caused by polarization of the ferroelectric under the effect of the internal field of the photoelectret have a complex nature and consist of two maxima, whose direction coincides and opposes the direction of the ordinary pyroelectric current (for brevity, we will call them the direct and reverse pyroelectric currents, respectively).

The maximum of the direct pyroelectric current is always much greater than the maximum of the inverse, while they both increase with increasing photoelectret charge.

After having produced the photoelectret state once in the crystal in the paraelectric region, one can observe the same pattern of the pyroelectric current corresponding to a fixed photoelectret charge with multiple repetition of the cooling and heating cycle and conversion of the crystal from one phase to the other (the time of these observation did not exceed several hours). Under these conditions, the ferroelectric photoelectret is essentially a pyroelectric. Depolarization of the photoelectret with illumination in the paraelectric region corresponds simultaneously to the ferroelectric depolarization: the pyroelectric charge decreases to the value corresponding to the unpolarized ferroelectric.

The data of quantitative investigation of this effect are presented in Figure 7.8. Figure 7.8a presents the dependence of the pyroelectric charge density Q_p on the external field intensity $\mathscr{E}$ (analogous to curve 2 in Figure 7.1). The dependence of the photoelectret charge density Q_f on the external field intensity $\mathscr{E}$ applied to the crystal in the paraelectric region is presented in Figure 7.8b. Figure 7.8c and d presents the charge densities Q_p^+ and Q_p^- corresponding to the direct and reverse pyroelectric current as a function of the magnitude and direction of the polarizing field. It is seen from Figure 7.8 that the direct and reverse pyroelectric currents increase with intensity of the polarizing field. Two sections are observed in the dependence of Q_p^+ on $\mathscr{E}$ — a rapid and slow growth. The presence of a significant direct pyroelectric current and the correspondingly high pyroelectric charge density $Q_p^+ > Q_p^-$ uniquely indicates the significant role played by the electrode photoelectret field $\mathscr{E}_1$ in the process of ferroelectric polarization [cf. (7.1) and (7.2)]. The reverse pyroelectric current (characterized by the corresponding charge Q_p^-) can be caused by ferroelectric polarization under the effect of the photoelectret field $\tilde{\mathscr{E}}_2$. To explain this process, it is necessary to investigate the relation between the distribution of the internal photoelectret field and the related magnitude and nature of the ferroelectric polarization.

This relation is of fundamental interest for the theory of ferroelectric polarization. The possibility of changing the nature of the distribution of the internal field in the photoelectret in an arbitrary manner and thereby affecting the possible mechanism of ferroelectric polarization is of significant interest.

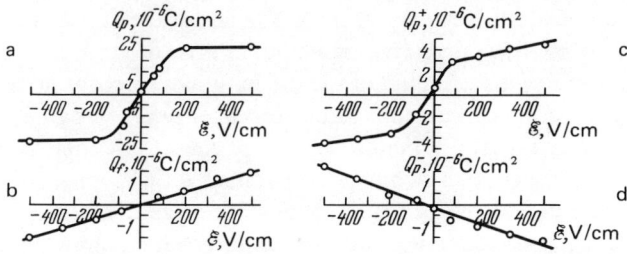

Figure 7.8. Dependence of the densities of the direct Q_p^+ and reverse Q_p^- pyroelectric charges on the photoelectret charge density Q_f.

We recall that the problem of the existence and relative role of two basic mechanisms has not been solved in the theory of ferroelectric polarization. The first of these is related to the formation of nucleation centers and their intergrowth along the polarizing field, and the second is related to lateral motions of the 180° domain walls [15].

A change in the distribution of the internal field in the photoelectret was carried out in [314] with the help of the probe technique, and the effect of this change on the ferroelectric polarization was investigated. The measurements were carried out in the following manner. A crystal of the solid solution $SbSI_{0.35}Br_{0.65}$ in the paraelectric region at a temperature of $\sim 175°K$ was illuminated in the region of the photoconduction maximum with an external electric field applied in the direction of the c axis of the crystal. Thus the direction of the photoelectret field coincided with the direction of the ferroelectric axis of the crystal. Investigation of the distribution of the internal field of the photoelectret produced in this manner was carried out by means of its local depolarization with the help of a light probe. This technique is described in detail in [273]. The nature of the internal field distribution in the photoelectret was compared with data on the ferroelectric polarization. For this, the photoelectret with short-circuited electrodes was converted to the ferroelectric region by lowering the temperature, held for strictly identical times at a temperature of $\sim 113°K$, after which it was again converted to the paraelectric region by heating. The pyroelectric current, whose direction and magnitude depended on the internal field distribution in the photoelectret, was then recorded in the region of the phase transition. Since, as has already been emphasized above, the ferroelectric is a photoelectret and a stable pyroelectric, observation of the pyroelectric current for the same photoelectret can be carried out repeatedly.

Figure 7.9 gives an idea about the typical nature of the internal field distribution in the photoelectret in the paraelectric region (a) and the corresponding temperature dependence of the pyroelectric current (b), which repeat curves 1 and 2 in Figure 7.7. The direction of the direct and reverse pyroelectric currents changes to the opposite with a change in direction of the internal field in the photoelectret (Figure 7.9, curve 2′). Under the effect of the photoelectret field, the pyroelectric current maximum shifts toward higher temperatures as compared to the maximum of the ordinary pyroelectric current.

The complex nature of the temperature dependence of the pyroelectric current presented in Figure 7.9b indicates the presence in the short-circuited photoelectret of a region with two oppositely directed fields, direct and reverse. As in [273], here and below by the direct field we mean the internal field of the photoelectret, whose direction coincides with the direction of the external field applied to the crystal with the formation of its photoelectret polarization, and accordingly by the reverse field—the field in the opposite direction.

Meanwhile, the probe technique displays the distribution of only the reverse

FERROELECTRIC PHOTOELECTRETS

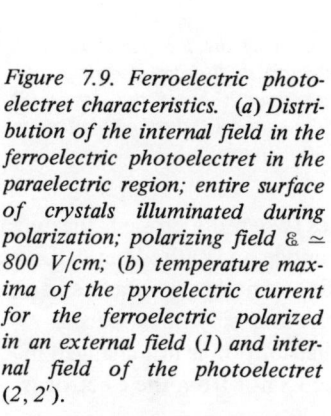
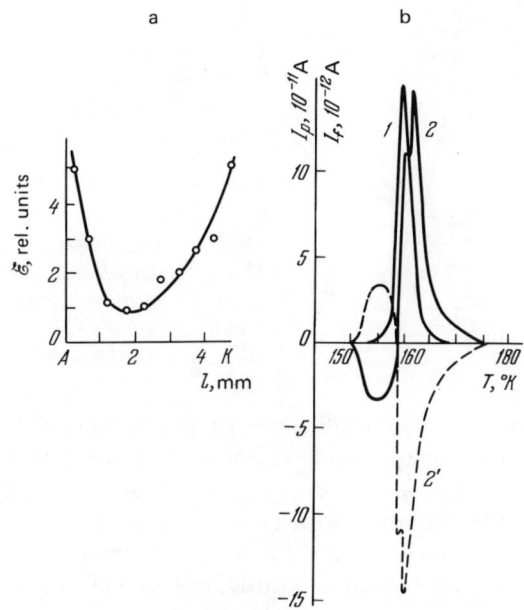

Figure 7.9. Ferroelectric photoelectret characteristics. (a) Distribution of the internal field in the ferroelectric photoelectret in the paraelectric region; entire surface of crystals illuminated during polarization; polarizing field $\mathcal{E} \simeq 800$ V/cm; (b) temperature maxima of the pyroelectric current for the ferroelectric polarized in an external field (1) and internal field of the photoelectret (2, 2').

field, since the field in Figure 7.9a does not change sign anywhere. Keeping in mind the electric neutrality of the short-circuited photoelectret, we are led to the conclusion that there exists a compensating direct field in the narrow electrode region of the crystal (narrow at least as compared to the thickness of the probe). According to the data of Figure 7.9b, the electrode direct field of the photoelectret plays the decisive role in the process of ferroelectric polarization. In this case, the maximum of the reverse pyroelectric current (the minor maximum) is naturally related to the existence of the reverse field in the volume of the photoelectret.

The following observations were directed to relate the change in nature of the distribution of the internal field of the photoelectret to a possible change in the pyroelectric current. For this purpose, with the formation of the photoelectret in the paraelectric region, the crystal was illuminated not completely, but with the help of a light probe, which could be set at an arbitrary position between the anode and cathode. As is well known from [273], this changes the relation between the direct and reverse internal fields in the photoelectret. Figure 7.10 presents the curves of the distribution of the internal field corresponding to different positions of the light probe with formation of the photoelectret. The obtained results permit us to reach two conclusions. First of all, the curves of the distribution correspond to the reverse field of the photoelectret, and, consequently, the compensating direct field must be concentrated in the narrow electrode region. Second, formation of the photoelectret in the crystal

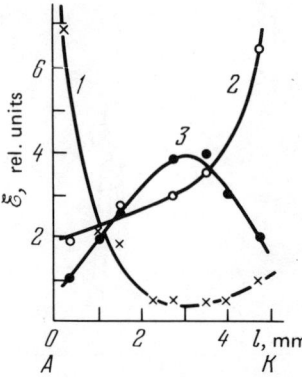

Figure 7.10. Distribution of the internal field in the ferroelectric photoelectret in the paraelectric region. Light probe located respectively at the anode (1), cathode (2), and at the center (3) during polarization; polarizing field $\mathcal{E} \simeq 800$ V/cm.

is related both to the internal distribution of charges (when the probe is located at the center) and to injection of carriers from the anode and cathode into the crystal (when the probe is located near the electrode). Analogous measurements were performed with a crystal insulated from the electrodes by mica spacers. They showed that the nature of the internal field distribution is not changed in this case and, consequently, injection of carriers occurs not from the electrodes, but from some surface layer.

Figure 7.11 presents the dependence of the pyroelectric current on the position of the light probe with formation of the photoelectret. It was obtained by integrating each of the maxima of curve 2 in Figure 7.9b. An interesting

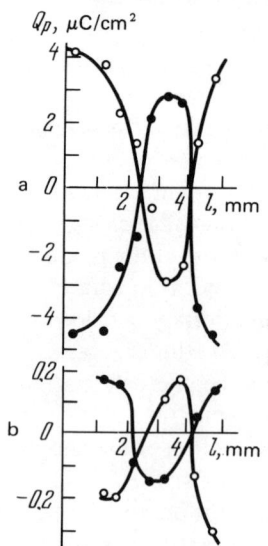

Figure 7.11. Dependence of the major (a) and minor (b) pyroelectric currents on the position of the light probe with formation of the photoelectret.

result of these measurements is that the change in nature of the internal field distribution of the photoelectret leads to a change not only in magnitude but also in the direction of the corresponding pyroelectric currents. It is seen by comparing Figures 7.10 and 7.11 that with displacement of the light probe from the electrode region to the center of the crystal and the corresponding replacement of the injection mechanism of the photoelectret polarization by the mechanism of internal separation of charges, the direction of ferroelectric polarization changes to the opposite. In the first case (probe near the electrodes), the major maximum of the pyroelectric current corresponds to the direct pyroelectric current, and, consequently, the direct field of the photoelectret localized in the surface layer near the electrode is decisive in the process of ferroelectric polarization. In the second case (probe at the center of the crystal), the major maximum of the pyroelectric current corresponds to the reverse pyroelectric current, and the reverse field in the volume of the photoelectret is accordingly responsible for the ferroelectric polarization. As is seen from Figure 7.11b, the minor maximum of the pyroelectric current changes in the opposite phase with the major maximum and, consequently, it is related to the reverse field of the photoelectret in the first case and to the direct field in the second case.

It was found possible to evaluate the role of the direct and reverse field of the photoelectret in the process of ferroelectric polarization with the help of measurements of the temperature dependence of the dielectric constant. It was found that the latter also depends on the nature of the distribution of the internal field of the electret. Figure 7.12, curve 1, corresponds to the unpolarized sample; curve 2 corresponds to the photoelectret produced by polarization with the light probe at the center of the crystal. These results indicate that the strong direct field of the photoelectret in the electrode region decreases the

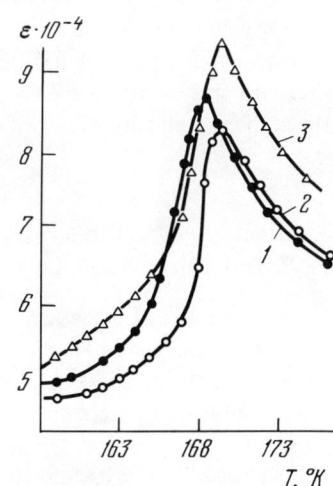

Figure 7.12. Temperature dependence of the dielectric constant.

dielectric constant of the crystal and, on the contrary, the reverse field in the volume of the photoelectret leads to an increase of the dielectric constant.

The observed peculiarities in the distribution of the internal field of the ferroelectric photoelectret and its effect on the ferroelectric polarization are naturally related to the existence of surface layers with low dielectric constant, which play the role of dielectric gaps (cf. Section 3.5).

It is natural to assume that the direct field of the photoelectret is concentrated in the surface layer and, moreover, the latter plays the role of injector of current carriers into the volume. The formation of the photoelectret is caused by the application of two polarization mechanisms: injection of carriers from the surface layer into the volume of the crystal and the volume separation of charges. When the probe is located close to the electrode in the process of polarization, the injector mechanism prevails, which leads to the formation of a strong direct field in the surface layer. This is also confirmed, in particular, by the effect of the photoelectret field on the dielectric constant ϵ (Figure 7.12, curve 2). As in the case of the reversible dielectric constant, the observed increase in ϵ is related to the formation of a strong field in the surface layer of the ferroelectric [15]. The data of Figure 7.11 indicate that the field in the surface layer in this case determines the ferroelectric polarization mechanism.

When the light probe is placed at the center of the crystal in the polarization process, the volume separation of charges should be considered as the predominating photoelectret polarization mechanism. In this case, the direct field in the surface layer screening the reverse field distribution in the volume of the crystal is relatively small and does not determine the ferroelectric polarization mechanism. The latter is determined by the field in the volume of the photoelectret in the case under consideration. This is also confirmed by measurement of the dielectric constant (Figure 7.12, curve 3). In accordance with thermodynamics, the increase of the field in the volume of the crystal leads to an increase of ϵ [15].

Thus, by changing the nature of the internal field distribution in the crystal, one can not only separate the two basic mechanisms of ferroelectric polarization, but also observe both the negative and the positive field effect on the dielectric constant.

7.3. Electron Mechanism of the Effect of Radiation on Ferroelectric Polarization Reversal

In investigating the coexistence of the two forms of polarization (ferroelectric and photoelectret) in solid solutions of $SbSI_{0.35}Br_{0.65}$, it was found possible to reach several conclusions with respect to the mechanism of the effect of nonequilibrium carriers on the process of ferroelectric reversal [315] (cf. Chapter 6).

Photoelectret polarization formed with screening of the ferroelectric polarization by nonequilibrium carriers was investigated. The crystal was polarized by cooling and transition from the paraelectric to the ferroelectric phase in an external field. After removal of the field and short-circuiting the electrodes, the ferroelectric was illuminated in the natural spectral range. As has already been noted in Section 7.1, this led to screening of P_0 by nonequilibrium carriers and the formation of the photoelectret charge caused by localization of the screening charge at deep levels. The effect of the photoelectret polarization thus produced on the ferroelectric polarization of the crystal was investigated in [315]. For this, the crystal, previously polarized in a field above the coercive field and illuminated, was transformed by heating to the paraelectric region; the pyroelectric current corresponding to a pyroelectric charge $Q_p \simeq 20\ \mu C/cm^2$ (saturation pyroelectric charge) was recorded at the phase-transition temperature. The crystal was then again cooled without an external field and transformed to the ferroelectric phase, after which it was again heated, where the pyroelectric current corresponding to $Q_p \simeq 18\ \mu C/cm^2$ was observed at the phase-transition temperature. This is close to the value of P_0 (we note that the unpolarized crystal displays $Q_p \simeq 0.4\ \mu C/cm^2$ because of unipolarity). The results obtained indicated that the internal photoelectret field below the Curie point T_1 causes the ferroelectric polarization of the crystal. The photoelectret produced with screening of spontaneous polarization is stable in the temperature region under investigation and is thus also the analog of a pyroelectric. An investigation of this phenomenon with screening of P_0 shows a number of peculiarities.

1. The dependence of the magnitude of P_0 evaluated from the pyroelectric charge on Q_f was investigated. As is seen from Figure 7.5, Q_f in the crystals under investigation can increase from 0 to $\sim 2\ \mu C/cm^2$ with increasing light intensity. Table 11 presents the dependence of the pyroelectric charge density on the transmission of neutral light filters used for changing the light intensity and, accordingly, Q_f. The maximum light intensity corresponded to a density $Q_f \simeq 2\ \mu C/cm^2$. As is seen from the data of the table and as has already been noted above, the presence of $Q_f \simeq 2\ \mu C/cm^2$ caused a direct pyroelectric current and, accordingly, $P_0 \simeq 18\ \mu C/cm^2$, which corresponds to almost complete saturation of the polarization. Moreover, the same magnitude of the photoelectret charge produced in the paraelectric phase by illumination of the crystal in an external field leads to a value of P_0 which was an order of magnitude smaller (Section 7.2). As is seen from the data of the table, not only did the pyroelectric charge decrease with a decrease of Q_f, but the nature of the pyroelectric current also changed; a second maximum of the reverse pyroelectric current appeared along with the direct pyroelectric current. We recall that an analogous structure of the pyroelectric current was observed for the photoelectret produced in the paraelectric phase (Section 7.2).

2. The effect of the internal photoelectret field on the process of ferroelectric polarization reversal was investigated with the help of the dielectric

TABLE 11. Effect of the Light Intensity on the Pyroelectric Charge

Transmission of filters, %	Q_p in direct direction, $\mu C/cm^2$	Q_p in reverse direction, $\mu C/cm^2$	Transmission of filters, %	Q_p in direct direction, $\mu C/cm^2$	Q_p in reverse direction, $\mu C/cm^2$
0	2	0.15	2	8.4	0.6
0.6	3.3	0.2	3	10	0
1	6.2	1	100	18	0

hysteresis loop. The formation of the photoelectret was carried out according to the technique described above. The crystal with short-circuited electrodes, previously polarized in an external field, was illuminated with natural light at a temperature below T_1. After this, the crystal was transformed to the paraelectric phase by heating and again cooled below T_1. The dielectric hysteresis loops were then observed with the help of the usual technique. Figure 7.13a

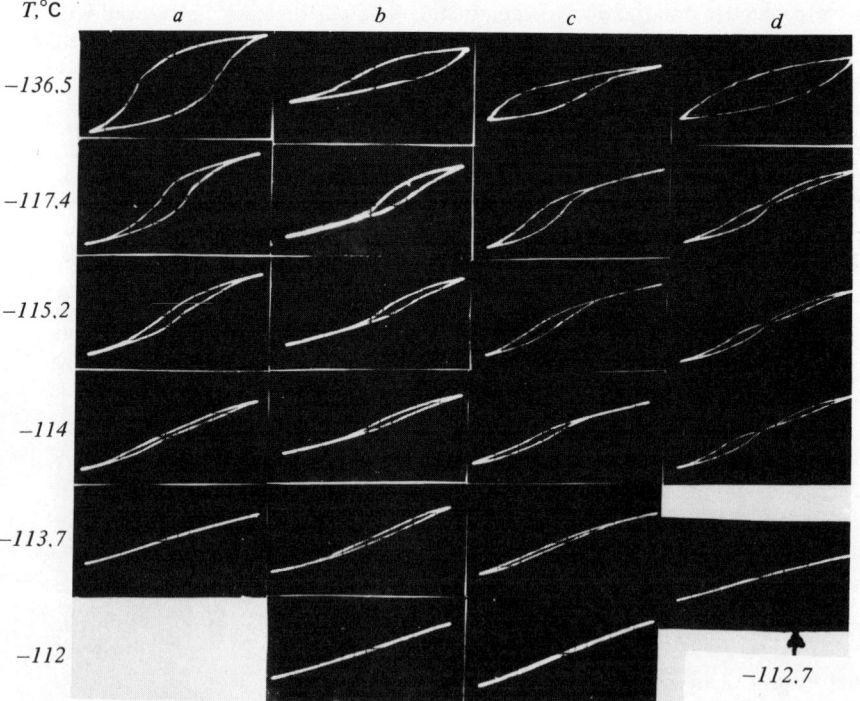

Figure 7.13. Effect of illumination of the ferroelectric on the process of ferroelectric polarization reversal.

presents the hysteresis loops for a crystal in which there is no photoelectret charge (without preliminary illumination in the polarized state). Figure 7.13b and c presents the hysteresis loops for the same crystal with $Q_f \simeq 2 \ \mu C/cm^2$ for two opposite directions of photoelectret polarization. The effect of the photoelectret field on the polarization reversal appeared in the asymmetry of the loops. Moreover, the shift of T_1 toward higher temperatures, which is caused by the photoelectret field and observed with the help of pyroelectric measurements (Section 7.1), is clearly seen in Figure 7.13. Depolarization of the photoelectret by repeated illumination of the crystal with natural light in the paraelectric phase reproduces the symmetric form of the hysteresis loops (transition from the loops of Figure 7.13b or c to the loops of Figure 7.13a) and returns T_1 to the initial equilibrium value. The symmetric form of the hysteresis loop is also recovered if the crystal is illuminated with natural light in the process of observing the hysteresis loop. This is illustrated by Figure 7.14. The measurements showed that the recovery time for the hysteresis loop is close to Maxwellian time $\tau = \epsilon/4\pi\sigma$. Thus, at a temperature of $-119°C$ (corresponding to a dielectric constant $\epsilon \simeq 4 \cdot 10^4$ and photoconductivity $\sigma \simeq 1.4 \cdot 10^{-10} \Omega^{-1} \cdot cm^{-1}$), $\tau \simeq 25$ sec, which agrees well with the data of Figure 7.14. The loop recovery time decreases with increasing light intensity and, consequently, photoconductivity.

The same magnitude of photoelectret charge produced in the paraelectric region by illumination of the crystal in an external electric field did not change the form of the hysteresis loop, which agrees with the data of the pyroelectric measurements presented above.

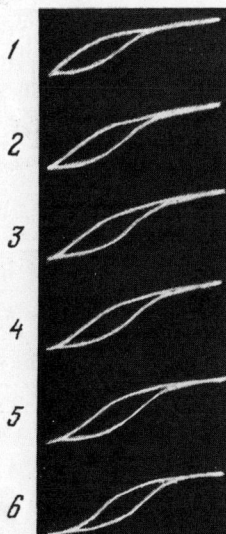

Figure 7.14. Effect of natural illumination on the asymmetric hysteresis loops and their recovery during polarization reversal. (1) 0; (2) 3; (3) 7; (4) 10; (5) 30; (6) 50 sec; $\sigma = 14 \cdot 10^{-11}$ $\Omega^{-1} \cdot cm^{-1}$.

3. It follows from the preceding that the internal field of the photoelectret affects the ferroelectric polarization differently depending on the conditions for formation of the photoelectret polarization in the crystal. This means that the process of ferroelectric polarization is determined not only by the magnitude of the photoelectret charge but also by the distribution of its internal field. The distribution of the internal field of the photoelectret formed with screening of P_0 by nonequilibrium carrier was investigated by the probe method in [315]. The obtained distribution is presented in Figure 7.15, and actually displays a different nature than in Figures 7.9 and 7.10. The distribution in Figure 7.15 displays a section of direct field evidently determining the ferroelectric polarization and the corresponding direct pyroelectric current. The other difference consists of the asymmetric nature of the internal field distribution.

4. Illumination of the unpolarized ferroelectric in the ferroelectric phase with natural light also strongly affects the process of ferroelectric polarization and polarization reversal. The unpolarized ferroelectric at a temperature below T_1 was illuminated in the region of its natural photosensitivity. The dielectric hysteresis loops presented in Figure 7.13d were measured after this in darkness. Comparison of the loops for the unilluminated (Figure 7.13a) and illuminated (Figure 7.13d) crystals indicates that the previous illumination of the unpolarized ferroelectric hinders the process of ferroelectric polarization reversal, which is equivalent to a sharp increase in the coercive field with illumination. The data of the pyroelectric measurements also indicate this. The unpolarized and illuminated ferroelectric was transformed to the paraelectric phase and then cooled below T_1 in a field above the dark coercive field. With subsequent heating, the crystal displayed $Q_p \simeq 1\ \mu C/cm^2$ (whereas the saturation pyroelectric charge $Q_p \simeq 20\ \mu C/cm^2$ was always measured for the unilluminated crystal under the same conditions). The pyroelectric charge increased somewhat with increasing polarizing field, but was much less than the saturation pyroelectric charge in the region of accessible fields. Moreover, as is seen from Figure 7.13d, illumina-

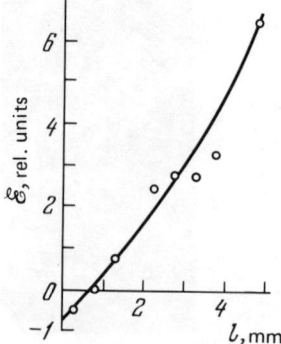

Figure 7.15. Distribution in the paraelectric region of the crystal of the internal field formed as a result of screening of the spontaneous polarization by nonequilibrium carriers.

FERROELECTRIC PHOTOELECTRETS

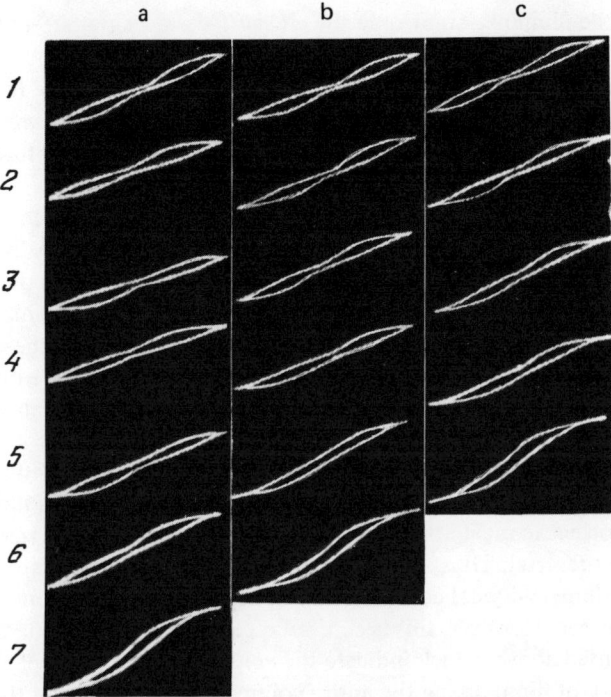

Figure 7.16. Effect of natural illumination on the double hysteresis loops and their recovery during polarization reversal: a, b, and c correspond to three different light intensities in order of increase; 1-7 denote different observation times.

tion of the crystal in the ferroelectric phase leads to the generation of a double hysteresis loop near T_1. If the unpolarized crystal previously illuminated in the ferroelectric phase is again illuminated in the paraelectric phase, its coercive field decreases to the dark value: the saturated form of the hysteresis loop is recovered, the pyroelectric charge increases and reaches saturation, and the double hysteresis loops disappear. Illumination of the crystal with natural light in the process of observing the hysteresis loop also re-establishes the form of the hysteresis loop. This is illustrated by Figure 7.16, which shows the kinetics of the hysteresis-loop recovery for three different light intensities and the corresponding three values of photoconductivity σ at a temperature of $-119°C$ (the value of σ was measured near the phase transition in the paraelectric phase for the same light intensities). The Maxwellian time for the three values of σ are 25, 75, and 180 sec, respectively.

The individual observations showed that the distortion and recovery of the hysteresis loop is produced only with illumination of the entire volume of the crystal, while illumination of only the end surfaces does not affect the form of the loops and does not lead to any sort of significant unipolarity.

The results presented above indicate that illumination of the ferroelectric with natural light in the ferroelectric phase strongly affects the process of ferroelectric polarization and polarization reversal, where the nature of this effect differs for the polarized and unpolarized states.

In the unpolarized state, curvature of the bands at the surface of the ferroelectric is determined by the charge of the surface levels. This leads to the existence of oppositely directed fields in the screening length at the opposite end surfaces of the crystal, perpendicular to the direction of spontaneous polarization. According to the assumptions discussed in Section 6.6, these fields lead to the generation of nucleation centers of antiparallel domains, while the difference in charge of the opposite surfaces and the corresponding fields can be one of the causes of unipolarity of the ferroelectric.

When the ferroelectric polarization process is determined by nucleation-center formation, the change in charge of the surface levels with illumination of the ferroelectric must affect the process of ferroelectric polarization and polarization reversal. Thus, the effect of illumination on polarization reversal of a multiple-domain crystal could have been explained starting from the model discussed above. However, this is contradicted by the entire set of experimental facts presented above, which indicate the volume nature of this phenomenon: the necessity of illuminating the entire volume of the crystal and the coincidence of the recovery time of the loop form with the Maxwellian time.

Thus, it is natural to conclude that the effect of illumination on the polarization reversal of a multiple-domain crystal is related to the process of nonequilibrium carrier screening of the domain boundaries in the volume of the crystal and their corresponding immobilization. Domain boundaries should then be understood both as boundaries of tapered domains as well as other possible domain boundaries not parallel to the c axis and observed in SbSI. The recovery of the form of the hysteresis loop with illumination of the crystal in the paraelectric phase or in the process of polarization reversal is then understood since the disappearance of the domain structure or its restructuring in these cases makes possible the disappearance of the screening charges localized at trapping levels due to the photoconductivity. Of course, the kinetics of hysteresis-loop recovery must then be described by a time close to the Maxwellian time.

We note in conclusion that the stable nature of the phenomena described above is related to the large value of u/kT and these phenomena are thus weakly expressed or not observed at all in SbSI at room temperatures (although the effect of illumination on the "molding" of the hysteresis loops in SbSI was observed in [295]). On the other hand, in such wide-band ferroelectrics at $BaTiO_3$, these phenomena are also observed in the region of the high-temperature phase transition. One can assume that the previously observed analogous effect of hard radiation on the hysteresis loop and the process of polarization reversal in Rochelle salt, triglycine sulfate [316], and barium titanate [5] are related not only to the generation of radiation defects, as previously assumed, but also to screening of the domain boundaries by nonequilibrium carriers. This does not contradict the possible absence of the significant photoconductivity for Rochelle salts and for triglycine sulfate because of the short lifetime and low carrier mobility in these crystals.

Finally, the formation of photoelectret polarization can be related not only to screening of spontaneous polarization but also to the presence of photovoltaic currents I_{sc} whose nature has not yet been investigated to a sufficient degree [232, 232a, 239] (cf. Section 5.5).

8
Ferroelectrics as Nonlinear Semiconductors

Investigation of the semiconductor properties of ferroelectrics is of fundamental interest. For example, the development of work on the photoconductivity of ferroelectrics is stimulated by at least two factors. First, the photoconductivity of ferroelectrics is the basis of a number of phenomena involving the effect of nonequilibrium carriers on the ferroelectric properties and the phase transition in the crystal. The first and second chapters were devoted respectively to the thermodynamic and microscopic theories of these phenomena. Second, in investigating photoferroelectric phenomena, it is necessary to know such characteristics of the photoconductivity as the spectral and temperature dependences, relaxation times, nature, parameters, and filling of local energy levels in the forbidden band since the photoferroelectric phenomena are related generally to optical recharging of the levels. As another example one can indicate the peculiarities in the transport phenomena in ferroelectrics, which are semiconductors with strongly expressed electrical nonlinearity. These peculiarities must appear in the temperature and field dependences of the current limited by the space charge, in differential resistance, in instability phenomena, etc. Unusual behavior of such microscopic parameters as the mobility, cross section of capture and recombination centers, and lifetime of nonequilibrium carriers can be expected in ferroelectrics near the Curie temperature.

Some of these problems are considered in this chapter. We give a few qualifying remarks with respect to its content. It has already become traditional in dividing the semiconductor properties of ferroelectrics to include such problems as the nature of the carriers in $BaTiO_3$, scattering mechanism, transport phenomena in reduced barium titanate, the positive temperature coefficient of resistance, etc. Although many of these problems are far from solved, much work has been devoted to them, which has been generalized by Bursian in his detailed monograph [7]. Thus, the author has limited this chapter to some relatively new phenomena, to whose investigation he and co-workers have made a definite contribution.

8.1. Electrical Conductivity and Photoconductivity of Ferroelectrics near the Phase Transition

The effect of phase transitions on the electrical conductivity σ_e and photoconductivity σ_{ph} of semiconductors is well known [196]. Ferroelectric semiconductors are not exceptional in this sense.

According to published data, the electrical conductivity of "pure" $BaTiO_3$ [317, 318] and $SrTiO_3$ [319] does not experience an anomaly in the region of the phase transition. At the same time, alloyed $BaTiO_3$ or, more precisely, a crystal with oxidized (and generally defect) surface, as a rule, has a positive temperature coefficient of resistance in this region [7]. In Section 3.7 an assumption was made about the relation of the positive temperature coefficient to the behavior of the Schottky barrier near the phase transition. An increase of the electrical conductivity discontinuity at the transition through the Curie point has been noted in SbSI, and also in the change in the activation energy of the electrical conductivity with transition from the ferroelectric to the paraelectric phase. According to the data of Volk et al. [210], there is an anomaly of the electrical conductivity in pure $BaTiO_3$ in the region of the phase transition, although the activation energy does not change with the transition through the Curie point and is 1.3 eV (Figure 8.1). Figure 8.2 presents the anomaly of the dark conductivity for some crystals of SbSI near the phase transition [323]. It is seen that this anomaly has a complex form. The decrease in activation energy of the electrical conductivity with the transition to the paraelectric phase by several tenths of an electron volt (from 0.6–1.2 eV in the ferroelectric phase to 0.3–0.6 eV in the paraelectric phase) is typical for all the crystals.

Peculiarities of the steady-state photoconductivity near the Curie point are observed for SbSI [320–322], $BaTiO_3$ [20, 210], and $SrTiO_3$ [324].

A sharp maximum of the steady-state photoconductivity σ_{ph} is observed in $SrTiO_3$ crystals at the phase-transition temperature 47°K. Figure 8.3 represents the temperature dependences of the photoconductivity in the [100] direction and spontaneous polarization P_0, illustrating the anomaly of σ_{ph} at this point

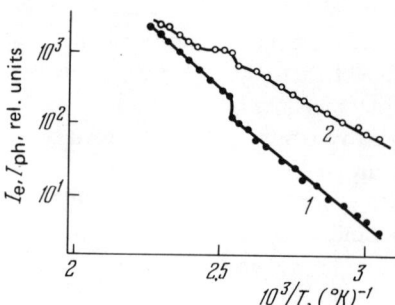

Figure 8.1. Temperature dependence of the dark current (1) and photocurrent (2) in $BaTiO_3$.

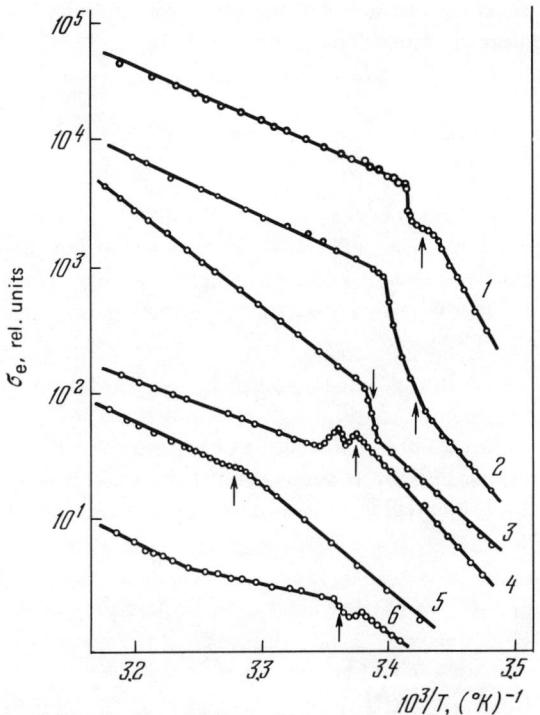

Figure 8.2. Temperature dependence of the dark current along the c axis in various SbSI crystals. The arrows denote the phase transition.

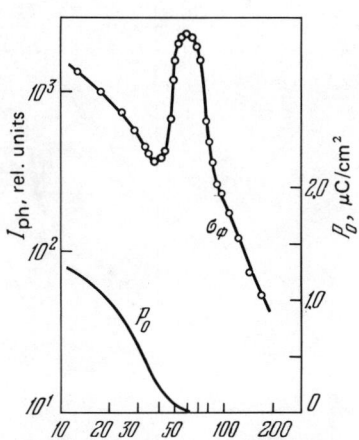

Figure 8.3. Temperature dependence of the photoconductivity and spontaneous polarization in $SrTiO_3$ in the region of the phase transition at $T = 47°K$ [324].

[327]. We will discuss the anomalies of the photoconductivity of BaTiO$_3$ and SbSI at the Curie point in more detail below.

The photoconductivity of a ferroelectric can be expressed in the following manner:

$$\sigma_{ph} = k\beta I \tau q \mu. \qquad (8.1)$$

Here, k is the light absorption coefficient, β is the quantum yield of the process of photocarrier generation, τ is the lifetime of nonequilibrium carriers, μ is the mobility, q is the electron charge, and I is the light intensity. The anomaly of the photoconductivity can obviously have a more complex nature than the anomaly of the electrical conductivity,

$$\sigma_e = qn\mu, \qquad (8.2)$$

because of the peculiarities in the behavior of k, β, and τ.

Both in BaTiO$_3$ and in SbSI, the temperature dependence $\sigma_{ph} = \sigma_{ph}(T)$ displays a maximum in the region of the ferroelectric phase transition. The data

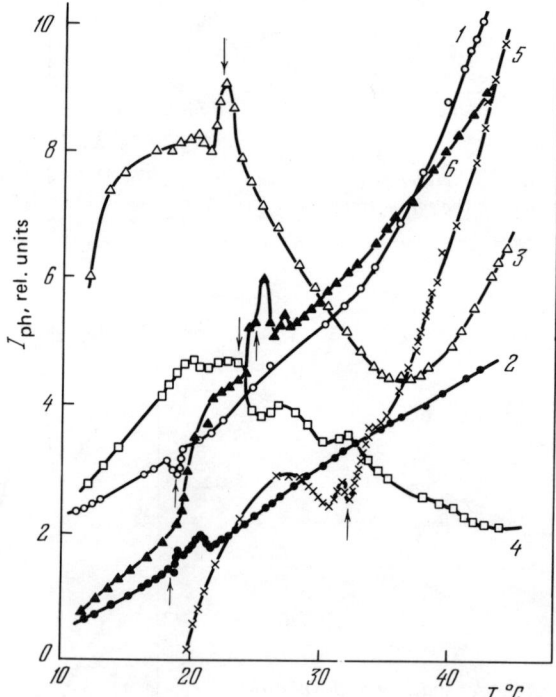

Figure 8.4. Temperature dependence of the photocurrent along the c axis in various SbSI crystals. The arrows denote temperature of the phase transition.

for BaTiO$_3$ are illustrated by Figure 8.1. Figure 8.4 presents the results of more detailed investigations of the temperature variation of the photoconductivity of SbSI for the same crystals for which the temperature dependence $\sigma_e = \sigma_e(T)$ is presented in Figure 8.2. Besides the complex anomaly of the photocurrent in the region of the phase transition, there is a section in the curve $I_{ph} = I_{ph}(T)$ for a major fraction of the investigated crystals where the photocurrent drops with increasing temperature, while this section is outside the region of the phase transition and is obviously not related to it (curves 3-5). The nature of the temperature dependence of the photocurrent is completely preserved with darkening of the contact regions. With excitation of the sample by strongly absorbed light (350-580 nm band), the dropping section disappears, and only the peculiarity near the Curie point is preserved. Hence, it follows that the dropping section is related to processes in the volume of the crystal and is not determined by the contacts or by the surface.

It is interesting to compare the temperature dependence of the photoconductivity with the temperature dependence of the parameters k, β, τ, and μ appearing in (8.1). Unfortunately, the temperature dependence of the mobility $\mu = \mu(T)$ has not been determined for SbSI because of experimental difficulties. The dependence $k = k(T)$ is significant only near the Curie point (cf. Chapter 4). It is by the change in absorption coefficient that Besdetnyi et al. [325] explain the decrease in photocurrent with the transition of SbSI from the ferroelectric to the paraelectric phase. A wide spectral region was used by Fridkin et al. [323] for excitation of the crystal in order to eliminate the possible dependence $k = k(T)$.

The phenomenological lifetime τ_{phen} [46] of the nonequilibrium carriers in SbSI was determined from the initial drop of the photocurrent after a long weak light pulse [321]. There is a relation between $\sigma_{ph} = \sigma_{ph}(T)$ and $\tau_{phen} = \tau_{phen}(T)$ near the phase transition, although it is absent beyond the vicinity of the Curie point. Moreover, appearing in (8.1) are the values of the true lifetime and quantum yield under conditions where the trapping levels do not affect the relaxation time of the nonequilibrium carriers [46]. The relaxation of the photocurrent was measured in [323] by the method of a short light pulse on the background of constant illumination [326, 327] to determine the temperature variation of the true lifetime and quantum yield. The pulse length in these measurements must be significantly less than the lifetime of the free carrier; the curve of the photocurrent decay is then an exponent with a relaxation time equal to the true lifetime. The amplitude of the photoresponse under these conditions does not depend on τ and is proportional for a constant absorption coefficient to the product of the quantum yield and the mobility $A \sim \beta\mu$. The constant illumination eliminates the effect of trapping levels and, moreover, decreases the Maxwellian relaxation time and thereby the time for establishing diffusion-drift equilibrium with the passage of the photocurrent pulse, which

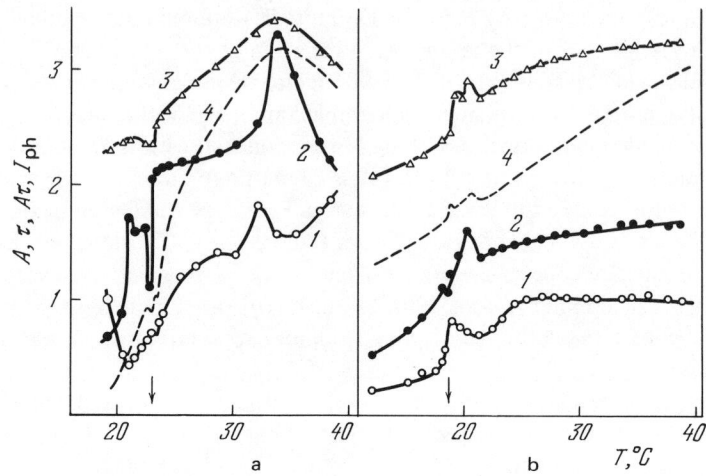

Figure 8.5. Temperature dependences of the photoresponse amplitude A (1), lifetime τ (2), product Aτ (3), and photocurrent I_{ph} (4) for two SbSI crystals.

permits one to observe an undistorted pattern of the relaxation of the photoexcited carriers [46].

Typical results for two crystals, for one of which the steady-state photocurrent decreases with temperature (a) and increases for the other (b), are presented in Figure 8.5. The curves of the photocurrent $I_{ph} = I_{ph}(T)$ agree well with the product $A(T) \cdot \tau(T)$, which speaks in favor of the applied technique. As a rule, the dependence $\tau = \tau(T)$ correlates better than $A = A(T)$ with $I_{ph} = I_{ph}(T)$ in the region of the phase transition and especially beyond it. It is also seen that the drop in photocurrent with increasing temperature is related to the decrease in the lifetime of the carriers, since the quantity $A \sim \beta\mu$ increases in this section. This conclusion is confirmed by the results for all the other crystals, although the dependence $A = A(T)$ has a different form for the various crystals.

Let us consider the results obtained in [323]. The dependence of the quantum yield of the photocurrent on temperature can hardly be significant in the temperature region of the phase transition. Thus, the anomaly of the photoconductivity is most likely determined by the behavior of the product of the mobility and the lifetime of the nonequilibrium carriers $\mu\tau$.

The scattering mechanisms are most often considered at present in studying the temperature dependence of the carrier mobility in ferroelectrics. Low and Pines [328] assume that the electron mobility in perovskite ferroelectrics is mainly determined by scattering of polarons at longitudinal optical phonons. In this case, the temperature dependence of the mobility has the form

$$\mu \sim f(\alpha) \exp(\hbar\omega/kT), \qquad (8.3)$$

where $\hbar\omega$ is the energy of the longitudinal optical phonon and α is the electron-phonon coupling constant. Comparison of this theory with experiment was carried out in [329]. A number of works have appeared recently, for example [329, 330], where the authors are led to the conclusion that scattering at transverse optical vibrations of the "soft" mode is decisive in ferroelectrics, especially near the Curie point. In this case [330],

$$\mu = \frac{f(T)}{\varepsilon_0 - \varepsilon_\infty}, \qquad (8.4)$$

where ε_0 and ε_∞ are the static and optical dielectric constants; $f(T)$ depends weakly on temperature [for example, $f(T) \sim T^{-1/2}$ for BaTiO$_3$]. According to [329], in the region where the Curie-Weiss law is valid, we have

$$\mu \sim T^{-3.5}(T - T_1 + T^*). \qquad (8.5)$$

where T_1 is the Curie temperature and $T^* = $ const ($T^* = 85°$ for BaTiO$_3$). The presence of the constant term $T = T^*$ and the factor $T^{-3.5}$ in (8.5) is related to the fact that there is a "background" of scattering at longitudinal optical vibrations along with scattering at the "soft" mode. According to [329, 330], the temperature variation of the mobility must display a peculiarity at $T = T_1$ in the form of a sharp minimum, although in the approximation (8.5) and in particular for $T^* \gg T - T_1$, this peculiarity is weakly expressed and $\mu \sim T^{-3.5}$ in the region $T > T_1$. The conclusions of the theory [329, 330] agree in general with the behavior of the mobility in BaTiO$_3$ in the region of the phase transition and in the presence of the positive temperature coefficient of resistance. To explain the temperature dependence of the mobility in BaTiO$_3$, Bursian successfully applied the small-radius polaron model, for which the dependence $\mu = \mu(T)$ is determined by an activation mechanism [7]. In the family of perovskite ferroelectric semiconductors, a transition from the polaron to the ordinary band model accompanied by an increase in mobility occurs with decreasing lattice constant and a corresponding increase in the overlap integral.

The scattering mechanism noted above [328-330] cannot explain the abrupt increase in the electrical conductivity at the phase transition in SbSI. It is natural to assume in this case that this peculiarity of the electrical conductivity is caused by the anomalous increase in the carrier concentration near the Curie point. Within the framework of the band model, the impurity electrical conductivity σ_e of the ferroelectric can display an anomaly in the region of the phase transition caused by a change in the activation energy of the donor levels u, the mobility μ, and the effective electron mass m^*. Substituting expression (1.74) for the carrier concentration n into (8.2),

$$n = (N_c N)^{1/2} \exp(-u/2kT),$$

and making use of the relations of [35],

$$m^* \sim E_g, \quad \mu \sim (m^*)^{-\alpha}, \quad \frac{3}{2} \leqslant \alpha \leqslant \frac{5}{2},$$

we arrive at the following expressing under the assumption that the phase transition does not affect the concentration of donor levels N:

$$\frac{d}{dT} \ln \sigma_e \simeq \left(\frac{3}{4} - \alpha\right) \frac{1}{E_g} \frac{dE_g}{dT} + \frac{1}{2kT^2} u - \frac{1}{2kT} \frac{du}{dT}. \quad (8.6)$$

Here, E_g is the width of the forbidden band, N_c is the density of states, u is the activation energy, and α is a coefficient determined by the nature of the carrier scattering.

Thus, as follows from (8.6), the temperature anomaly of the electrical conductivity can be caused by the change in activation energy of the donor levels u at the phase transitions as a result of a strong electron–phonon interaction. The discontinuity in the activation energy of the electrical conductivity of SbSI has already been discussed above. Nonetheless, the activation energy must be determined with care from the experimental values of $(d/dT) \ln \sigma_e$ since the mobility can depend strongly on temperature according to (8.4) and (8.5), while the terms $u/2kT^2$ and $(d/dT) \ln \mu$ are quantities of the same magnitude at $T = 300°$K and $u = 0.6$ eV. In the presence of the polaron mechanism of electrical conductivity in the ferroelectric, the anomaly of the electrical conductivity at the Curie point can be caused by the sharp decrease in the polaron activation energy at the phase transition [331]. Finally, as has already been noted in Section 1.6 and 5.6, the anomaly of the electrical conductivity σ_e near the Curie point can be related to the formation of new quasi-particles, for example, localized phasons, which leads to a change in both the activation energy u and in the concentrations of donor centers N.

As has already been noted, the temperature dependence of the photocurrent in SbSI usually agrees satisfactorily with the temperature dependence of the lifetime of nonequilibrium carriers. The anomaly of $\tau = \tau(T)$ can be related to restructuring of the system of recombination levels at the phase transition, also including the formation of phason levels. On the other hand, with recombination at charged centers, we have

$$\tau \sim \varepsilon^2, \quad (8.7)$$

since according to [196], $\tau = (SvN)^{-1}$ and

$$S = \left(\frac{q^2}{2\pi kT\varepsilon}\right)^2. \quad (8.8)$$

Here, N and S are the concentration and capture cross section of recombination centers. According to (8.7), the temperature anomaly of the photoconductivity at the Curie point is determined by the Curie-Weiss law. The model of Coulomb centers was assumed, in particular, in [324] to explain the temperature anomaly of the photoconductivity in $SrTiO_3$ (Figure 8.3). Nonetheless, in connection with the application of the Coulomb-center model and equation (8.8) to ferroelectric semiconductors, it is necessary to make the following comment.

A characteristic feature of Coulomb centers in ferroelectrics must be their sharp anisotropy caused by the anisotropy of the dielectric constant. Thus, the temperature dependence of the effective cross section S must be more gradual than $(\epsilon T)^{-2}$, where ϵ is the dielectric constant measured along the ferroelectric axis. The divergence between the experimental and theoretical values of S will also increase with increasing ϵT for another reason. In the region of temperatures near the Curie-Weiss temperature, the value of S calculated for the Coulomb-center model is less than the atomic cross section, i.e., the model itself loses physical meaning. This means exactly a change in the nature of capture at the centers under consideration in the region of temperatures near the phase-transition temperature, possibly because of a change in the nature of the centers themselves. As for the possible anomaly of S in the region of ferroelectric phase transition, equation (8.8) is suitable for its description only in the case of low-temperature phase transitions and not too large values of ϵ. This case is evidently realized in $SrTiO_3$ according to [324]. For large ϵT_1, as in the case of SbSI, the anomaly of $\tau(T_1)$ is more likely related to the formation of new levels (for example, phason) or coupled electron-phonon states.

The drop in photocurrent with increasing temperature observed in a significant fraction of SbSI crystals is also typical for linear semiconductors and can be explained by the mechanism of temperature quenching of the photocurrent. According to the Rose model [196], temperature quenching of the photoconductivity is explained by a decrease in the lifetime of nonequilibrium carriers because of "switching" of the recombination channel. In accordance with this model, the quenching section must be shifted toward higher temperatures with increasing light intensity, which was actually observed for SbSI in [323]. It also follows from the Rose model that the intensity-current characteristic of the photocurrent must be ultralinear in the range of temperatures where quenching is observed. However, the ultralinear section was not observed for the investigated crystals of SbSI. If one takes into account the narrowness of the temperature region of quenching ($\sim 10°C$) and the relatively small magnitude of the quenching (cf. Figure 8.4), this can be explained by the specifics of the "sensitizing" levels in this semiconductor, in particular, that their concentration is small and they are localized in a narrow energy band. In any case, the results of [323] indicate that this drop in the photocurrent is not related to the phase transition. The independent measurements of the thermostimulated currents carried out in [323] indicated that in crystals where quenching is observed, there are several maxima of the thermostimulated current corresponding to impurity levels in the band 0.1–0.6 eV. In crystals for which quenching was not observed, there was only one maximum of the thermostimulated current corresponding to the level 0.5 eV.

As for the phase-transition region itself, the dependences $\tau = \tau(T)$ and $A = A(T)$ have a complex nature there for crystals with temperature quenching. This does not yet make it possible to uniquely explain the temperature anomaly of the photocurrent in the region of the phase transition for these crystals. It is possible that the region of temperature quenching of the photoconductivity in some of these crystals is superimposed onto the region of the phase tranisition, which significantly complicates the observed experimental curves.

Relaxation of the photocurrent in the ferroelectric phase is accompanied by attenuated vibrations, which reach the amplitude of the photoresponse signal itself in a number of crystals. These vibrations have a spectrum consisting of two or three frequencies in the range from 30 to 200 kHz, similar to the vibration spectrum arising in the same circuit with delivery of rectangular voltage pulses to the samples in darkness. The frequencies of the vibrations are inversely proportional to the length of the samples. The amplitude has a maximum near the Curie point and displays temperature hysteresis correlated with the piezoelectric modulus. All this indicates the piezoelectric nature of the vibrations. Tatsuzaki et al. [214] reported on the deformation of SbSI crystals with illumination in a constant field. In these experiments, the light flash evidently causes pulses of deformation of the sample, after which follow natural mechanical vibrations and current oscillations in the circuit because of the piezoelectric effect. Constant "natural" illumination completely suppresses the vibrations, which is in agreement with the phenomenon of deformation saturation at comparatively low light intensities observed in [215].

8.2. Effect of Screening of Spontaneous Polarization on the Photoconductivity of Ferroelectrics

It is well known that the state of the surface, to which recombination of nonequilibrium carriers is related, significantly affects the photoconductivity of semiconductors and causes, in particular, the appearance of a maximum in its spectral characteristics [332]. Qualitative changes in the spectral distribution of the photoconductivity occur with large surface curvatures of the bands. The spatial separation of electron–hole pairs by the field of the surface barrier leads to a decrease in the recombination rate in the surface layer and accordingly to an increase in the lifetime of nonequilibrium carriers here as compared to the lifetime in the volume. The increase in photoconductivity observed in the short-wavelength region of the spectrum with illumination at the barrier can be explained by the increase in lifetime with the approach of the nonequilibrium carrier generation region toward the surface. The subsequent drop in the photocurrent is evidently caused by the effect of surface recombination, which begins to predominate over the factor of spatial separation of electron-hole pairs with contraction of the region of their generation directly toward the surface [196].

Investigation of the spectral distribution of the photoconductivity of the ferroelectric semiconductors SbSI [17, 333] and $SrTiO_3$ [334] reveals a maximum located near the intrinsic absorption edge, i.e., in the region of photon energy of 2 and 3.2 eV, respectively. The intrinsic absorption edge of $BaTiO_3$ is located in the region 3.0–3.2 eV at room temperature, whereas the spectral maxima of the photoconductivity correspond to the energies of 3.55 and 3.8 eV, respectively, according to [335]. Besides this, photoconductivity was discovered in $BaTiO_3$ [336] and SbSI [337] in the short-wavelength region, caused by generation of nonequilibrium carriers in the region of Schottky barrier arising with contact of metal with the ferroelectric. It is obvious that ferroelectric semiconductors must display yet another anomaly of the photoconductivity related to the existence of surface screening layers (Section 3.5). According to generally accepted assumptions, the surface layers in ferroelectrics are layers of space charge screening spontaneous polarization and producing a field $\sim 10^5$ V/cm. The strong electric field causes dielectric saturation and a low dielectric constant of the surface layer. The assumption has been made in a number of works that the Schottky barrier model is applicable to the surface layer of a ferroelectric [15]. Thus, the surface screening layer must determine the photosensitivity in the region of strong absorption of the crystal. A feature of this photoconductivity must be the effect on it of the domain structure and ferroelectric polarization reversal. The surface layer can also be observed in the paraelectric region, although its parameters can sharply change at the phase transition [15, 338, 339]. According to this, the singular photoconductivity of a ferroelectric can also experience a change in this temperature region. Of course, this type of photoconductivity can be observed only with illumination of those faces of the crystal where the surface screening layer is located, and must be absent with illumination of faces parallel to the ferroelectric axis. The features enumerated above were the basis of observation and investigation of the singular photoconductivity of $BaTiO_3$ and SbSI related to surface layers [19, 340].

The spectral distribution of both the longitudinal and transverse photoconductivities was investigated for single crystals of $BaTiO_3$. The longitudinal photoconductivity was measured in the direction of the c axis, for which semitransparent gold electrodes, through which illumination was carried out, were applied to the (001) faces. The spectral distribution of the transverse photoconductivity was measured with the application of the field in the [100] direction and illumination by a light probe in the [001] direction, while the electrodes were not illuminated. In both cases, the spectral distribution had one maximum at $\lambda \simeq 360$ nm without any sort of peculiarity in the region of the intrinsic absorption edge, i.e., at $\lambda \simeq 400$ nm (Figure 8.6). In the case of trans-

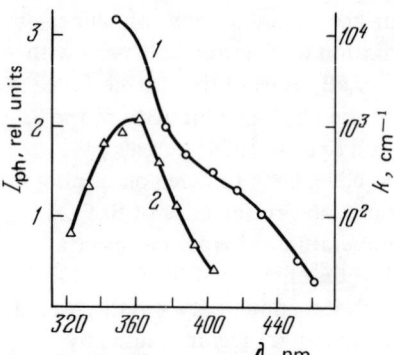

Figure 8.6. Spectral distribution of the absorption (1) [111] and photoconductivity (2) for $BaTiO_3$.

verse photoconductivity, the position of the maximum did not depend on the amplified field, electrode material, and illumination of the contacts. This maximum was also stably observed in the paraelectric region of the crystal, which corresponds to the well-known observation of a surface level above the Curie temperature for $BaTiO_3$ [15, 339].

It was of the greatest interest to compare for single crystals of SbSI the curves of the spectral distribution of the transverse photoconductivity with illumination of the faces of the {001} zone, where the surface layer is absent for reasons of symmetry, with the corresponding curves obtained with illumination of the (101) and (001) faces, where the surface screening layer is located (we note that analogous observations for $BaTiO_3$ are made difficult by the presence of a and c domains).

Figure 8.7 shows the spectral distribution of the photoconductivity for the (100) and (101) faces. The short-wavelength shift of the maximum of curve

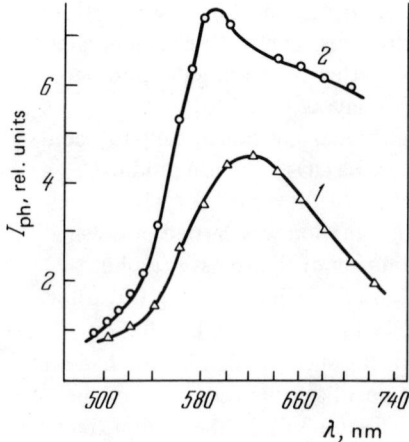

Figure 8.7. Spectral distribution of the transverse photoconductivity in the ferroelectric region of SbSI. Curve 1 corresponds to the (100) face, and curve 2 to the (101) face.

2 with respect to the maximum of the "natural" photoconductivity was observed for all the investigated crystals, although its magnitude was not constant. We will show below that this shift is actually caused by the singular photoconductivity of the surface layer of SbSI.

Figure 8.8 illustrates the effect of the direction of the spontaneous polarization on the spectral distribution of the photoconductivity. Before measurement of the photoconductivity, the crystal was polarized by a constant field greater than the coercive field applied along the ferroelectric c axis. The spectral distribution of the transverse photoconductivity for the (100) face does not change with polarization of the crystal and with a change in direction of polarization (curve 1). For the (101) face, on the contrary, the direction of the preliminary polarization significantly affects the spectral distribution of the photoconductivity (curves 2 and 3). This indicates that there is an asymmetry of the surface layers in SbSI, where the anode surface layer is significantly thinner than the cathode layer and the electric field in it is greater in magnitude.

The temperature dependences of the spectral distribution of the photoconductivity for the (100) and (101) faces are also very different. The spectral maximum of the photoconductivity for the (100) face increases with increasing temperature and shifts toward lower energies, while the shift coefficient in both the ferroelectric and paraelectric regions is close to the temperature coefficient of the width of the forbidden band dE_g/dT in SbSI. The maximum of the spectral distribution of the photoconductivity for the (101) face remains in place in the ferroelectric region in the investigated temperature interval from

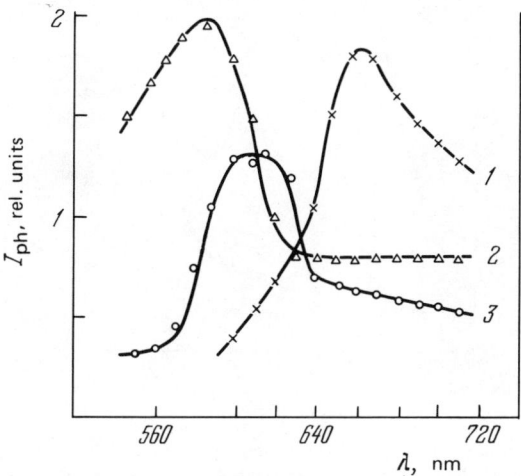

Figure 8.8. Effect of the ferroelectric polarization on the spectral distribution of the transverse photoconductivity of SbSI in the ferroelectric region. Curve 1 corresponds to the (100) face; curves 2 and 3, to the (101) face; (2) anode region; (3) cathode region.

0°C to the Curie temperature. The spectral distributions of the photoconductivity for the (100) and (101) faces coincide in the paraelectric region after holding the crystal at a temperature above the Curie temperature for a long time. Thus, in contrast to $BaTiO_3$, the ferroelectric SbSI does not display a singular photoconductivity related to surface screening layers in the paraelectric region.

Figure 8.9 presents the results of observations of the kinetics of the disappearance of the singular photoconductivity in the paraelectric region of SbSI, caused by the decay of the surface layer. Measurements of the spectral distribution of the photoconductivity, which were carried out successively in time after transition of the crystal to the paraelectric phase, indicate that the process of disruption of the surface layer is accompanied by the disappearance of the short-wavelength maximum and by the formation of the long-wavelength maximum. With the reverse transition from the paraelectric to the ferroelectric region, an analogous pattern is observed: a gradual disappearance of the long-wavelength maximum and the formation of the short-wavelength maximum. The kinetics of this process depend on the short-circuiting of the ferroelectric in the direction of the c axis: the formation time increases with short-circuiting.

The experimental results presented above lead to the conclusion that the singular photoconductivity of ferroelectrics in the region of strong optical

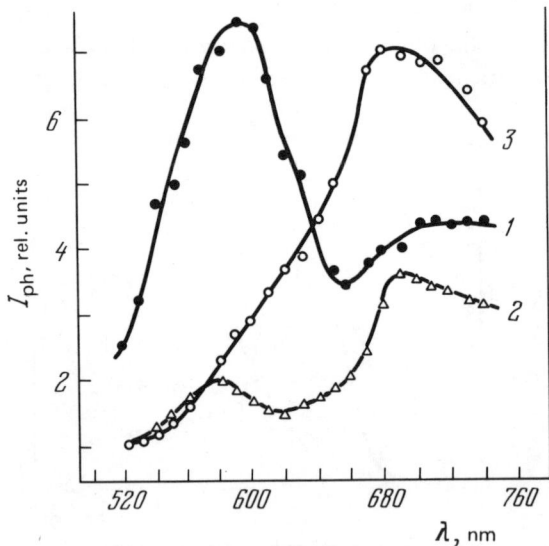

Figure 8.9. Relaxation of the singular photoconductivity of SbSI at the phase transition. Spectral distribution of the photocurrent. (1) Directly after the transition from the ferroelectric to the paraelectric region; (2) after a single heating to +50°C and subsequent cooling to the temperature of the measurement; (3) after a long time in the paraelectric phase. Temperature of the measurements 19°C, $T_1 = 18°C$.

absorption is related to the surface screening layer. The results obstained indicate that this layer is maintained in the paraelectric region of $BaTiO_3$, whereas the surface layer is absent in the nonpolar base of SbSI. However, it is possible that a surface layer exists in the paraelectric phase of SbSI, but its field is not great enough to generate the singular photoconduction band. The photoconductivity of the surface layer can, as has already been indicated, be caused by the effect of the strong electric field on the lifetime of nonequilibrium carriers. On the other hand, the short-wavelength shift of the maximum of the spectral distribution of the photoconductivity can also be determined by the short-wavelength shift of the intrinsic absorption edge in the ferroelectric under the effect of the strong electric field concentrated in the surface layer. Actually, Nosov and Lyakhovitskaya [341] observed a change in the spectral distribution of the photoconductivity of SbSI due to this effect. Comparison of the spectral distributions of the photoconductivity and absorption coefficients for $BaTiO_3$ (Figure 8.6) and SbSI shows that the maximum of the singular photoconductivity corresponds to an effective thickness of the surface layer of $\sim 10^{-4}$ cm for $BaTiO_3$ and $\sim 10^{-5}$ cm for SbSI, which is a reasonable estimate (Section 3.5). If the surface layer is identified with the Schottky barrier and the surface levels are neglected, then the thickness of the anode and cathode layers can be evaluated with the help of (3.98). Thus, $d_+ > d_-$ for SbSI and, consequently, the concentration of donors is greater than the concentration of acceptors. The values of $P_0 \simeq 20 \ \mu C/cm^2$ and $d \simeq 10^{-5}$ cm in SbSI correspond in order of magnitude to the concentration of donors (acceptors) of $N_d \simeq 10^{13}$ cm^{-3}. This value agrees satisfactorily with the electrical conductivity of SbSI.

Direct measurements of the recombination lifetime of nonequilibrium carriers in the surface layer of SbSI were carried out in [323]. The lifetime was measured from the photocurrent relaxation curve by the light shock method [326, 327] (cf. Section 8.1). The measurements were carried out at the (101) and (100) faces of the crystal in the ferroelectric and paraelectric phases, respectively. The lifetime of nonequilibrium carriers was measured with excitation of the crystal by light pulses with wavelengths corresponding to the maximum of the intrinsic photoconductivity (λ_1 = 620 nm), the photoconductivity maximum in the surface layer (λ_2 = 580 nm), and also with short-wavelength light $\lambda_3 < \lambda_2$ (λ_3 = 400 nm). Typical photocurrent relaxation curves on a semilogarithmic scale consisted of three linear sections, which possibly correspond to three recombination channels. The time constants determined from the relaxation curves are summarized in Table 12 [the measurements were carried out on the three (1, 2, 3) isolated sections of the two faces]. As is seen from the table, for the (101) face of the crystal in the ferroelectric phase, the carrier lifetime with excitation by light with wavelength corresponding to the maximum of the intrinsic photoconductivity is actually maximum. In all the remaining cases, that is, with illumination of the (101) face of the crystal in the paraelectric phase

TABLE 12. Time Constants (μsec) of the Photocurrent Relaxation Process at Various Sections of the Relaxation Curve

	Ferroelectric phase					
	(101) face			(100) face		
λ, nm	1	2	3	1	2	3
620	19	39	82	27.5	74	74
580	25	51	142	20.5	43	43
400	9.5	32	68	16.5	34	34
	Paraelectric phase					
	(101) face			(100) face		
λ, nm	1	2	3	1	2	3
620	15	40	100	20	44	160
580	10	23	36	17.5	24	39
400	6.5	22	22	8.5	18	34

or with illumination of the (100) face, the lifetime decreases monotonically with decreasing wavelength of the exciting light.

8.3. Some Photoelectric Phenomena in SbSI

The combined study of photoelectric phenomena in SbSI leads to the conclusion of the presence in the forbidden band of this ferroelectric of a complex system of levels determining the photoelectric properties. The spectral characteristics of some of the photoelectric phenomena in SbSI are presented in Figure 8.10 [342, 343].

The maximum of the photoconductivity at 90°K is in the region of 540 nm, which corresponds to a width of the forbidden band of 2.3 eV. Photosensitivity in the region of $\lambda = 1$ μm is practically absent. Sensitivity in the long-wavelength region appears after preliminary excitation of the crystal by "natural" light. With successive excitation by "natural" and long-wavelength light, an induced impurity photoconductivity is observed [46] which has the usual nature of "flashes," transforming to a quasi-stationary section where the photocurrent decays with a time constant equal to about 1 h. Steady-state impurity photoconductivity is observed in this section, which obviously corresponds to the case of nonlinear recombination [46]. Curve 4 in Figure 8.10 was obtained with successive excitation of the crystal by "natural" and then long-wavelength lights, and its points correspond to the maximum values of the "flash" photocurrent. Curves 5 and 6 give, respectively, the spectral distribution of the photo-

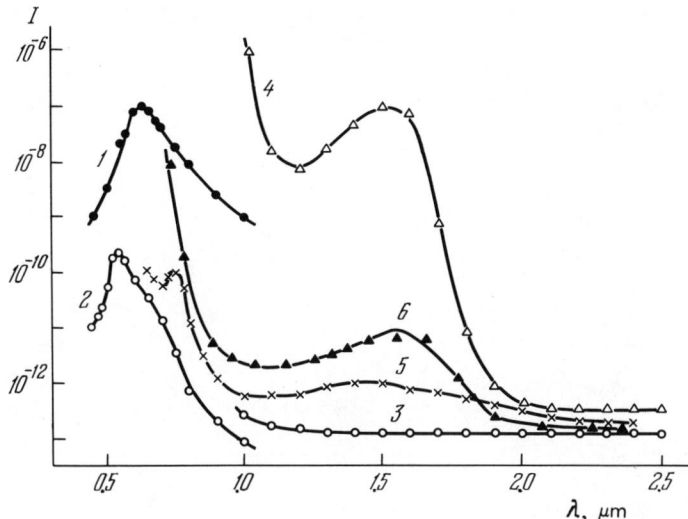

Figure 8.10. Spectral distribution of the photoelectric effect in SbSI. (1) Photoconductivity, T = 290°K; (2) photoconductivity, T = 90°K; (3) photoconductivity of the undisturbed crystal, T = 90°K; (4) induced impurity photoconductivity of the excited crystal, T = 90°K; (5) formation of the photoelectret polarization, T = 90°K; (6) depolarization of the photoelectret, T = 90°K.

electret polarization formation and the spectral distribution of the photoelectret depolarization current [273]. Good coincidence of the long-wavelength maxima is seen in curves 4-6, which correspond to excitation of the electrons from levels at 0.7-0.8 eV below the bottom of the conduction band. The long-wavelength maximum in curve 5 in the region of ~1.6 eV may correspond to a transition of electrons from the valence band to the levels under consideration with subsequent photoelectret formation by capture of holes at the other levels.

Quenching of the photoconductivity in the region $\lambda \simeq 0.75$ nm (which corresponds to ~ 1.6 eV) is detected in SbSI crystals. The quenching effect is small and is 3-4% at most of the background current value. The quenching region coincides with one of the maxima of the spectral distribution of the photoelectret formation (curve 5).

The thermostimulated currents were measured for the same crystals. The energy of the trapping levels, which is calculated from the single maximum of the thermostimulated conductivity, is ~ 0.55 eV, and the concentration of levels is $N \simeq 10^{18}$ cm^{-3}. An attempt was undertaken to observe the decrease in area under the thermostimulated current curve after preliminary excitation by impurity light. Quenching of the thermostimulated conductivity was not observed when a flash of induced photoconductivity was generated.

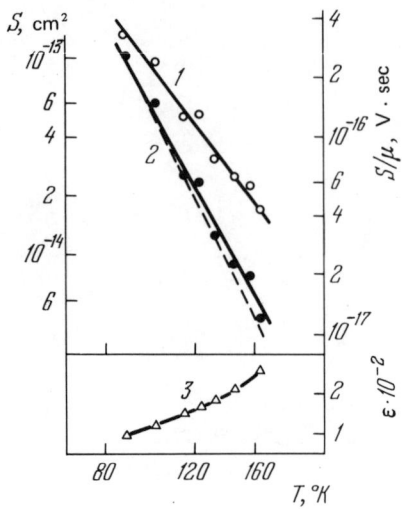

Figure 8.11. Temperature dependence of S/μ (1), the electron capture cross section S (2), and the dielectric constant (3) for SbSI. The dashed line shows the course of the function $f(T) = \text{const}/(\epsilon T)^2$.

One can determine from the kinetics of the induced photoconductivity the interaction cross section with photons (q) and electrons (S) for the impurity levels M responsible for the induced photoconductivity [46]. The value of the electron mobility μ is unknown at low temperatures in SbSI, and thus the value of S/μ (Figure 8.11, curve 1) was calculated from the experimental data in [342, 343]. It was assumed in [342, 343] that the mobility in SbSI in the temperature interval under consideration is determined by scattering at transverse optical vibrations (the "soft" mode) and $\mu \sim 1/\epsilon$ for perovskite ferroelectrics [329, 330]. The value of S (Figure 8.11, curve 2) was calculated by extrapolation according to this law from the value $\mu = 50 \text{ cm}^2/(\text{V} \cdot \text{sec})$ at $T = 290°\text{K}$ [199].

The large value of the capture cross section and also the nature of the dependence $S = S(T)$ are the basis for assuming that capture occurs at Coulomb centers, for which the cross section satisfies (8.8). This is also indicated by the close coincidence of the curves $S = S(T)$ and $f(T) = (\epsilon T)^{-2}$ (Figure 8.11). The strong dependence of the capture cross section on temperature in this case is determined by the temperature dependence of the dielectric constant of the ferroelectric semiconductor far from the Curie point (Figure 8.11, curve 3).

The possibility of using the Coulomb-center model in the region of the phase transition was already considered in Section 8.1.

Analysis of the kinetics of induced impurity photoconductivity led to an estimate of the interaction cross section q of photons with M levels of approximately 10^{-15} cm^2 for $\lambda = 1.4$ μm and indicated that q is practically independent of temperature. The concentration of M levels of the order of 10^{11}–10^{12} cm^{-3} and the capture cross section of electrons at recombination levels of $\sim 10^{-14}$ cm^2 for $T = 90°\text{K}$ are determined from the kinetics.

Fine trapping levels with energy of ~0.04 eV and concentration of 10^{18}-10^{19} cm^{-3} are also characteristic in the investigated crystals.

The results of investigation of the intensity-current characteristics of the photoconductivity for SbSI at room temperature are presented in Figure 8.12. Each characteristic consists of two linear sections with different slopes, which corresponds to the power dependence of the photocurrent on the light intensity $I_{ph} \sim I^{\alpha}$. The value of α on the second section is less than on the first and is ≈ 0.5, which indicates the nonlinear nature of the recombination. The reverse path of the characteristic does not repeat the direct path. The initial direct path of the curve is reproduced after holding the sample in darkness for a long time (about a day). The holding time can be reduced to 1 h with heating to 50-60°C. A significant fraction of the crystals display the following feature of the intensity-current characteristics (curves 2 and 3 in Figure 8.12). At the end of the first section having a slope close to unity, the photocurrent increases by 1.5 orders of magnitude, being established during several hours, after which the second section begins with a slope of ~0.5. The reverse path does not reproduce this feature. The discontinuity in the photocurrent decreases in magnitude with increasing temperature and then disappears. The temperature at which the discontinuity disappears does not usually coincide with the phase-transition temperature. The discontinuity is not observed with excitation by strongly absorbed light (curve 1, Figure 8.12) and is thus related to absorption in the volume.

The phenomena considered can be explained on the basis of ideas about light-stimulated conductivity [344, 345]. Tyler and Woodburry [345] have expressed the assumption that the cause of stimulation is the creation of centers, which has a still insignificantly small electron capture cross section after the capture of a hole. These centers can be, for example, acceptor levels, which have

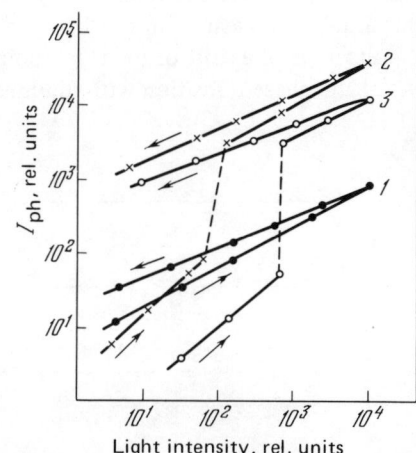

Figure 8.12. Intensity-current characteristics of SbSI at $T = 10°C$, illumination in various spectral regions. (1) 540 nm; (2) 630 nm; (3) 720 nm.

a double negative charge in equilibrium. According to another point of view [344], the stimulated conductivity is related to the formation of a barrier at the recombination centers. There is no basis at present for preferring one model or the other in the application to SbSI. If, for example, one assumes the model of [345], then, by assuming that the quasi-Fermi level for the principal carriers converges to the level responsible for stimulated conductivity on the "ultralinear" section of the intensity-current characteristic, one can evaluate roughly the position of the latter in the forbidden band of SbSI. The energy of this level is 0.5-0.65 eV under the assumption that the carrier mobility is $\mu = 50$ cm^2/(V · sec).

Figure 8.13 shows the assumed energy level diagram for SbSI, constructed on the basis of the results of photoelectric measurements at $T \simeq 100°$K. The data presented above indicate the presence in SbSI of two different types of electron capture levels close in energy (0.7-0.8 eV) and differing not only in concentration ($N \simeq 10^{18}$ cm^{-3}, $M \simeq 10^{11}$-10^{12} cm^{-3}), but also in the interaction cross section with photons ($q_N \ll q_M$). The latter is explained by the absence of significant quenching of the thermostimulated conductivity and the simultaneously significant emptying of M levels at the flash of induced impurity photoconductivity. If it is taken into account that investigation of the flashes was carried out at 90°K, while the maximum of the thermostimulated current corresponds to 220°K, then it is clear that the difference in energies of the M and N levels is insignificant. We note that the existence in the ferroelectric of two local levels, close in energy, related to the 180° domain boundaries was predicted in [346]. The hole trapping levels and recombination centers are denoted respectively by O and L in the diagram of Figure 8.13. The photoconductivity quenching mechanism is naturally related to the transition of electrons from the valence band to the M level (~1.6 eV) with the subsequent capture of holes by the recombination centers L. The formation of photoelectret polarization in the region of long-wavelength photocurrent quenching is also related to this mechanism. It is also clear from this that the photoferroelectric phenomena in SbSI (for example, the shift of the Curie point and the change in temperature hysteresis of the phase transition with illumination) are related to optical recharging of

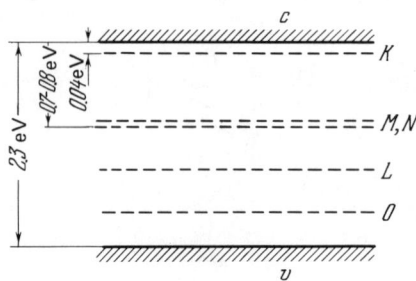

Figure 8.13. Energy level diagram of SbSI at $T = 90°$K.

the N and O levels since the concentrations of the other levels can be neglected, while recharging of the fine levels K is insignificant at room temperature. We also recall that the maximum of the optical quenching of the photohysteresis effect is located near ~ 0.6 eV (cf. Section 5.2), which agrees with the energy of the N level determined from the maximum of the thermostimulated conductivity.

8.4. Currents Limited by the Space Charge in Ferroelectrics

The theory of currents limited by the space charge (CLSC) in linear semiconductors was developed by Mott and Gurney and also by Adirovich [39, 347]. Rose and Lampert took the effect of traps into account within the framework of this theory [348]. A number of interesting peculiarities in the voltage–current characteristics of CLSC and also in the temperature dependence of CLSC near the Curie point are governed by the nonlinear dependence of the induction on the field. The theory of CLSC for ferroelectrics was developed in [349, 350].

In the one-dimensional case for a single-domain ferroelectric not containing traps, the expression for the steady-state current without consideration of diffusion, the Poisson equation, and the relation between the field and polarization can be written in the following form:

$$j = |q|\mu n \mathscr{E}, \quad dj/dx = 0, \qquad (3.43)$$

$$dP/dx = qn, \qquad (3.42)$$

$$\mathscr{E} = dF/dP, \qquad (3.1)$$

$$F = \frac{\alpha}{2} P^2 + \frac{\beta}{4} P^4 + \frac{\gamma}{6} P^6. \qquad (3.26)$$

Here, j is the current density, n is the carrier concentration, $\mathscr{E}$ is the field intensity, F is the free energy, and α, β, and γ are the Devonshire coefficients. We have neglected the elastic stresses and gradient energy in (3.26). Since we are interested in the region of high fields where the dielectric nonlinearity appears, we have neglected the contribution of the diffusion current in (3.43). Equations (3.43), (3.42), (3.1), and (3.26) must be supplemented by the boundary conditions

$$\begin{aligned} \mathscr{E}(x=0) &= 0, \\ P(x=0) &= P_0, \\ n(x=0) &= \infty, \end{aligned} \qquad (8.9)$$

which correspond to the case of the so-called virtual injection electrode at $x = 0$. The solution of the system (3.43), (3.42), (3.1) and (3.26) with consideration of the boundary conditions (8.9) has the form

$$j = \frac{q}{|q|}\frac{\mu}{L}\{F(P) - F(P_0)\}, \tag{8.10}$$

$$U = \frac{q}{|q|}\frac{\mu}{j}\int_{P_0}^{P}\mathscr{E}^2 dP, \tag{8.11}$$

where U is the voltage applied to the crystal in the direction of spontaneous polarization and L is the thickness of the crystal in this direction.

We consider first the ferroelectric with a second-order phase transition:

$$F = \frac{\pi}{C}(T - T_0)P^2 + \frac{1}{4}\beta P^4, \quad \beta > 0, \tag{8.12}$$

$$\mathscr{E} = \frac{2\pi}{C}(T - T_0)P + \beta P^3, \tag{8.13}$$

$$P = \sqrt{\frac{2\pi}{C\beta}(T_0 - T)}, \quad T < T_0. \tag{8.14}$$

Expressing β in terms of the spontaneous polarization $P_0 = (2\pi T_0/C\beta)^{1/2}$ at $T = 0$ and using the normalized notation

$$\eta = \frac{\mathscr{E}}{\mathscr{E}_0}, \quad \tau = \frac{T}{T_0}, \quad \pi = \frac{P}{P_0}, \quad \varphi = \frac{U}{\mathscr{E}_0 L},$$

$$f = \frac{F}{\mathscr{E}_0 P_0}, \quad \delta = \frac{jL}{\mu P_0 \mathscr{E}_0},$$

where $\mathscr{E}_0 = 2\pi T_0 P_0/C$ is the characteristic field close to the coercive field, we arrive at the solution in the following form:

$$\delta = f(\pi) - f(\pi_s), \tag{8.15}$$

$$\delta\varphi = \int_{\pi_s}^{\pi}\eta^2 d\pi, \tag{8.16}$$

$$f(\pi) = \frac{1}{4}[2(\tau - 1)\pi^2 + \pi^4], \tag{8.17}$$

$$\eta = \pi[\tau - 1 + \pi^2], \tag{8.18}$$

$$\pi_s = \begin{cases} 0 & \text{for } \tau \geq 1, \\ \sqrt{1 - \tau} & \text{for } \tau \leq 1. \end{cases} \tag{8.19}$$

At the Curie point ($\tau = 1$), the expression for the current has the form

$$\delta = \frac{7}{16}\left(\frac{7}{4}\right)^{1/3}\varphi^{4/3} = 0.527\varphi^{4/3} \tag{8.20}$$

or

$$j = 0.527\mu\beta^{-1/3}\frac{U^{4/3}}{L^{7/3}} = 0.527\mu\frac{P_0}{\mathscr{E}_0^{1/3}}\frac{U^{4/3}}{L^{7/3}}. \quad (8.21)$$

In the region of weak fields ($|\pi - \pi_s| \ll 1$), expansion of $f(\pi)$ at the point $\pi = \pi_s$ gives

$$f(\pi) = f(\pi_s) + \frac{1}{2}\frac{\partial \eta}{\partial \pi}\bigg|_{\pi_s}(\pi - \pi_s)^2 + \ldots,$$

$$\eta(\pi) = \frac{\partial \eta}{\partial \pi}\bigg|_{\pi_s}(\pi - \pi_s) + \ldots.$$

Hence

$$\delta = \frac{9}{8}\left(\frac{\partial \eta}{\partial \pi}\bigg|_{\pi_s}\right)^{-1}\varphi^2 \quad (8.22)$$

or

$$j = \frac{9}{8}\mu\left(\frac{d\mathscr{E}}{dP}\bigg|_P\right)^{-1}\frac{U^2}{L^3}, \quad (8.23)$$

where

$$\left(\frac{d\mathscr{E}}{dP}\bigg|_P\right)^{-1} = \varepsilon_0\varepsilon = \begin{cases} \dfrac{\varepsilon_0 C}{T - T_0} & \text{in the paraelectric region,} \\[1ex] \dfrac{\varepsilon_0 C}{2(T_0 - T)} & \text{in the ferroelectric region} \end{cases}$$

(ϵ_0 is the vacuum dielectric constant). Thus, in the region of weak fields, the CLSC in the ferroelectric is subject to the Child law known for linear semiconductors [39].

Nonlinearity of the ferroelectric appears in the region of strong fields where

$$\delta = \frac{1}{4}\pi^4 \quad \text{and} \quad \eta = \pi^3,$$

which leads to the voltage–current characteristic (8.21). Figure 8.14 presents the voltage–current characteristics corresponding to the solution (8.15)–(8.19) and different temperatures. The dashed line is the voltage–current characteristic corresponding to the case of strong fields and satisfying (8.20) or (8.21). It is seen from Figure 8.14 that the characteristics corresponding to satisfaction of the Child law and corresponding to different temperatures intersect the dashed line of (8.20) in the paraelectric region ($\tau > 1$) at

$$\varphi_g^p = 0.320(\tau - 1)^{3/2} \quad (8.24)$$

and in the ferroelectric region ($\tau < 1$) at

$$\varphi_g^s = 0.9(1 - \tau)^{3/2}. \quad (8.25)$$

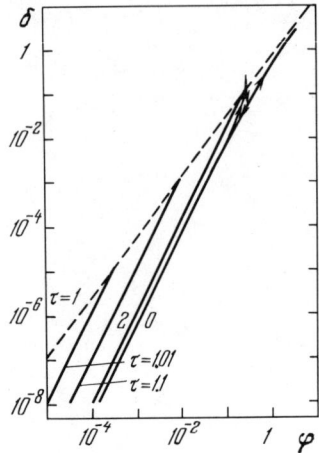

Figure 8.14. Voltage-current characteristics of a ferroelectric with a second-order phase transition.

If one neglects the temperature dependence of the mobility, one can determine the temperature dependence of the CLSC from Figure 8.14. If the normalized voltage φ is fixed, then the normalized current δ is proportional to $\epsilon \sim (|\tau - 1|)^{-1}$ for $\varphi < \varphi_g$, while it does not depend on temperature for $\varphi \geqslant \varphi_g$. Since the parameter φ_g itself depends on temperature, the dependence of the normalized current δ on temperature can be represented by the curves in Figure 8.15. The transition from $\delta = $ const to $\delta \sim |\tau - 1|^{-1}$ occurs at

$$\delta_g^p = \left(\frac{147}{432}\right)^2 (\tau - 1)^2 \quad \text{in the paraelectric region,} \tag{8.26}$$

$$\delta_g^s = \left(\frac{49}{72}\right)^2 (1 - \tau)^2 \quad \text{in the ferroelectric region.} \tag{8.27}$$

The dashed line in Figure 8.15 corresponds to (8.26) and (8.27). If we measure

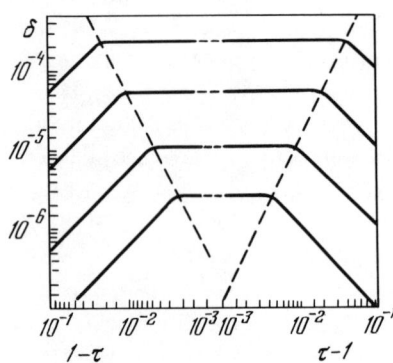

Figure 8.15. Temperature dependence of the CLSC in a ferroelectric. Dashed lines correspond to (8.26) and (8.27).

the voltage–current characteristic beginning with small voltages and the direction of the applied field opposite to the direction of spontaneous polarization ($\pi_s < 0$), we can obtain a characteristic with a slope greater than that according to the Child law (8.23), since the dielectric constant increases with increasing field. When the field reaches the value of the coercive field, the spontaneous polarization changes sign because of polarization reversal, and the CLSC drops discontinuously, which leads to hysteresis effects (cf. the vertical arrow in Figure 8.14).

We now consider a ferroelectric with a first-order phase transition. In the region of strong fields

$$F = \frac{1}{6}\gamma P^6 \quad \text{and} \quad \mathscr{E} = \gamma P^5,$$

and the voltage–current characteristic is accordingly described by the equation

$$j = \frac{11}{36}\left(\frac{11}{6}\right)^{1/5} \gamma^{-1/5} \mu \frac{U^{6/5}}{L^{11/5}}. \tag{8.28}$$

The current (8.28) does not depend on the temperature and thus corresponds to the limiting solution of (8.21).

For weak fields, one can obtain a solution analogous to (8.23) and depending on temperature:

$$j = \frac{9}{8} \varepsilon_0 \bar{\varepsilon} \mu \frac{U^2}{L^3}. \tag{8.29}$$

The family of curves of (8.29) satisfying the Child law intersects the limiting characteristic (8.28) at values of φ_g, depending on the temperature, where φ_g decreases with approach toward the Curie point as for the case of the second-order phase transition. There are regions near the Curie point where $\bar{\varepsilon}$ decreases or increases with increasing field. According to this, voltage–current characteristics are obtained with a slope smaller or greater than the Child law. A hysteresis variation of the voltage–current characteristics is observed in the region of temperatures where the double dielectric hysteresis loops occur.

The CLSC were investigated experimentally in crystals of SbSI [351, 352] and BaTiO$_3$ [213]. Both dark currents and photocurrents limited by the space charge are observed in SbSI with $j \sim U^\alpha$, where $\frac{3}{2} \leq \alpha \leq 2$. Investigation of the potential distribution by the probe method showed that a current limited by the space charge occurs in fields of $\mathscr{E} \gtrsim 10^3$ V/cm in SbSI, where the index of the voltage–current characteristic and the temperature dependence of the CLSC depend in a complex manner on the electrode material. The effect of the phase transition on the CLSC also could not be traced for SbSI. The voltage–current characteristics of BaTiO$_3$ crystals obtained in [213] with the photocurrent limited by the space charge agree well with the CLSC theory discussed above.

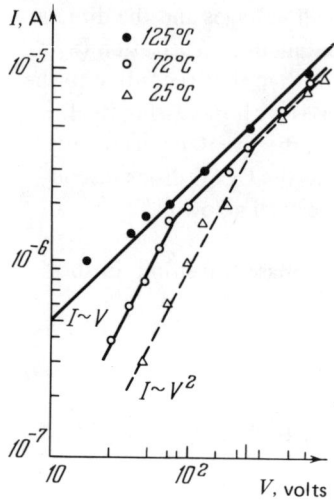

Figure 8.16. Voltage–current characteristics of the CLSC in $BaTiO_3$ [213].

Measurements were carried out with pulsed illumination, and a transient photocurrent signal caused by injection of holes into the crystal was observed. Figure 8.16 presents the voltage–current characteristics for three temperatures including the region of the high-temperature phase transition ($T_1 \simeq 131°C$). It is seen that the point of intersection of the quadratic voltage–current characteristic with the almost linear limiting characteristic at $T \to T_1$ is shifted toward lower voltages.

8.5. Induction Waves and Space Charge in Ferroelectrics

As is well known, the generation of the negative differential resistance in a linear semiconductor leads to the appearance of electrical instabilities, for which sections are formed in the semiconductor with increased concentrations of space charge and field (electric domains) [353]. The motion of the electric domains along the field leads to the generation of oscillations in the external circuit of the current source. Different mechanisms can be the basis for the formation of the negative differential resistance in linear semiconductors [353]. As an example, one can indicate the mechanism of recombination instability, when the capture cross section of charge centers depends on the field. The negative differential resistance can also be related to the scattering mechanism, when the mobility decreases with the field.

The nonlinear relation between the induction and field in ferroelectrics leads to a new form of electrical instability, which can also arise with positive differential resistance. Under specific conditions, an inhomogeneous distribution of the induction is formed in the crystal in the direction of the ferroelectric axis, where the inhomogeneity can move along the crystal under the effect of the external field. Since the change in the induction is screened by the space charge at the boundaries of the inhomogeneity, waves of induction and space charge arise in the ferroelectric under these conditions. These waves could have been called domains by analogy with linear semiconductors if this term had not already been used in the physics of ferroelectrics.

A number of works by Shur and Chenskii [354–358] have been devoted to the theory of electrical instability in ferroelectrics. We will show below, following [357], that the appearance of induction and charge waves is a consequence of the solution of the problem on the nonstationary distribution of induction and space charge in the "single-domain" ferroelectric which moves in the external field. For this, we will start from the equation for the current (3.43), the continuity equation, the Poisson equation (3.42), and the equation of state of the ferroelectric (3.1), neglecting elastic stresses and the gradiant energy:

$$j(z, t) = \mu q \mathscr{E} n(z, t) - qD^* \frac{\partial n(z, t)}{\partial z}, \qquad (3.43)$$

$$\frac{\partial j}{\partial z} + q \frac{\partial n}{\partial t} = 0, \qquad (8.30)$$

$$\frac{\partial D}{\partial z} = 4\pi q [n(z, t) - n_0], \qquad (3.42)$$

$$\frac{1}{4\pi} \mathscr{E} = \alpha D + \beta D^3 + \gamma D^5. \qquad (3.1)$$

If (3.42) is differentiated with respect to time and (8.30) is taken into account, we obtain

$$\frac{\partial^2 D}{\partial z\, \partial t} = -4\pi \frac{\partial j}{\partial z}. \qquad (8.31)$$

Integrating (8.31) with respect to z, we have

$$\frac{\partial D}{\partial t} = 4\pi [-j(z, t) + I(t)], \qquad (8.32)$$

here $I(t)$ is the charge supplied per unit time from the external circuit. We obtain the following equation from (8.32), (3.43), and (3.42):

$$-D^* \frac{\partial^2 D}{\partial z^2} + \mu \mathscr{E} \frac{\partial D}{\partial z} + \frac{\partial D}{\partial t} = 4\pi [I(t) - \mu \mathscr{E} n_0]. \qquad (8.33)$$

By setting $D = \epsilon \mathscr{E}$ for the linear semiconductor, we obtain the equation of Knight and Petersen [357] from (8.33), which leads to the domain instability under specific conditions imposed on the dependence $\mu = \mu(\mathscr{E})$ (the negative differential resistance).

Keeping in mind that the relation between D and $\mathscr{E}$ is given by the equation of state (3.1) for the ferroelectric and assuming $I = $ const, we seek a solution of (8.33) in the form of waves:

$$D(z, t) = D(z - ut) = D(y). \tag{8.34}$$

Substitution of (8.34) into (8.33) brings this equation to a form analogous to the form of the equation used in the theory of the Jahn effect [357]:

$$\frac{d^2 D}{dy^2} + \frac{u - 4\pi\mu (\alpha D + \beta D^3 + \gamma D^5)}{D^*} \frac{dD}{dy} +$$
$$+ \frac{4\pi \left[I - n_0^{-1} 4\pi\mu (\alpha D + \beta D^3 + \gamma D^5) \right]}{D^*} = 0. \tag{8.35}$$

Equation (8.35) was analyzed in [354-357] with the help of the phase plane method. The solution of (8.35) in the ferroelectric region corresponds to the case of propagating induction waves. In the paraelectric region, where the first-order phase transition is induced by the field, the solution of (8.35) corresponds to layers of the paraelectric phase propagating into the ferroelectric phase. Since the change in induction at the boundaries of the propagating "domain" does not exceed $2D_0$ and must be screened by the space charge with density qn_0, the width of the "domain" is

$$d \leqslant \frac{2D_0}{4\pi q n_0}. \tag{8.36}$$

Since the instability in the form of oscillations is possible only under the condition $d < L$ (L is the length of the crystal in the direction of propagation of the "domain"), then

$$n_0 L \gtrsim \frac{2D_0}{4\pi q}. \tag{8.37}$$

For $BaTiO_3$, for example, this condition corresponds to $n_0 L \gtrsim 2\text{-}5 \cdot 10^{13}$ cm^{-2}. However, n_0 should be understood as the concentration of both free carriers as well as carriers captured at fine traps ("slave" levels). Thus, condition (8.37) is not necessarily related to a high concentration of carriers in the band. The frequency of the oscillations caused by the motion of the "domain" is $\sim \mu\mathcal{E}/L$. For $BaTiO_3$, ($\mu \simeq 10^{-1}$ cm^2/(V · sec), this mechanism leads to frequencies of $10^2\text{-}10^3$ Hz in the interval of fields $10^2\text{-}10^3$ V/cm for $L = 10^{-1}$ cm. For SbSI, these frequencies must be an order of magnitude greater if one assumes $\mu \simeq 50$ cm^2/(V · sec) [199].

As was shown by Shur, an instability related simultaneously to a change in mobility and to a field shift of the Curie point can arise in a ferroelectric near the temperature of the first-order phase transition. However, this mechanism leads in essence to the formation of the negative differential resistance.

Of the experimental works devoted to electrical instability in ferroelectrics, we mention two [289, 359]. Low-frequency electric oscillations (several hertz) were observed in SbSI crystals near the Curie point in [298]. A parallel optical observation showed that these oscillations are caused by the motion of sections of the paraelectric phase into the ferroelectric phase. Electrical oscillations with a frequency of the order of several kilohertz in a field of ~500 V/cm were observed in SbSI crystals at a temperature of +15°C in [359]. The generation of oscillations was accompanied by the appearance between the electrodes of a band structure with a period of ~10^{-3} cm, which was observed with the help of an optical microscope. If one takes the period of this structure as the width of the "domain," then $n_0 \lesssim 10^{16}$ cm^{-3} for SbSI according to (8.36), which agrees in general with the high concentration of fine traps (Section 8.3).

References

References for Preface

1. B. M. Vul, *Dokl. Akad. Nauk SSSR* **43**, 308 (1944).
2. B. M. Vul and I. M. Gol'dman, *Dokl. Akad. Nauk SSSR* **46**, 154 (1945).
3. E. Fatuzzo, G. Harbeke, W. I. Merz, R. Nitsche, H. Roetschi, and W. Ruppel, *Phys. Rev.* **127**, 2036 (1962).
4. R. Nitsche, H. Roetschi, and P. Wild, *Appl. Phys. Lett.* **4**, 210 (1964).
5. G. A. Smolenskii, V. A. Bokov, V. A. Isupov, N. N. Krainik, R. E. Pasynkov, and M. S. Shur, *Ferroelectrics and Antiferroelectrics*, Nauka (1971).
6. V. G. Vaks, *Introduction to the Microscopic Theory of Ferroelectrics,* Nauka (1973).
7. E. V. Bursian, *Nonlinear Crystals. Barium Titanate,* Nauka (1974).
8. I. S. Zheludev, *Physics of Crystalline Dielectrics*, Nauka (1968).
9. J. Barfoot, *Introduction to the Physics of Ferroelectric Phenomena*, Mir (1970).
10. A. S. Sonin and B. A. Strukov, *Introduction to Ferroelectricity*, Vysshaya Shkola (1970).
11. *Barium Titanate*, E. G. Fesenko and O. I. Prokopalo, eds., Rostov State University Press (1971).
12. L. P. Kholodenko, *Thermodynamic Theory of Ferroelectrics of the $BaTiO_3$ Type*, B. N. Rolov, ed., Znanie, Riga (1971).
13. G. S. Zhdanov, *Solid State Physics*, Moscow State University Press (1961).
14. W. Kanzig, *Ferroelectrics and Antiferroelectrics*, Academic Press (1964).
15. F. Jona and G. Shirane, *Ferroelectric Crystals,* Pergamon Press (1962).
16. *Barium Titanate*, Nauka (1973), p. 6.

References for Chapter 1

17. V. M. Fridkin, L. M. Belyaev, A. A. Grekov, and A. I. Rodin, *J. Phys. Soc. Jpn.* **28**, Suppl. 448 (1970).
18. V. M. Fridkin, L. M. Belyaev, A. A. Grekov, N. A. Kosonogov, and A. I. Rodin, *J. Phys. (Paris)* **33**, Suppl. 123 (1972).
19. V. M. Fridkin, A. A. Grekov, E. A. Savchenko, and T. R. Volk, *J. Phys. (Paris)* **33**, Suppl. 127 (1972).
20. V. M. Fridkin, A. A. Grekov, N. A. Kosonogov, B. F. Proskuryakov, and T. R. Volk, *Ferroelectrics* **8**, 429 (1974).
21. V. M. Fridkin, A. A. Grekov, P. V. Ionov, A. I. Rodin, E. A. Savchenko, and K. A. Mikhailina, *Ferroelectrics* **8**, 433 (1974).
22. V. M. Fridkin, L. M. Belyaev, A. A. Grekov, I. I. Groshik, V. A. Lyakhovitskaya, and

A. I. Rodin, Proceedings of the IX International Conference on Semiconductor Physics, Nauka (1969), p. 1328.
23. *Ferroelectrics* **6**, $\frac{1}{2}$ (1973).
24. *Semiconductor Ferroelectrics*, A. A. Grekov, ed., Rostov State University Press (1973).
25. V. M. Fridkin. *Pis'ma Zh. Eksp. Teor. Fiz.* **3**, 252 (1956).
26. R. E. Pasynkov, *Izv. Akad. Nauk SSSR, Ser. Fiz.* **34**, 2466 (1970).
27. R. E. Pasynkov, *Ferroelectrics* **6**, 19 (1973).
28. L. D. Landau and E. M. Lifshits, *Statistical Physics*, Addison-Wesley (1969).
29. V. L. Ginzburg, *Usp. Fiz. Nauk* **38**, 490 (1949).
30. V. L. Ginzburg, *Usp. Fiz. Nauk* **77**, 621 (1962).
31. R. Kern, *J. Phys. Chem. Solids* **23**, 249 (1962).
32. G. Harbeke, *J. Phys. Chem. Solids* **24**, 957 (1963).
33. A. P. Levanyuk and D. G. Sannikov, *Zh. Eksp. Teor. Fiz.* **55**, 256 (1968).
34. V. Dvořak, *Czech. J. Phys.* **B21**, 1250 (1971).
35. H. M. James, Photoconductivity Conference, Atlantic City, New Jersey (1954).
36. H. Brooks, *Adv. Electron. Electron Phys.* **7**, 117 (1960).
37. R. W. Keyes, *Solid State Phys.* **11**, 179 (1960).
38. J. C. Slater, *Introduction to Chemical Physics*, McGraw-Hill (1939), p. 216.
39. N. Mott and R. Gurney, *Electronic Processes in Ionic Crystals*, Dover (1948).
40. I. B. Levinson and E. I. Rashba, *Usp. Fiz. Nauk* **108**, 385 (1972).

References for Chapter 2

41. W. Cochran, *Adv. Phys.* **9**, 387 (1960).
42. J. J. Hallers and W. T. Caspers, *Phys. Status Solidi* **36**, 587 (1969).
43. K. Kreher, *Phys. Lett. A* **30**, 384 (1969).
44. K. Kreher, *Wiss. Z. Karl-Marx-Univ. Leipzig, Math.-Naturwiss. Reihe* **20**, 287 (1971).
45. T. Natterman, *Phys. Status Solidi B* **51**, 395 (1972).
46. S. M. Ryvkin, *Photoelectric Phenomena in Semiconductors*, Fizmatgiz (1963).
47. J. Ziman, *Principles of the Theory of Solids*, Cambridge University Press (1972).
48. I. B. Bersuker, *Phys. Lett.* **20**, 589 (1966).
49. I. B. Bersuker and B. G. Vekhter, *Fiz. Tverd. Tela* **9**, 2652 (1967).
50. B. G. Vekhter and I. B. Bersuker, *Ferroelectrics* **6**, 13 (1973).
51. N. Kristoffel and P. Konsin, *Phys. Status Solidi* **21**, K39 (1967).
52. N. Kristoffel and P. Konsin, *Phys. Status Solidi* **28**, 731 (1968).
53. N. N. Kristofel' and P. I. Konsin, *Izv. Akad. Nauk SSSR, Ser. Fiz.* **33**, 187 (1969).
54. N. N. Kristofel' and P. I. Konsin, *Fiz. Tverd. Tela* **13**, 3513 (1971).
55. P. I. Konsin and N. N. Kristofel', *Kristallografiya* **17**, 712 (1972).
56. N. Kristoffel and P. Konsin, *Ferroelectrics* **6**, 3 (1973).
57. G. Chanussot, *Ferroelectrics* **8**, 671 (1974).

References for Chapter 3

58. I. I. Ivanchik, *Fiz. Tverd, Tela* **3**, 3731 (1961).
59. G. M. Guro, I. I. Ivanchik, and N. F. Kovtonyuk, *Pis'ma Zh. Eksp. Teor. Fiz.* **1**, 9 (1967).
60. G. M. Guro, I. I. Ivanchik, and N. F. Kovtonyuk, *Fiz. Tverd. Tela* **10**, 135 (1968).

61. E. V. Chenskii, *Fiz. Tverd. Tela* **11**, 666 (1969).
62. E. V. Chenskii, *Fiz. Tverd. Tela* **12**, 586 (1970).
63. V. A. Zhirnov, *Zh. Eksp. Teor. Fiz.* **35**, 1175 (1958).
64. M. DiDomenico, Jr., and S. H. Wemple, *Phys. Rev.* **155**, 539 (1967).
65. E. V. Chenskii and V. B. Sandomirskii, *Fiz. Tekh. Poluprovodn.* **3**, 857 (1969).
66. F. F. Vol'kenshtein, *Physical Chemistry of the Surface of Semiconductors*, Nauka (1973).
67. B. M. Vul, G. M. Guro, and I. I. Ivanchik, *Fiz. Tekh. Poluprovodn.* **4**, 162 (1970).
68. G. M. Guro and N. F. Kovtonyuk, *Fiz. Tekh. Poluprovodn.* **2**, 300 (1968).
69. G. M. Guro, I. I. Ivanchik, and N. F. Kovtonyuk, *Fiz. Tverd. Tela* **11**, 1956 (1969).
70. B. M. Vul, G. M. Guro, and I. I. Ivanchik, *Ferroelectrics* **6**, 29 (1973).
71. B. M. Vul, G. M. Guro, and I. I. Ivanchik, *Fiz. Tverd. Tela* **14**, 707 (1972).
72. S. Triebwasser, *Phys. Rev.* **118**, 100 (1960).
73. E. V. Bursian and V. V. Shapkin, *Fiz. Tverd. Tela* **9**, 283 (1967).
74. V. G. Poshin, V. K. Novik, B. V. Selyuk, V. A. Koptsik, N. D. Gavrilova, and V. A. Meleshina, *Kristallografiya* **19**, 809 (1974).
75. V. P. Konstantinova, in *Growth of Crystals*, N. N. Sheftal', ed., Vol. 7, Consultants Bureau (1969).
76. B. V. Selyuk, *Kristallografiya* **19**, 221 (1974).
77. B. V. Selyuk, *Izv. Akad. Nauk SSSR, Ser. Fiz.* **35**, 1798 (1971).
78. B. V. Selyuk, *Kristallografiya* **16**, 356 (1971).
79. B. V. Selyuk, *Ferroelectrics* **6**, 37 (1973).
80. W. Schottky, *Z. Phys.* **118**, 539 (1942).
81. J. Bardeen, *Phys. Rev.* **71**, 717 (1947).
82. D. Kahng and S. H. Wemple, *J. Appl. Phys.* **36**, 2925 (1965).
83. S. H. Wemple, D. Kahng, and H. J. Braun, *J. Appl. Phys.* **38**, 353 (1967).
84. T. Murakami, *J. Phys. Soc. Jpn.* **23**, 457 (1967).
85. T. Murakami, *J. Phys. Soc. Jpn.* **24**, 282 (1968).
86. A. M. Cowley, *J. Appl. Phys.* **36**, 3212 (1965).
87. A. G. Zhdan, E. V. Chenskii, E. S. Artobolevskaya, R. S. Gvozdover, É. I. Rau, and V. I. Petrov, *Pis'ma Zh. Eksp. Teor. Fiz.* **3**, 161 (1971).
88. L. D. Landau and E. M. Lifshits, *Electrodynamics of Continuous Media*, Addison-Wesley (1960).
89. I. E. Tamm, *Sov. Phys.* **1**, 731 (1932).
90. A. I. Larkin and D. E. Khmel'nitskii, *Zh. Eksp. Teor. Fiz.* **55**, 2345 (1968).

References for Chapter 4

91. K. Gulyamov, V. A. Lyakhovitskaya, N. A. Tikhomirova, and V. M. Fridkin, *Dokl. Akad. Nauk SSSR*, **161**, 1060 (1965).
92. V. M. Fridkin, K. Gulyamov, V. A. Lyakhovitskaya, V. N. Nosov, and N. A. Tikhomirova, *Fiz. Tverd. Tela* **8**, 1907 (1966).
93. K. A. Verkhovskaya and V. M. Fridkin, *Fiz. Tverd. Tela* **8**, 1620 (1966).
94. K. A. Verkhovskaya and V. M. Fridkin, *Fiz. Tverd. Tela* **8**, 3129 (1966).
95. K. A. Verkhovskaya and A. S. Sonin, *Zh. Eksp. Teor. Fiz.* **52**, 383 (1967).
96. K. A. Verkhovskaya and V. M. Fridkin, *Izv. Akad. Nauk SSSR, Ser. Fiz.* **31**, 1156 (1967).
97. N. R. Ivanov and K. A. Verkhovskaya, *Fiz. Tverd. Tela* **10**, 2974 (1967).
98. V. M. Fridkin and K. A. Verkhovskaya, *Appl. Opt.* **6**, 1825 (1967).

99. K. A. Verkhovskaya, I. P. Grigas, and V. M. Fridkin, *Fiz. Tverd Tela* **10**, 2015 (1968).
100. V. M. Fridkin, E. I. Gerzanich, I. I. Groshik, and V. A. Lyakhovitskaya, *Pis'ma Zh. Eksp. Teor. Fiz.* **4**, 201 (1966).
101. N. R. Ivanov and K. A. Verkhovskaya, *Izv. Akad. Nauk SSSR, Ser. Fiz.* **34**, 2563 (1970).
102. V. M. Fridkin, K. A. Verkhovskaya, and B. G. Bochkov, *Phys. Status Solidi A* **22**, 759 (1974).
103. A. I. C. Wilson, *Phys. Rev.* **54**, 1103 (1938).
104. F. Herlach, *Helv. Phys. Acta* **34**, 305 (1961).
105. A. P. Kasatkin, T. G. Petrov, and E. B. Treizus, *Kristallografiya* **7**, 952 (1962).
106. R. P. Ozerov, N. V. Rapnev, V. I. Pakhomov, I. S. Rez, and G. S. Zhdanov, *Kristallografiya* **7**, 620 (1962).
107. S. Savada, S. Nomura, S. Fujii, and I. Yoshika, *Phys. Rev. Lett.* **1**, 320 (1958).
108. R. Pepinsky and K. Vedam, *Phys. Rev.* **114**, 1217 (1959).
109. L. A. Shuvalov, N. R. Ivanov, and G. K. Sitnik, *Kristallografiya* **12**, 366 (1967).
110. T. Horie, K. Kawabe, and S. Sawada, *J. Phys. Soc. Jpn.* **9**, 823 (1954).
111. C. Gähwiller, *Phys. Kond. Mater.* **6**, 269 (1967).
112. T. Horie, K. Kawabe, M. Tachiki, and S. Sawada, *J. Phys. Soc. Jpn,* **10**, 541 (1955).
113. Y. Kaifu and T. Komatsu, *J. Phys. Soc. Jpn.* **23**, 903 (1967).
114. T. R. Brews, *Phys. Rev. Lett.* **18**, 662 (1967).
115. G. A. Cox, G. G. Roberts, and R. H. Tredgold, *Br. J. Appl. Phys.* **17**, 743 (1966).
116. S. H. Wemple, *Phys. Rev. B* **2**, 2679 (1970).
117. A. Frova. *Nuovo Cimento B* **55**, 1 (1968).
118. E. V. Burtsev, *Ferroelectrics* **6**, 15 (1973).
119. H. Y. Fan, *Rep. Prog. Phys.* **19**, 107 (1956).
120. K. A. Verkhovskaya, Author's abstract of candidate's dissertation, Moscow (1968).
121. M. I. Cohen and R. F. Blunt, *Phys. Rev.* **168**, 929 (1968).
122. M. DiDomenico, Jr., and S. H. Wemple, *Phys. Rev.* **166**, 565 (1968).
123. E. Doenges, *Z. Anorg. Allg. Chem.* **263**, 112, 280 (1950).
124. R. Arndt and A. Niggli, *Naturwissenschaften* **51**, 158 (1964).
125. V. N. Nosov, *Zh. Strukt. Khim.* **9**, 338 (1968).
126. T. A. Pikka and V. M. Fridkin, *Fiz. Tverd. Tela* **10**, 3378 (1968).
127. D. M. Bercha, I. V. Bercha, V. Yu. Slivka, I. D. Turyanitsa, and D. V. Chepur, *Fiz. Tverd. Tela* **11**, 1677 (1969).
128. K. Ishikawa, R. Tanaka, and K. Toyoda, *Phys. Lett. A* **42**, 289 (1972).
129. E. I. Gerzanich, *Fiz. Tverd. Tela* **9**, 2995 (1967).
130. E. I. Gerzanich, A. N. Borets, and D. Sh. Kovach, *Opt. Spektrosk.* **32**, 1141 (1972).
131. E. I. Gerzanich, A. N. Borets, and D. Sh. Kovach, *Izv. Vyssh. Uchebn. Zaved., Fiz.* **7**, 85 (1972).
132. T. P. McLean, *Prog. Semicond.* **5**, 53 (1960).
133. K. Ohi, *J. Phys. Soc. Jpn.* **25**, 1369 (1968).
134. V. Riede, *Phys. Lett. A* **29**, 715 (1969).
135. D. M. Bercha, V. Yu. Slivka, N. N. Syrbu, I. D. Turyanitsa, and D. V. Chepur, *Fiz. Tverd. Tela* **13**, 276 (1971).
136. B. P. Grigas, *Fiz. Tekh. Poluprovodn.* **2**, 481 (1969).
137. I. Horak, I. D. Turjanica, and K. Nejezchleb, *Krist. Tech.* **3**, 231 (1968).
138. M. Barbe, D. Bruebois, M. Dimani, and M. Laurent, *C. R. Acad. Sci. Paris* **268**, 2053 (1969).
139. K. Toyodo and K. Ishikawa, *J. Phys. Soc. Jpn.* **28**, Suppl. 451 (1970).
140. L. L. Golik and M. I. Elinson, *Fiz. Tverd. Tela* **12**, 2895 (1970).
141. H. Kamimura, S. M. Shapiro, and B. Balkanski, *Phys. Lett. A* **33**, 277 (1970).

REFERENCES

142. A. Kh. Zeinally, A. M. Mamedov, and Sh. M. Efendiev, *Fiz. Tekh. Poluprovodn.* 7, 383 (1973).
143. É. V. Chisler, I. T. Savatinova, and V. M. Fridkin, *Fiz. Tverd. Tela* 12, 2882 (1970).
144. M. Balkanski, M. K. Teng, S. M. Shapiro, and M. K. Ziolkiewicz, *Phys. Status Solidi* 44, 355 (1971).
145. I. P. Grigas and A. S. Karpus, *Kristallografiya* 12, 719 (1967).
146. I. P. Grigas and A. S. Karpus, *Fiz. Tverd. Tela* 9, 2882 (1967).
147. I. P. Grigas and A. S. Karpus, *Fiz. Tverd. Tela* 9, 2887 (1967).
148. V. S. Dinkyavichus and M. P. Mikalkevichus, *Fiz. Tverd. Tela* 9, 2997 (1967).
149. A. Orlyukas and I. P. Grigas, *Litov. Fiz. Sb.* 14, 313 (1974).
150. I. Petzelt and I. Grigas, *Ferroelectrics* 5, 59 (1973).
151. H. Frölich, *Adv. Phys.* 3, 325 (1954).
152. V. L. Bonch-Bruevich, *Phys. Status Solidi* 42, 35 (1970).
153. S. H. Wemple and M. DiDomenico, Jr., *Phys. Rev. Lett.* 23, 1156 (1969).
154. S. H. Wemple and M. DiDomenico, Jr., *Phys. Rev. B* 1, 193 (1970).
155. Y. Yamada and G. Shirane, *Phys. Rev.* 117, 848 (1969).
156. M. G. Cohen, M. DiDomenico, Jr., and S. H. Wemple, *Phys. Rev. B* 1, 4334 (1970).
157. L. V. Keldysh, *Zh. Eksp. Teor. Fiz.* 34, 38 (1958).
158. W. Franz, *Z. Naturforsch. Teil. A* 13, 484 (1958).
159. E. I. Gerzanich and V. M. Fridkin, *Zh. Eksp. Teor. Fiz.* 56, 780 (1969).
160. E. I. Gerzanich and V. M. Fridkin, *Fiz. Tverd. Tela* 10, 3111 (1968).
161. E. I. Gerzanich and V. M. Fridkin, *Kristallografiya* 14, 3 (1969).
162. E. I. Gerzanich and V. M. Fridkin, *Pis'ma Zh. Eksp. Teor. Fiz.* 8, 553 (1968).
163. C. Gähwiller, *Helv. Phys. Acta* 38, 361 (1965).
164. O. V. Kovalev and G. Ya. Lyubarskii, *Zh. Tekh. Fiz.* 28, 1151 (1958).
165. F. M. Gashimzade, *Fiz. Tverd. Tela* 2, 2070 (1960).
166. A. S. Karpus and I. V. Batarunas, *Litov. Fiz. Sb.* 1, 315 (1961).
167. K. D. Tovstyuk and D. M. Gemus, *Fiz. Tverd. Tela* 5, 142 (1963).
168. K. D. Tovstyuk and D. M. Bercha, *Fiz. Tverd. Tela* 6, 662 (1964).
169. E. N. Rashba, *Ukr. Fiz. Zh.* 8, 1064 (1963).
170. A. N. Borets, *Fiz. Tverd. Tela* 8, 2440 (1966).
171. D. V. Chepur, D. M. Bercha, I. D. Turianitsa, and V. Yu. Slivka, *Phys. Status Solidi* 30, 461 (1968).
172. S. D. Shutov, V. V. Sobolev, V. Yu. Popov, and S. N. Shestatskii, in *The Chemical Bond in Semiconductors,* Nauka i Tekhnika, Minsk (1968), p. 58.
173. Y. Yamada and H. Chihara, *J. Phys. Soc. Jpn.* 21, 2085 (1966).
174. A. G. Khasabov and I. Ya. Nikiforov, *Izv. Akad. Nauk SSSR, Ser. Fiz.* 34, 2480 (1970).
175. A. G. Khasabov and I. Ya. Nikiforov, *Kristallografiya* 16, 41 (1971).
176. F. Michel-Calendini and G. Mesnard, *Phys. Status Solidi B* 44, K117 (1971).
177. F. Michel-Calendini and G. Mesnard, *J. Phys. C* 6, 1709 (1973).
178. L. F. Mattheiss, *Phys. Rev. B* 2, 3918 (1970).
179. T. Soules, E. Kelly, D. Vaught, and I. Richardson, *Phys. Rev. B* 6, 1519 (1972).
180. L. F. Mattheiss, *Phys. Rev.* 181, 987 (1969).
181. A. Kahn and A. Leyendecker, *Phys. Rev. A* 135, 1321 (1964).
182. I. Zook and T. Casselman, *Surf. Sci.* 37, 244 (1973).
183. A. G. Khasabov and I. V. Grekova, in *Semiconductor Ferroelectrics,* Rostov State University, Rostov-on-Don (1973), p. 85.
184. A. G. Khasabov and I. Ya. Nikiforov, in *Semiconductor Ferroelectrics,* Rostov State University, Rostov-on-Don (1973), p. 79.
185. S. Scavničar, *Z. Kristallogr.* 114, 85 (1960).

186. I. Ya. Nikiforov and A. G. Khasabov, in *The Chemical Bond in Semiconductors and Semimetals*, Nauka i Tekhnika, Minsk (1972), p. 101.
187. I. Ya. Nikiforov and A. G. Khasabov, *Fiz. Tverd. Tela* **13**, 3589 (1971).
188. K. Nakao and M. Balkanski, *Phys. Rev. B* **8**, 5759 (1973).
189. C. Fong, J. Petroff, S. Kohn, and J. Shen, *Solid State Commun.* **14**, 681 (1974).

References for Chapter 5

190. A. A. Grekov, V. A. Lyakhovitskaya, A. I. Rodin, and V. M. Fridkin, *Fiz. Tverd. Tela* **8**, 3092 (1966).
191. L. M. Belyayev, I. I. Groshik, V. A. Lyakhovitskaya, V. N. Nosov, and V. M. Fridkin, *Pis'ma Zh. Eksp. Teor. Fiz.* **6**, 481 (1967).
192. B. P. Grigas, I. P. Grigas, and R. P. Belyatskas, *Fiz Tverd. Tela* **9**, 1532 (1967).
193. I. I. Groshik, P. V. Ionov, and V. M. Fridkin, *Fiz. Tekh. Poluprovodn.* **2**, 1630 (1968).
194. S. Ueda, I. Tatsuzaki, and J. Shindo, *Phys. Rev. Lett.* **18**, 453 (1967).
195. M. D. Volnyanskii, A. Yu. Kudzin, and A. N. Sukhinskii, *Fiz. Tverd. Tela* **15**, 2819 (1973).
196. R. Bube, *Photoconductivity of Solids*, Wiley (1960).
197. B. P. Grigas, *Fiz. Tekh. Poluprovodn.* **2**, 481, 585 (1968).
198. R. Bube and S. Thomsen, *J. Chem. Phys.* **23**, 15 (1955).
199. V. A. Alekseeva and E. G. Landsberg, *Fiz. Tverd. Tela* **8**, 3138 (1966).
200. M. K. Sheinkman and I. E. Korsunskaya, *Ukr. Fiz. Zh.* **12**, 2042 (1967).
201. V. N. Nosov, *Kristallografiya* **12**, 359 (1967).
202. A. I. Rodin, Author's abstract of candidate's dissertation, Rostov-on-Don (1971).
203. I. I. Groshik and V. M. Fridkin, *Fiz. Tverd. Tela* **10**, 2878 (1968).
204. M. DiDomenico, Jr., and S. H. Wemple, *Phys. Rev.* **155**, 545 (1967).
205. A. A. Grekov, A. I. Rodin, and V. M. Fridkin, *Fiz. Tverd. Tela* **12**, 3643 (1970).
205a. V. M. Fridkin, A. A. Grekov, and A. I. Rodin, *Appl. Phys. Lett.* **14**, 119 (1969).
206. G. I. F. Garlich and A. F. Gibson, *Proc. R. Soc.* **60**, 574 (1948).
207. I. B. Newkirk, *Acta Metall.* **6**, 316 (1956).
208. R. H. Bube, *Phys. Rev.* **106**, 103 (1957).
209. T. R. Volk, N. A. Tikhomirova, I. D. Turyanitsa, and V. M. Fridkin, *Fiz. Tverd. Tela* **12**, 3645 (1970)
210. T. R. Volk, A. A. Grekov, N. A. Kosonogov, A. I. Rodin, and V. M. Fridkin, *Kristallografiya* **16**, 241 (1971).
211. T. R. Volk, A. A. Grekov, N. A. Kosonogov, and V. M. Fridkin, *Fiz. Tverd. Tela* **14**, 3216 (1972).
212. A. Yu. Kudzin, A. N. Sukhinskii, and R. V. Oginov, *Fiz. Tverd. Tela* **10**, 1577 (1968).
213. L. Benguigui, *Solid State Commun.* **7**, 1245 (1969).
214. I. Tatsuzaki, K. Itoh, S. Ueda, and I. Shindo, *Phys. Rev. Lett.* **17**, 198 (1966).
215. A. A. Grekov, E. D. Rogach, and A. G. Sukiyazov, *Fiz. Tverd. Tela* **12**, 3559 (1970).
216. A. A. Grekov, R. E. Pasynkov, and E. D. Rogach, *Fiz. Tverd. Tela* **14**, 2216 (1972).
217. A. A. Grekov, E. D. Rogach, and L. N. Syrkin, *Fiz. Tverd. Tela* **14**, 2768 (1972).
218. A. M. Zav'yalova, P. L. Zaks, and L. N. Syrkin, *Fiz. Tverd. Tela* **12**, 1580 (1970).
219. A. Ashkin, C. D. Boyd, T. M. Dziedzic, R. G. Smith, A. A. Ballmann, I. I. Levinstein, and K. Nassau, *Appl. Phys. Lett.* **9**, 72 (1966).
220. A. Ashkin, B. Tell, and I. M. Dziedzic, *J. Quantum Electron.* **3**, 400 (1967).
221. F. S. Chen, *J. Appl. Phys.* **8**, 3418 (1967).
221a. F. S. Chen, *J. Appl. Phys.* **40**, 3389 (1969).
222. R. L. Townsend and I. T. LaMacchia, *J. Appl. Phys.* **41**, 5188 (1970).

REFERENCES

223. W. D. Johnston, *J. Appl. Phys.* **41**, 3279 (1970).
224. I. I. Amodei, *RCA Rev.* **32**, 185 (1971).
225. I. I. Amodei, D. L. Staebler, and W. Stephens, *Appl. Phys. Lett.* **18**, 507 (1971).
226. I. I. Amodei D. L. Staebler, and W. Phillips, *Appl. Opt.* **11**, 390 (1972).
227. D. L. Staebler and I. I. Amodei, *Ferroelectrics* **3**, 107 (1972).
228. N. B. Angerb, V. A. Pashkov, and I. M. Solov'eva, *Zh. Eksp. Teor. Fiz.* **62**, 1666 (1972).
229. H. Tsuya and Y Fujiuo, *Jpn. J. Appl. Phys.* **12**, 1896 (1973).
230. I. B. Thaxter and M. Kestigian. *Appl. Opt.* **13**, 913 (1974).
231. F. Micheron, C. Mayeux, and I. F. Trotier, *Appl. Opt.* **13**, 784 (1974).
232. P. V. Ionov, K. A. Verkhovskaya, L. I. Ivleva, Yu. S. Kuz'minov, and V. M. Fridkin, *Kratk. Soobshch. Fiz.* **10**, 24 (1973).
232a. P. V. Ionov, *Fiz. Tverd. Tela* **15**, 2827 (1973).
233. T. R. Volk, K. D. Kochev, and Yu. S. Kuz'minov, *Kristallografiya* **20**, 583 (1975).
234. A. V. Guinzberg, K. D. Kochev, Yu. S. Kusminov, and T. R. Volk, *Phys. Status Solidi A* **29**, 309 (1975).
235. A. G. Chynoweth, *Phys. Rev.* **102**, 705 (1956).
236. N. M. Bezdetnyi, A. Kh. Zeinally, N. N. Lebedeva, and M. K. Sheinkman, *Fiz. Tekh. Poluprovodn.* **5**, 1061 (1970).
237. N. M. Bezdetnyi, G. Z. Gorbatov, A. Kh. Zeinally, N. N. Lebedeva, and M. K. Sheinkman, *Fiz. Tverd. Tela* **14**, 574 (1972).
238. B. S. Agaronov, N. M. Bezdetnyi, A. Kh. Zeinally, and N. N. Lebedeva, *Kristallografiya* **17**, 677 (1972).
239. A. M. Glass, D. von der Linde, and T. I. Negran, *Appl. Phys. Lett.* **25**, 233 (1974).
240. A. M. Glass, *J. Appl. Phys.* **40**, 4699 (1969).
241. I. P. Kaminov and E. H. Turner, *Proc. IEEE* **54**, 1374 (1966).
242. P. V. Lenzo, E. G. Spencer, and A. A. Bollmann, *Appl. Phys. Lett.* **11**, 33 (1967).
243. M. G. Clark, F. S. DiSalvo, A. M. Glass, and G. E. Peterson, *J. Chem. Phys.* **59**, 6209 (1973).
244. D. von der Linde, A. M. Glass, and K. F. Rodgers, *Appl. Phys. Lett.* **25**, 155 (1974).
245. A. M. Glass and D. H. Auston, *Opt. Commun.* **5**, 45 (1972).
246. A. M. Glass and D. H. Auston, *Phys. Rev. Lett.* **28**, 897 (1972).
247. D. von der Linde, D. H. Auston, A. M. Glass, and K. F. Rodgers, *Solid State Commun.* **14**, 137 (1974).
248. I. F. Kanaev and V. K. Malinovskii, *Fiz. Tverd. Tela* **16**, 3694 (1974).
249. T. F. Moss, *Optical Properties of Semiconductors*, Foreign Literature Press (1961).
250. I. M. Lifshits, *Usp. Fiz. Nauk* **83**, 617 (1964).
251. M. A. Krivoglaz, *Fiz. Tverd. Tela* **11**, 2230 (1969).
252. M. A. Krivoglaz, *Fiz. Tverd. Tela* **12**, 1705 (1970).
253. E. A. Galashin, *Dokl. Akad. Nauk SSSR* **171**, 366 (1966).
254. I. Tyndall, *Proc. R. Soc.* **17**, 92 (1869).
255. C. T. R. Wilson, *Philos. Trans. R. Soc. London, Ser. A* **192**, 403 (1899).
256. M. T. Mulcachy and E. Kuffel, *Proc. Phys. Soc.* **80**, 1333 (1962).
257. E. A. Galashin, *Vestn. Mosk. Gos. Univ., Khim.* **2**, 28 (1969).
258. E. A. Galashin, *Dokl. Akad. Nauk SSSR* **198**, 1360 (1971).
259. I. Förster and K. Kasper, *Z. Electrochem.* **59**, 975 (1955).
260. E. A. Galashin and S. S. Dorofeev, in *Mechanism and Kinetics of Crystallization*, Nauka i Tekhnika, Minsk (1969), p. 483.
261. A. P. Kasatkin, *Kristallografiya* **11**, 328 (1966).
262. H. M. Papee, A. C. Montefinale, and T. U. Zanidzki, *Nature* **203**, 1343 (1964).
263. I. A. Paribok-Aleksandrovich, *Fiz. Tverd. Tela* **11**, 2018 (1969).

264. A. V. Ginzberg, A. D. Sablin-Yavorskii, and V. M. Fridkin, *Pis'ma Zh. Eksp. Teor. Fiz.* **12**, 22 (1970).
265. C. B. Van den Berg, I. E. Van Delden, and I. Bouman, *Phys. Status Solidi* **36**, K89 (1969).
266. P. Wachter and P. Weber, *Solid State Commun.* **8**, 1133 (1970).
267. V. A. Izvozchikov and V. A. Bordovskii, in *Semiconductor Ferroelectrics*, Rostov State University, Rostov-on-Don (1973), p. 152.

References for Chapter 6

268. I. C. Crawford, *Ferroelectrics* **1**, 23 (1970).
269. P. Würfel and W. Ruppel, *Proceedings of the Third International Conference on Photoconductivity*, Stanford, 1969, Pergamon Press, Oxford (1971), p. 379.
270. V. V. Bogatko and N. F. Kovtonyuk, *Fiz. Tverd. Tela* **12**, 605 (1970).
271. A. D. Goncharov, A. F. Mal'tsev, and I. A. Primachenko, *Fiz. Tekh. Poluprovodn.* **3**, 102 (1969).
272. G. M. Guro, I. I. Ivanchik, and N. F. Kovtonyuk, in Barium Titanate, Nauka (1973), p. 71.
273. V. M. Fridkin, *Physical Principles of the Electrophotographic Process*, Energiya Press (1966).
274. G. G. Harman, *Phys. Rev.* **111**, 27 (1958).
275. L. Godefroy, P. Jullien, B. Morlon, and G. Godefroy, *Ferroelectrics* **8**, 421 (1974).
276. E. G. Fesenko, V. G. Gavrilyachenko, M. A. Martinenko, A. F. Semenenko, and I. P. Lapin, *Ferroelectrics* **6**, 61 (1973).
277. A. A. Adonin and A. A. Grekov, *Fiz. Tverd. Tela* **16**, 568 (1974).
278. A. G. Zhdan and E. S. Artobolevskaya, *Fiz. Tverd. Tela* **13**, 1247 (1971).
279. K. Ohi and K. Irie, *J. Phys. Soc. Jpn.* **28**, 1379 (1970).
280. A. A. Grekov, V. A. Lyakhovitskaya, A. I. Rodin, and V. M. Fridkin, *Dokl. Akad. Nauk SSSR* **169**, 810 (1966).
281. T. Mori, H. Tamura, and E. Sawaguchi, *J. Phys. Soc. Jpn.* **20**, 1294 (1965).
282. S. Kawada and M. Ida, *J. Phys. Soc. Jpn.* **20**, 1287 (1965).
283. D. S. Lieberman, M. S. Wechsler, and T. H. Read, *J. Appl. Phys.* **26**, 473 (1955).
284. T. Mori and E. Sawaguchi, *Jpn. J. Appl. Phys.* **6**, 792 (1967).
285. T. Mori and E. Sawaguchi, *J. Phys. Soc. Jpn.* **25**, 1195 (1968).
286. T. S. Travina, L. L. Golik, E. I. Gerzanich, I. I. Groshik, M. I. Elinson, P. V. Ionov, V. A. Lyakhovitskaya, and V. M. Fridkin, *Fiz. Tverd. Tela* **9**, 3664 (1967).
287. L. L. Golik, I. I. Groshik, M. I. Elinson, P. V. Ionov, V. A. Lyakhovitskaya, T. S. Travina, and V. M. Fridkin, *Kristallografiya* **13**, 1014 (1968).
288. D. Meyerhoffer, *Phys. Rev.* **112**, 413 (1958).
289. V. G. Kuznetsov and V. Z. Borodin, *Ukr. Fiz. Zh.* **13**, 1907 (1968).
290. V. M. Rudyak and A. A. Bogomolov, *Fiz. Tverd. Tela* **9**, 3336 (1967).
291. V. M. Rudyak, A. A. Bogomolov, and V. V. Ivanov, *Fiz. Tverd. Tela* **11**, 1700 (1969).
292. V. V. Ivanov, A. A. Bogomolov, and V. M. Rudyak, *Fiz. Tverd. Tela* **11**, 3394 (1969).
293. A. A. Bogomolov, V. V. Ivanov, and V. M. Rudyak, *Kristallografiya* **14**, 1033 (1969).
294. A. A. Grekov, V. A. Lyakhovatskaya, A. I. Rodin, and V. M. Fridkin, *Fiz. Tverd. Tela* **10**, 2239 (1968).
295. A. Yu. Kudzin and A. N. Sukhinskii, *Fiz. Tverd. Tela* **11**, 1389 (1969).
296. A. Yu. Kudzin, A. N. Sukhinskii, and M. D. Volnyanskii, *Fiz. Tverd. Tela* **11**, 2679 (1969).
297. A. Yu. Kudzin, A. N. Sukhinskii, and V. A. Butsychenko, *Fiz. Tverd. Tela* **12**, 2180 (1970).

298. V. M. Fridkin, M. I. Gorelov, A. A. Grekov, V. A. Lyakhovitskaya, and A. I. Rodin, *Pis'ma Zh. Eksp. Teor. Fiz.* **4**, 461 (1966).
299. E. Fatuzzo, *J. Appl. Phys.* **33**, 2588 (1962).
300. V. M. Fridkin and I. I. Groshik, *Appl. Phys. Lett.* **10**, 354 (1967).
301. M. P. Michailov, *C. R. Acad. Bulg. Sci.* **21**, 1029 (1968).
302. B. P. Grigas, *Fiz. Tverd. Tela* **9**, 2430 (1967).
303. V. P. Bender and V. M. Fridkin, *Fiz. Tverd. Tela* **13**, 614 (1971).
304. V. M. Fridkin, A. A. Grekov, N. A. Kosonogov, and T. R. Volk, *Ferroelectrics* **4**, 169 (1972).
305. P. W. Forsberg, *Phys. Rev.* **76**, 1187 (1949).
306. V. M. Fridkin, *Ferroelectrics* **2**, 119 (1971).
307. K. Imai, S. Kawada, and M. Ida, *J. Phys. Soc. Jpn.* **21**, 1855 (1966).
308. M. K. Sheinkman, A. Kh. Zeinally, N. M. Bezdetnyi, N. N. Lebedeva, and G. Z. Gorbatov, *Ukr. Fiz. Zh.* **11**, 1914 (1970).
309. B. S. Agaronov, N. M. Bezdetnyi, A. Kh. Zeinally, N. N. Lebedeva, and M. K. Sheinkman, *Fiz. Tverd. Tela* **14**, 254 (1972).
310. N. M. Bezdetnyi, G. Z. Gorbatov, A. Kh. Zeinally, N. N. Lebedeva, and M. K. Sheinkman, *Fiz. Tverd. Tela* **14**, 924 (1972).
311. D. Berlincourt, H. Jaffe, W. I. Merz, and R. Nitsche, *Appl. Phys. Lett.* **4**, 61, (1964).
312. N. M. Bezdetnyi, G. Z. Gorbatov, A. Kh. Zeinally, N. N. Lebedeva, and M. K. Sheinkman, *Fiz. Tekh. Poluprovodn.* **6**, 1189 (1972).

References for Chapter 7

313. A. A. Grekov, M. A. Malitskaya, V. D. Spitsyna, and V. M. Fridkin, *Kristallografiya* **15**, 500 (1970).
314. A. A. Grekov, M. A. Malitskaya, and V. M. Fridkin, *Kristallografiya* **17**, 581 (1972).
315. A. A. Grekov, M. A. Malitskaya, and V. M. Fridkin, *Kristallografiya* **18**, 788 (1973).
316. V. A. Yurin, A. S. Baberkin, E. N. Kornienko, and I. V. Gavrilova, *Izv. Akad. Nauk SSSR, Ser. Fiz.* **24**, 1334 (1960).

References for Chapter 8

317. M. S. Kosman and E. V. Bursian, *Izv. Akad. Nauk SSSR, Ser. Fiz.* **22**, 1459 (1958).
318. G. Godefroy and B. Couson, *J. Phys. (Paris)* **33**, Suppl. C2-120 (1972).
319. H. P. R. Frederikse, W. R. Thurber, and W. R. Hosler, *Phys. Rev. A* **134**, 442 (1964).
320. J. Sasaki, *Jpn. J. Appl. Phys.* **3**, 558 (1964); **4**, 228 (1965).
321. V. N. Nosov and V. M. Fridkin, *Fiz. Tverd. Tela* **8**, 148 (1966).
322. A. G. Zhdan and E. S. Artobolevskaya, *Prib. Tekh. Eksp.* **2**, 215 (1972).
323. V. M. Fridkin, A. A. Grekov, A. I. Rodin, E. A. Savchenko, and T. R. Volk, *Ferroelectrics* **6**, 71 (1973).
324. H. Jasunaga, *J. Phys. Soc. Jpn.* **28**, Suppl. 454 (1970).
325. N. M. Bezdetnyi, A. Kh. Zeinally, N. N. Lebedeva, and M. K. Sheinkman, *Fiz. Tverd. Tela* **12**, 2480 (1970).
326. E. A. Sal'kov and M. K. Sheinkman, *Fiz. Tverd. Tela* **5**, 397 (1963).
327. V. E. Lashkarev, A. V. Lyubchenko, and M. K. Sheinkman, *Fiz. Tverd. Tela* **7**, 1717 (1965).
328. P. L. Low and D. P. Pines, *Phys. Rev.* **98**, 414 (1955).
329. S. H. Wemple, M. DiDomenico, Jr., and A. Jayaraman, *Phys. Rev.* **180**, 547 (1969).
330. V. L. Vinetskii, M. A. Itskovskii, and L. S. Kukushkin, *Phys. Status Solidi* **39**, K23

(1970); V. L. Vinetskii, M. A. Itskovskii, and L. S. Kukushkin, *Fiz. Tverd. Tela* **13**, 76 (1971).
331. A. I. Gubanov and M. S. Shur, *Fiz. Tverd. Tela* **12**, 664 (1970).
332. H. B. De Vore, *Phys. Rev.* **102**, 86 (1956).
333. R. Nitsche and W. Merz, *J. Phys. Chem. Solids* **13**, 1954 (1960).
334. H. Jasunaga and L. Nakada, *J. Phys. Soc. Jpn.* **22**, 338 (1967).
335. R. Farrel, *Proceedings of the Third International Conference on Photoconductivity,* Stanford, 1969, Pergamon Press, Oxford (1971), p. 93.
336. D. E. Sawer, *Appl. Phys. Lett.* **13**, 392 (1968).
337. A. G. Zhdan and E. S. Artobolevskaya, *Fiz. Tverd. Tela* **13**, 1242 (1971).
338. F. Prokert and G. Schmidt, *Izv. Akad. Nauk SSSR, Ser. Fiz.* **33**, 1090 (1969).
339. H. Motegi and S. Hoshino, *J. Phys. Soc. Jpn.* **29**, 524 (1970).
340. V. M. Fridkin, A. A. Grekov, E. A. Savchenko, and T. R. Volk, *Phys. Status Solidi* **8**, K55 (1971).
341. V. N. Nosov and V. A. Lyakhovitskaya, *Kristallografiya* **11**, 322 (1966).
342. A. A. Grekov, A. I. Rodin, and V. M. Fridkin, *Fiz. Tekh. Poluprovodn.* **5**, 1287 (1971).
343. V. M. Fridkin, A. A. Grekov, and A. I. Rodin, *Phys. Lett. A* **35**, 59 (1971).
344. M. K. Sheinkman, I. V. Markevich, and V. A. Khvostov, *Fiz. Tekh. Poluprovodn.* **5**, 1904 (1971).
345. W. Tyler and H. H. Woodburry, *Phys. Rev.* **96**, 874 (1954).
346. D. M. Bercha, A. A. Kikineshi, O. G. Semak, and D. V. Chepur, *Fiz. Tverd. Tela* **14**, 573 (1972).
347. E. I. Adirovich, *Fiz. Tverd. Tela* **2**, 1410 (1960).
348. M. Lampert and P. Mark, *Injection Currents in Solids,* Academic Press (1970).
349. V. M. Fridkin and K. Kreher, *Phys. Status Solidi A* **2**, 281 (1970).
350. V. F. Krapivin and E. V. Chenskii *Fiz. Tverd. Tela* **12**, 597 (1970).
351. Yu. M. Popov and P. S. Nesterenko, *Fiz. Tekh. Poluprovodn.* **2**, 1373 (1968).
352. H. Neumann, K. Kreher, and W. Schmitz, *Phys. Lett. A* **28**, 743 (1969).
353. V. L. Bonch-Bruevich, I. P. Zvyagin, and A. G. Mironov, *Domain Electrical Instability in Semiconductors,* Consultants Bureau (1975).
354. M. S. Shur, *Fiz. Tverd. Tela* **10**, 2652 (1968).
355. M. S. Shur, *Fiz. Tverd. Tela* **10**, 3560 (1968).
356. M. S. Shur, *Fiz. Tverd. Tela* **10**, 3684 (1968).
357. M. S. Shur, *Izv. Akad. Nauk SSSR, Ser. Fiz.* **33**, 207 (1969).
358. E. V. Chenskii, *Fiz. Tverd. Tela* **11**, 666 (1969).
359. E. Sawaguchi and T. Mori, *J. Phys. Soc. Jpn.* **21**, 2077 (1966).